Blue and lesser snow geese at the Squaw Creek National Wildlife Refuge. Note loess hills in background. Vegetation in foreground is giant cord grass (*Spartina michauxiana*). Refuge lies in the valley of the Missouri River near Mound City, Missouri.

The
BIOGEOCHEMISTRY
of Blue, Snow,
and Ross'
GEESE

By Harold C. Hanson and Robert L. Jones

Published by
SOUTHERN ILLINOIS UNIVERSITY PRESS
Carbondale and Edwardsville
for
THE ILLINOIS NATURAL HISTORY SURVEY
Urbana, Illinois

Published by Southern Illinois University Press for the State of Illinois Department of Registration and Education, Natural History Survey Division, Urbana, Illinois.

Illinois Natural History Survey Special Publication No. 1

Publication of this book has been made possible by generous grants from the Gaylord Donnelley Foundation, the Max McGraw Wildlife Foundation, and by funds provided by the Illinois Department of Conservation.

Manufactured in the United States of America

Designed by Dwight Agner

Library of Congress Cataloging in Publication Data

Hanson, Harold Carsten, 1917–
    The biogeochemistry of blue, snow, and Ross' geese.

    (Special publication—Illinois Natural History Survey; no. 1)
    Bibliography: p.
    Includes index.
    1. Blue goose. 2. Snow goose. 3. Ross's goose. 4. Biogeochemistry—Hudson Bay.
5. Feathers. 6. Birds—Racial analysis.
7. Waterfowl management—Hudson Bay. I. Jones, Robert L., 1936– joint author. II. Title.
III. Series: Illinois. Natural History Survey. Special publication—Illinois Natural History Survey; no. 1.
QL696.A52H36   598.4'1   76-46617
ISBN 0-8093-0751-0

Contents

List of
Maps

List of
Tables

Foreword

THE WORDS *ecology, ecosystems, renewable resources,* and *pollution* loom large in our vocabulary and concerns today. The studies described in this book encompass these concerns as the authors trace an array of twelve elements—principally minerals—from their geographical sources to incorporation in the keratin of the primary feathers of goose wings. Doctors Hanson and Jones determined that the quantitative pattern of minerals found in feather keratin can be employed usefully in determining the birthplaces of wild geese and, in their subsequent years of life, their molting or breeding areas, or both. This is possible because each nesting area appears to be unique with respect to its local geology or to the input of minerals it has received over thousands of years from adjacent areas from wind deposits, the action of glaciers, rivers, and streams, its contact with the oceans, or a combination of these factors. Thus, the patterns of minerals incorporated into the feathers reflect in varying degrees the nutrient chain that can be traced back through the plants to the soils and, ultimately, to their rock origins.

The original purpose of the research was to determine the usefulness of feather mineral patterns in determining the geographical origins of wild, migrant, and wintering goose populations. This technique promises to supplement or, in many cases, replace banding as a means of establishing such information. In the case of Canada geese, determination of sites of origin will also effectively identify racial or subspecies stocks and, hence,

permit improved management of the populations.

In order to understand the relationship between mineral levels in the environment and those in the feathers, the authors found it necessary to explore in detail current understandings of mineral metabolism. Their data suggest that the importance of the role of sodium has been underestimated in its influence on the absorption of other mineral elements. It would appear that certain of their findings have an important relationship to medical problems common to both man and domestic mammals and that feather keratin essentially serves as a monitor of background levels of minerals important to the health and welfare of birds.

George Sprugel, Jr.
Chief, Illinois Natural History Survey

Urbana, Illinois
June 25, 1975

Acknowledgments

THE PRESENT book is continent-wide in scope. It was made possible only through financial support and cooperation in the field received from many federal, state, provincial, and private agencies. But institutions consist of people; this book, therefore, also reflects the assistance of a host of personnel engaged in waterfowl research and management.

Grants for field studies on wild geese were received from the John Simon Guggenheim Foundation, the Arctic Institute of North America, the Wildlife Management Institute, and the Graduate College of the University of Illinois. We can only hope that this book and several to follow will have justified their deeply appreciated support.

Most wild geese nest in inaccessible and remote parts of the North. These areas can only be visited readily by airplane. Thanks to the direct or indirect support of the Ontario Ministry of Natural Resources, the Manitoba Department of Mines and Natural Resources, the Canadian Wildlife Service, and the United States Fish and Wildlife Service one or both of us were able to see and photograph from the air the major breeding areas of blue, snow, and Ross' geese. For this privilege, we are indeed indebted to Harry G. Lumsden, W. C. Currie, Eugene F. Bossenmaier, Richard H. Kerbes, and James G. King and their respective agencies.

Although we personally made feather collections during some of these northern field trips and during visits to key stopover and wintering areas of wild geese, we are largely obligated to others for the collections made in the field of wing feathers of geese shot by hunters. We therefore wish to acknowledge the contributions of these men, who made or implemented feather, and in some cases plant and soil, collections for us in the various provinces and states: *Quebec,* J. Douglas Heyland; *Ontario,* Harry G. Lumsden, James R. Bailey, W. C. Currie, and John Lessard; personnel of the Royal Canadian Mounted Police who monitored the goose hunting camps on James Bay; *Manitoba,* Eugene F. Bossenmaier and the late Al Pakulak; *Saskatchewan,* Alex Dzubin; *British Columbia,* late Hugh Monahan and William A. Morris; *District of Keewatin,* Charles D. MacInnes and B. Lief; *District of Franklin,* Richard H. Kerbes; *District of Mackenzie,* John P. Ryder and Tom Barry; *Virginia,* Donald R. Ambrosen; *South Dakota,* Lyle J. Schoonover and Lee Herzberger; *Missouri,* Harold H. Burgess; *Louisiana,* John J. Lynch, Ted Joanen, John Walther, and Jake Valentine; *Utah,* John E. Nagel; *Nevada,* Donald E. Lewis; *New Mexico,* Richard W. Rigby and Raymond J. Buller; *Washington,* Robert G. Jeffrey; *California,* Richard D. Bauer, Harry A. George, Ed McNeil, Newell B. Morgan, Walter T. Shannon, Mr. and Mrs. Chester Stoncypher, Jr., and Sanford R. Wilbur.

We also wish to express our special indebtedness to Harry G. Lumsden, John P. Ryder, and John Chattin without whose aid in getting specimens we would have had very little to report upon.

Few field studies permit such relatively great gains in knowledge to be made from the brief excursions into the field as represented by the present research approach. A striking instance was the occasion provided one of us (Hanson) to collect Canada goose wings on a brief visit to the Eskimo settlement on the Belcher Islands, N.W.T. Findings from the collection made on the Belcher Islands are probably the most significant that accrued from any sample of feathers we studied. For the opportunity to visit the Belcher Islands, the gracious hospitality of Rene Brunelle, former Minister of the Ontario Ministry of Natural Resources, is acknowledged.

For the privilege of studying several key museum specimens we are indebted to Earl Godfrey of the National Museum of Natural Sciences of Canada.

Three grants from the Graduate College of the University of Illinois provided funds for the greater portion of the chemical analyses which were carried out in the laboratories of the Agronomy Department of the University of Illinois College of Agriculture. Without this support, the study would have been indefinitely postponed. For making analytical facilities available to us we are indebted to M. D. Thorne, R. W. Howell, and T. R. Peck. The proficiency of Emil Marcusiu in operating the optical spectrograph was of vital importance to the study. Preparation of feather material for analyses is exacting and sometimes frustrating work; hence our abiding gratitude to student assistants Robert Sinclair, Eugene L. Ziegler, and Lee Herzberger.

The efficient use of feather mineral analyses to determine the geographical origins of wild geese is dependent upon the use of a computer. We were fortunate to have had the able assistance of Jeffrey Tyler in handling computer analyses of the data and in adapting the discriminant function program for use in determining the origins of geese.

The photograph of the Louisiana coast was taken from Skylab (negative number 640A0A4206000). We thank the National Aeronautics and Space Administration for permitting us to use this picture and the United States Geological Survey, Eros Data Center, for a negative reproduction of the original.

Maps of Canada, with the exception of Figure 90, are adaptations of official Canadian federal and provincial topographic maps and geologic maps prepared by the Geological Survey of Canada (as indicated in legends). We are indebted to the Arctic Institute of North America for permission to use the map of the surficial geology and soil configurations of Cape Churchill, Manitoba, from Ritchie (1962). The outline portion of the geologic map of Wrangel Island, U.S.S.R., is based on the World Aeronautical Chart CC–8, 1–1,000,000, second edition, March 1, 1973. We are appreciative of the skills of Lloyd LeMere and Richard M. Sheets, present and former technical illustrators, respectively, of the Illinois Natural History Survey, in preparing the maps presented. Feather mineral diagrams were drafted by Lloyd LeMere.

Professional reviewing prior to publication, an important requisite for publication and safeguard for authors, was especially needed and appreciated in the present study because of its cross-disciplinary nature. We are, therefore, particularly grateful to J. Mannery, University of Toronto; Robert E. Johnson, Knox College (formerly of the University of Illinois); James D. Jones, Mayo Clinic; and Jean Graber, Illinois Natural History Survey, for their helpful reviews of sections dealing with biochemistry and physiology. Also, throughout the study it has been most helpful to have been able to discuss biochemical problems with Robert C. Hiltibran of the Illinois Natural History Survey. Richard H. Kerbes made valuable suggestions in relation to our discussions of the blue and snow goose colonies. We have also benefited from discussions of ecological problems with William R. Edwards of the Illinois Natural History Survey.

The reader, like ourselves, will appreciate the lively and valuable contribution John J. Lynch (retired from the United States Fish and Wildlife Service) has made in Chapter 6 to an understanding of blue and lesser snow goose populations on the Gulf Coast.

Plant identifications were kindly made for us by Robert A. Evers of the Illinois Natural History Survey.

We would be remiss if we did not acknowledge the splended library facilities of the University of Illinois. Without its in-depth collections, our findings simply could not have been adequately integrated.

Freedom to pursue a research program down avenues and in ways judged best is the most precious and vital commodity an investigator can have. For this reason, the authors take pleasure in acknowledging respectively the stimulating support they have received from Glen C. Sanderson, Head, Section of Wildlife Research, and George Sprugel, Jr., Chief, Illinois Natural History Survey, and Robert W. Howell and Marlowe D. Thorne, present and former heads, respectively, of the Department of Agronomy, University of Illinois.

Manuscripts have a way of requiring many retypings. For their patience in such matters, we are most grateful to Eleanore Wilson and Elizabeth McConaha. Robert M. Zewadski, Technical Editor of the Illinois Natural History Survey, ably demonstrated that the process of having a major publication edited can be both a profitable and pleasurable experience. And our high personal and professional regard for Vernon A. Sternberg, Director of the Southern Illinois University Press, and the superior standards he has attained for the Press are reasons why we are pleased to achieve publication under his aegis.

The publication of academic books having limited distribution always poses a financial dilemma. The printing of this book was made possible largely by generous grants from the Gaylord Donnelly Foundation, the Max McGraw Wildlife Foundation, and funds provided by the Illinois Department of Conservation. We trust the findings will have merited their confidence.

Harold C. Hanson
Robert L. Jones

Urbana, Illinois
June 15, 1975

THE BIOGEOCHEMISTRY
OF BLUE, SNOW,
AND ROSS' GEESE

1. Geographic setting of the blue and lesser snow goose colonies in the Hudson Bay area.

OUR PRESENT fund of knowledge on waterfowl populations has been importantly derived, directly or indirectly, from the use of various banding and marking techniques. Yet, despite the notable successes achieved, individualized external marking systems have not yielded certain needed kinds of quantitative population data on dispersal, distribution, and losses from hunting—or at best have permitted only inferences or estimates of actual values. The basic handicap of leg banding is not the inherent defects of the technique itself, but that it is seldom possible to band all or nearly all of the individuals in a population or in discrete populations that subsequently mix—a theoretically desirable objective if patterns of distribution are to be determined accurately from banding alone.

One of the notable challenges in waterfowl management is the problem of accurately determining the winter distribution of blue and lesser snow geese (*Anser caerulescens caerulescens*) that originate from discrete breeding colonies scattered across the Canadian Arctic and on Wrangel Island, U.S.S.R. (Figures *1–4*). Regional groupings and flocks from these colonies intermingle in migration and subsequently use overlapping wintering grounds, particularly along the coast of the Gulf of Mexico. Most of these colonies are so large that only token bandings are possible. Recoveries from a single nesting colony or from several adjacent colonies banded in the same year cannot provide a reliable estimate of the origins of most of the individuals wintering in any one area. When such information is lacking, it is axiomatic that hunting losses from such mixed populations cannot be related satisfactorily to their respective breeding colonies. Yet such data are needed to place the management of this species on a better footing.

In a preliminary study (Hanson and Jones 1968) we reported on the apparent usefulness of quantitative data on twelve minerals incorporated in the vane portion of primary feathers grown on the breeding grounds (Figures *5–6*) as a means of differentiating the origins of blue and snow goose populations. Since the inception of the study in 1965, we have analyzed the mineral content of the primary feathers of over 3,000 wild geese. This series has included, in addition to blue and lesser snow geese (Figure *7*), the greater snow goose (*Anser caerulescens atlanticus*) (Figure *8*), Ross' goose (*Anser rossii*) (Figure *9*), the various races of the Canada goose (*Branta canadensis*), the black brant (*Branta bernicla*), and the white-fronted goose (*Anser albifrons*). However, the resolution of feather mineral patterns of blue and

CHAPTER ONE

Geographic Origins of Wild Geese— The Biogeochemical Approach

snow geese was given priority over the completion of similar studies on other species, because the individuals from the various colonies are look-alikes—in contrast to most Canada geese that are readily separable by morphology into regional taxonomic units. Hence, any hope of managing blue and snow geese on a colony basis, if only to a partial degree, appeared to rest on achieving success with the feather mineral technique. ("Feather mineral pattern" refers to the absolute and relative concentrations of the twelve chemical elements that were determined.)

The question will likely arise in the minds of many readers as to why the twelve elements were chosen. The answer is that it was partially a question of expediency; these are the elements for which routine analyses of plant materials are made in the Agronomy Department at the University of Illinois. However, had it been possible to select the key elements to be studied from throughout the periodic table, the choice would probably have been essentially the same. Calcium, magnesium, sodium, potassium, and phosphorus are essential macroelements, and their availability in parent rock, soils, and plants varies greatly; iron, copper, zinc, manganese, and silicon are included in the twenty trace elements essential for higher animals (Underwood 1971:2); and aluminum is ubiquitous in tissues of both plants and animals, although it is highly variable in concentration. Boron is not essential to animals, but it "plays a regulatory role in carbohydrate metabolism in plants" (Epstein 1972:293). Plants of the genus *Equisetum*, which are wetlands or aquatic plants having unusual capacity to concentrate silica, constitute one of the

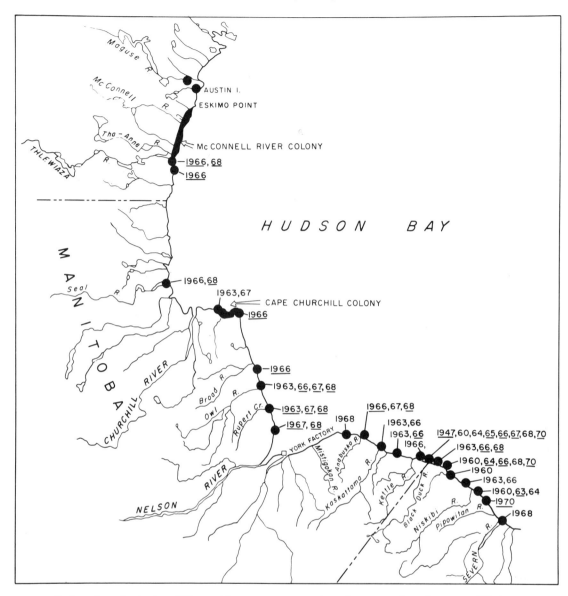

2. Location of colonies of blue and lesser snow geese along the west and south coasts of Hudson Bay. Also shown are locations of sporadic occurrences of molting and nesting birds. Underlined years indicate goslings were observed. (Reprinted from Hanson et al. 1972)

preferred foods of geese; consequently, the presence of relatively large amounts of silicon, recently shown to be an essential element for chickens (Carlisle 1974), in the feathers of geese may in part also be indicative of their diet and feeding habitat.

In respect to the scope of treatment and the breadth of the geographic area involved, this book probably represents the first study of its kind. We believe that the nature of our research required the kind of presentation given if an attempt at a synthesis of findings from many areas of inquiry that could contribute to an understanding of our data was to be made. For example, the significance of variations in the mineral contents of wild geese cannot be understood adequately without consideration of certain aspects of their life histories or put into perspective without an appreciation of the

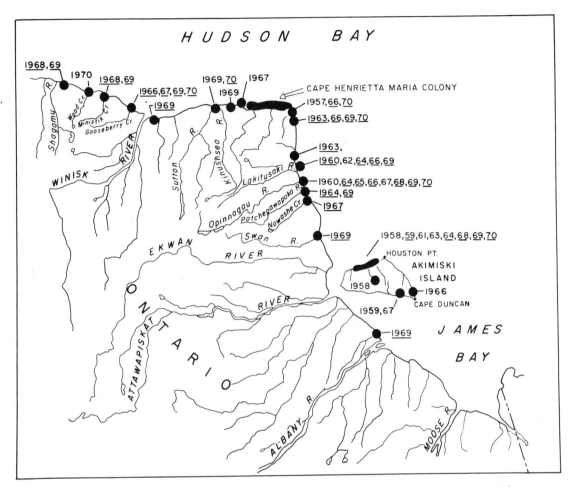

3. Location of the Cape Henrietta Maria colony and miscellaneous records of molting and nesting blue and lesser snow geese along the eastern sector of the south coast of Hudson Bay and along the upper portions of James Bay and Akimiski Island. Underlined years indicate goslings were observed. (Reprinted from Hanson et al. 1972)

local geology and soils that in large measure determine mineral levels in food plants. These plants in turn account partly for variations in mineral intake and absorption by geese.

Further, the feather mineral patterns of goose populations cannot be evaluated adequately unless the ranges and movements of these populations on their breeding grounds are taken into consideration. The immediate nesting areas of wild geese seldom afford the quality and, especially, the quantity of food required for large numbers of adults and their rapidly growing broods. These factors and others, such as the safety of the young and reduction of the spread of diseases and parasites, dictate that adults and their young disperse and feed over a wide area during the period of gosling

growth and molt of the adults. The mineralogy of these generally extensive feeding areas may vary considerably; variation in mineral input into the feeding areas is, in turn, related to the geologies of the drainage basins, whether small or large.

If game managers are to use the feather mineral technique, evaluations of their findings on the origins of bagged geese and the impact of the kill should be made in relation to the sizes of the breeding colonies which contribute to migrating and wintering populations—or in the case of Canada and other geese, in the light of knowledge of the sizes of more diffusely spread breeding populations. For these reasons we have included in our introductory remarks for each colony of blue and snow geese recent minimal estimates of its size.

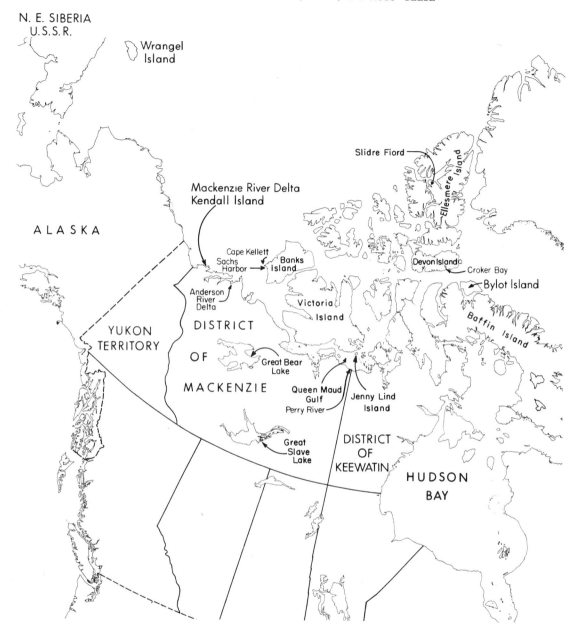

4. Geographic setting of blue, lesser and greater snow, and Ross' goose colonies in the central, western, and high Arctic.

The primary objective of this study was to define the limits and variability of the levels of twelve minerals in the primary feathers of blue, snow, and Ross' geese of known origins so that these data could be used to determine birthplaces and breeding or molting areas, or both, of birds of unknown origins. It should be emphasized, however, that populations of blue and lesser snow geese are less rewarding to work with than are some other species, notably Canada geese. Whereas most populations of the former spend the summer along coastal plains of similar geological origin—principally Paleozoic limestones that have a history of former marine submergence—the various races of Canada geese breed in areas of highly diversified geological structures present

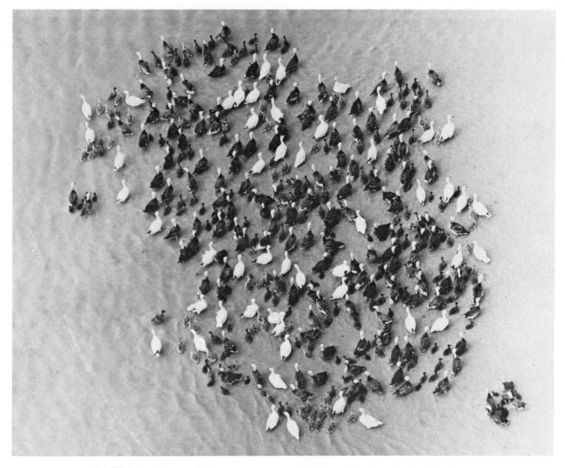

5. Flightless blue and lesser snow geese near Cape Henrietta Maria, Ontario.

over a great part of the continent north of Mexico. As might be expected, the feather mineral patterns derived from these latter populations are correspondingly diverse. From the standpoint of gaining firmer and more complete insights into the ''pure-science'' aspects of the biogeochemistry of feather minerals, it is unfortunate that our studies on the feather mincrals of Canada geese were not completed first. Because of the added insights that analyses of the feathers of Canada geese have provided, we have taken the liberty of citing data from a few of these geese that bolster points of view on mineral metabolism that are less strongly illustrated by our data for blue, snow, and Ross' geese.

The term *biogeochemistry* as defined by subject matter and usage in this report is a study (albeit limited in certain instances by inadequate size of samples) of the relationships of mineral concentrations in the various trophic levels of the ecosystem, the mode of transfer of these minerals from one trophic level to the next above, the mechanisms and factors involved in mineral transfers, the metabolic relationships of the concerned minerals to each other in each trophic level, and factors relating to their deposition in feather keratin.

This study was initially begun with the interests of game management foremost. Although we do not discount the continued need for banding for specific purposes, we hope that this report will constitute a substantial contribution toward the goal of determining the origins of wild geese without resort to the expensive procedure of banding in remote Arctic regions. Our findings and their interpretations, being of a cross-disciplinary nature, should also be at least of passing interest to a fairly wide range of professional workers, such as geologists, geochemists, plant physiologists, nutritionists, biochemists, physiologists, and ecologists. Specialists in each area will find some of our remarks such common knowledge that they

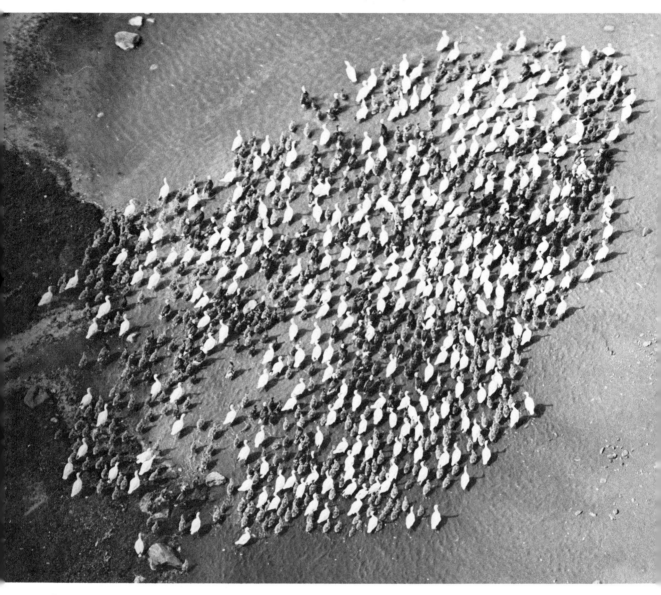

6. Flightless blue and lesser snow geese near the McConnell River delta, N.W.T. This flock contained 294 adult snow geese, 98 adult blue geese, and 661 goslings (total 1,053). Flocks containing as many as 1,200 adults and goslings have been photographed in the McConnell River area.

may be regarded as an affront, but we hope that for most readers our discussions provide the minimal number of links in a chain of information needed to follow the fate of a given element from its rock origins to its incorporation in feather keratin. Lastly, for ornithologists and the ecologists interested in the North, we hope that our report will also serve as a useful reference for the geography of the northern areas under review, brought into overall perspective by the inclusion of low-altitude oblique aerial photographs.

If a defense is needed for the cross-disciplinary nature of our consideration of our data, we can do no better than cite the parallel, perhaps somewhat exaggerated, viewpoint presented by Masoro (1973:vii): "Because of the arbitrary separation of these two disciplines [biochemistry and mammalian physiology] in most teaching programs and all

7. A captive lesser snow goose on Westham Island, Frazier River delta, B.C. Flocks stopping in this area originate on Wrangel Island, U.S.S.R. The small Canada shown with it is from the north slope of Alaska.

9. A captive Ross' goose at the Sather Game Farm near Round Lake, Minnesota.

8. Greater snow goose on the Fosheim Peninsula, Ellesmere Island, N.W.T. (Photograph courtesy of Dalton Muir)

textbooks, the vast majority of students do not see the intimate relationships between them. Consequently, the medical student and the beginning graduate student as well as the recently trained physician find it difficult, if not impossible, to utilize the principles of biochemistry as they apply to the physiological and pathological events they observe in man and other mammals.''

It will be apparent that because of the variability of the analytical techniques employed, the variable nature of avian metabolism, and the place-to-place variability of the mineralogy of the breeding ground environment of any given population, this report cannot be considered definitive or final. We are certain, however, that it provides a broader perspective for research in behalf of wildlife management as well as for basic inquiries into avian and mammalian metabolic responses to mineral environments.

CHAPTER TWO

Samples, Analytical Procedures, and Data Analyses

SOIL SAMPLES

As our soil samples, like many of the plant samples, were collected by several collaborators, no consistency in sampling technique was likely. In any event, the samples represented the uppermost organic material-rich portion of the soil profile—the A0, the organic-rich layer, or A1 horizon, the surficial mineral horizon bearing substantial organic matter, or both. Prior to analysis the samples were oven-dried at 110° C and then split to obtain an aliquot for pulverizing in a mill having tungsten carbide grinding faces. A representative sample of the pulverized soil, ground to less than 60-mesh screen size, was treated twice with hydrofluoric and hydrochloric acids and dried each time, and finally the salts were taken into solution with hydrochloric acid. Calcium, magnesium, iron, zinc, manganese, and copper were analyzed by the atomic absorption method. Phosphorus was determined by the vanadomolybdate-yellow method.

PLANT SAMPLES

Plant samples used in the analysis, with the exception of *Triglochin maritima,* were collected within the feeding ranges of the four colonies represented. Only the aerial portions of the plants were used in the analyses, and no effort was made to clean these samples of adventitious matter. We reasoned that any extraneous matter would also be consumed by feeding geese and become an integral part of the diet at an unknown rate of availability. Before

analysis the samples were dried at 60° C, finely ground in a Wiley mill, subsampled, and then ashed at 500° C. The ash was analyzed for the same mineral elements and by the same optical-emission spectroscopy method used for analyses of the primary feather vanes. Nitrogen was determined on another aliquot after acid digestion by the micro-Kjeldahl method.

PRIMARY FEATHER SAMPLES

We believe that the feathers of geese, particularly those of immatures, banded on the breeding grounds and subsequently shot by hunters constitute the ideal reference material (sometimes hereafter referred to as reference feathers) to which to relate feather mineral data from birds of unknown origin shot by hunters. These banded geese are presumed to have duplicated the environmental exposures of the unbanded geese. The most important source of wings from geese banded in the Hudson Bay colonies were birds killed on the coastal marshes of northern Ontario. In addition, incompletely developed primaries of geese in the molt were analyzed as well as molted primaries of the previous year. As will be pointed out later, neither of these two kinds of samples served ideally as basic reference material to portray the breeding colonies. We also sought assistance in obtaining random collections of wings of blue and snow geese from points on migration routes and the wintering grounds in the belief that if the feather mineral technique is to be useful, we should be the first to be confronted with the practical problem of relating geese of unknown origins to their respective breeding grounds.

In most mass banding operations of blue and lesser snow geese in the North, individuals have been classified only as to age. Obviously, if few or no geese in the sample banded in a colony have been sexed, the influence of sex on the mineral content of the feathers cannot be determined subsequently from wings of banded geese. Our samples were adequate in respect to sex identification for only one population, the Ross' geese breeding at Karrak Lake, District of Keewatin. Age data were available for most banded wings or could be determined from characteristics of the tips of the primaries (Hanson 1965), but in most cases too few wings from birds of known age were obtained from geese banded in the same areas of the various colonies to yield significant findings on the influence of age on the mineral content of feathers. Though not statistically significant, the array of elemental differences within extensive breeding ranges—e.g., the Baffin Island colony—decreased the likelihood that data samples from widely separated points within the colony could be meaningful when grouped accordingly to age only.

ANALYTICAL PROCEDURES AND RELATED STUDIES

As the portions of the primary feathers used and the analytical techniques employed have remained constant throughout the study, we reprint our procedures here with only slight changes (Hanson and Jones 1968).

In our studies only the vane portions of the primary feathers were used in the analyses because they were found to be more highly mineralized than the shaft. The basal quarter of the vane was also excluded because part of this portion of the feather is grown after the goose has regained the power of flight (Hanson 1962) and could possibly, but not probably, have left the breeding or molting area. In such a case it would be possible for a somewhat different mineral pool to be incorporated into the basal portion of the feathers.

To facilitate washing, the remaining portions of the primary feathers were cut transversely into 1.5-inch (38 mm) pieces. These were washed with about 175 ml of distilled water by shaking in 250-ml conical flasks on a reciprocating shaker for at least four hours with a change of water after each hour of washing. The feather pieces were finally rinsed several times with distilled water. The flasks and contents were then drained of excess water and dried in an oven held at 60° C for at least twenty-four hours. Approximately 1 gram of the dried vane was carefully trimmed free of the shaft and placed in a 50-ml Vycor crucible.

Analytical determinations were performed with a Jarrell-Ash direct reading emission spectrograph on an acid solution (containing 4.5 percent HCl and 1.5 percent HNO_3, both by volume, and 1 percent lithium as lithium chloride) of the dry-ashed (500° C) sample. A rotating-disc solution technique with alternating-current spark excitation was used with lithium as an internal standard. The amount of each element was estimated from working curves obtained from reference plant samples.

Feather ash color was determined by the Munsell color notation system.

Mineral Content

Some fragmentary data on shaft and vane mineralization were collected early in the evaluation of this feather method for determining goose birthplaces. We had anticipated lower levels of minerals in the shaft than in the vane after reading the reports of McCullough and Grant (1952 and 1953) on ruffed grouse in New Hampshire. As the analyses in Table 1 were considered to be indicative of the relative mineral concentrations to be found in shaft and vane parts, we settled on analyses of the vane portion of the primary feathers as the basis for this report, bearing in mind the improving accuracy of the optical-emission spectrograph with increasing concentrations of minerals. Also, from a small sample we determined that for the outermost 75 percent of the length of fully grown primaries the ratio of the weight of the shaft to the weight of the vane is 1.34 (V = 16.8 percent).

Because different amino acids have different metal binding capacities, the warning of Schroeder et al. (1955:3908) has particular relevance for those who contemplate studies of the origin of waterfowl based on feather mineral analyses: "It is apparent that in studies of so-called feather keratin, definite portions rather than the whole feather should be investigated if the results are to be meaningful."

Kelsall (1970) investigated the relative concentrations of zinc, copper, calcium, and manganese in vanes and whole flight feathers of the mallard

1. Distribution of selected elements in parts per million in the shaft and vane portions and in the whole primary feather of a male Canada goose collected at Horicon Marsh, Wisconsin

Part	n	Element							
		Ca	Mg	Na	Fe	Zn	Cu	Si	Al
Shaft	1	<500	130	28	15	90	4.6	<500	20
Vane	1	1,740	275	465	148	225	6.0	2,750	81
Whole	4	822	172	318	51	165	6.6	681	31
Standard error of the mean			20	6	8	4	8	0.2	61	1

(*Anas platyrhynchos*). He found the respective concentrations to be 183 and 124 ppm, 5.7 and 4.6 ppm, 350 and 209 ppm, and 9.6 and 5.6 ppm for vane and whole feather samples. Only copper concentrations were not significantly different between vanes and whole feathers. These ratios of vane to whole-feather mineral contents are not appreciably different from the data reported in Table 1 for a Canada goose. The shaft and vane data in Table 1 result from single analyses, whereas four replicates of the whole-feather analysis were performed. Kelsall concluded that the time involved in clipping vane from shaft was not warranted largely because he found less scatter in data for whole-feather analyses. Differences in mineral content between the shaft and vane are probably due to differences in the amino acid content of these structures (Schroeder et al. 1955:3908; Blackburn 1961:703).

Analytical Variation Due To Wash Time

Several experiments were conducted to evaluate the effects of the length of wash time on the elemental composition of the feather. In one of these experiments we cut to the same length and divided into ten lots the primaries of an immature Canada goose, sex unknown, collected at Union County, Illinois. Each feather lot weighed 0.7 g ± 2 percent. Four lots were washed in 150 ml of distilled

water for one hour, another four for four hours, and the remaining two lots were washed for eight hours. All washings were done on a reciprocating shaker. After each hour of washing the water was changed so that diffusion of ions out of the feather matrix would be enhanced. Each batch of wash water was retained for analysis. The feathers were dried at 60° C, and the vane was cut from the shaft for analysis by atomic absorption and flame-emission spectroscopy.

The data on the mineral composition of the feathers of this individual are given in Table 2. For this particular bird the one-hour wash appears sufficient. In the four- and eight-hour periods, the samples were rinsed at one-hour intervals and fresh water added. None of the means of elements in feathers washed for four hours differs significantly from those washed for one hour. Surprisingly, when we refer to the data on the mineral composition of the wash water (Table 3), the amounts of all of these elements that have been washed from the feather are impressive. None of the pairs of means of elements between one- and four-hour treatments, as shown in Table 3, is significantly different. Values for four- and eight-hour experiments in Table 3 are accumulated from four one-hour washings and eight one-hour washings, respectively. There is some suggestion that important losses of sodium occurred between four and eight hours.

2. Concentrations in micrograms per gram of total feather of selected elements and ash by percent in feathers washed in distilled water

Sample	Washing Time (hours)	Ca	Mg	Na	Fe	Zn	Cu	Feather Ash (%)
R–1	1	1,130	187	34	122	184	18.5	0.58
R–2	1	1,110	186	34	163	193	15.5	0.60
R–3	1	1,120	170	36	180	168	15.1	0.58
R–4	1	1,300	215	30	159	180	13.2	0.73
Mean		1,165	190	34	156	181	15.6	0.62
Standard error of the mean		45	9	1	12	5	1.1	0.04
R–5	4	1,230	223	37	110	220	9.3	0.53
R–6	4	1,450	182	34	182	232	14.2	0.75
R–7	4	1,380	166	33	122	203	13.2	0.65
R–8	4	1,160	148	38	231	173	13.5	0.70
Mean		1,305	180	36	161	207	12.6	0.66
Standard error of the mean		66	16	2	28	13	1.1	0.05
R–9	8	1,010	119	32	138	177	12.7	0.47
R–10	8	1,070	106	34	149	223	15.4	0.52
Mean		1,040	112	33	144	200	14.0	0.50

3. Concentrations in parts per million of selected elements found in water in which feathers were washed

Sample	Washing Time (hours)	Ca	Mg	Na	K
R–1	1	749	1,284	244	219
R–2	1	846	1,322	242	185
R–3	1	841	1,197	219	241
R–4	1	919	1,363	331	283
Mean		839	1,292	259	232
Standard error of the mean		35	35	25	20
R–5	4	927	1,234	282	248
R–6	4	1,050	1,288	396	347
R–7	4	590	1,815	240	228
R–8	4	1,063	1,254	450	431
Mean		908	1,398	342	314
Standard error of the mean		110	140	49	47
R–9	8	1,059	1,150	552	392
R–10	8	1,217	1,339	553	449
Mean		1,138	1,250	552	420

Effect of Exposure to Seawater

In another experiment, primary feathers from an adult female Canada goose collected at Horseshoe Lake in southern Illinois on 16 February 1943 were cut into short lengths, combined, and then divided into six lots. Artificial seawater was prepared, and one lot was washed in a reciprocating shaker for 0.75 hour. In a similar manner other lots were washed for 1.5, 3.0, 6.0, and 48 hours. After these exposures to seawater the feathers were washed with distilled water in a reciprocating shaker for 2 hours with a change of wash water every 0.5 hour. Unfortunately, the feather sample representing the control was lost. A sample of secondary wing feather from the same bird was analyzed in place of the lost control, and data for its chemical content are given in Table 4 for purposes of comparison in the event that subsequent research should indicate that the mineral composition of secondary feathers is not different from that of primary feathers.

Disregarding the control secondary feather data, the analyses of the secondary feathers (Table 2) suggest that the feather content of sodium and magnesium will be increased three times and calcium almost twofold by continuous contact with seawater for periods up to 48 hours (Table 4). These increases, of course, would have been larger if the feathers had not been washed with distilled water before being analyzed. The treatments were

not replicated, and the decrease in the amount of copper after 0.75 hour and the increase of zinc at 48 hours cannot be tested for significance although we can infer that these elements are firmly bound in the feather matrix. Notably, potassium did not undergo any appreciable change. Ash weight increased, reflecting the increases in sodium, calcium, and magnesium.

In a third experiment four lots of primary feathers from an immature Canada goose collected 9 December 1968 at the Union County, Illinois, public shooting area were placed in seawater for two hours and then washed for different lengths of time, in an unreplicated experiment. Data for calcium, magnesium, sodium, potassium, copper, zinc, and ash in the feathers are given in Table 5. These data provided evidence for the rapid loss of sodium, potassium, calcium, and magnesium in two hours of washing. Sodium required, perhaps,

4. Concentrations in parts per million of selected elements in primary feathers of a Canada goose (*Branta canadensis interior*) after exposure to artificial seawater

Seawater Contact Time (hours)	Ca	Mg	Na	K	Zn	Cu	Ash (%)
0	680	105	22	20	259	20	0.40
0.75	1,161	263	203	95	275	29	0.63
1.5	1,442	378	306	48	273	19	0.84
3.0	1,780	569	405	94	260	22	0.97
6.0	1,978	720	515	79	278	21	1.29
48.0	2,015	882	620	73	349	22	1.44
Percentages in seawater	0.040	0.127	1.05	0.38	na*	na	

* Not added to artificial seawater.

5. Concentrations in parts per million of selected elements in feathers exposed to artificial seawater for two hours and subsequently washed for different lengths of time

Wash Time (hours)	Ca	Mg	Na	K	Fe	Zn	Cu	Ash (%)
1	1,640	820	942	68	145	217	28	1.14
2	1,110	469	457	34	116	172	22	0.62
4	1,210	417	207	30	139	215	32	0.75
8	1,080	317	160	22	120	202	43	0.60
Percentages in seawater	0.040	.127	1.05	0.38	na*	na	na	

* Not added to artificial seawater.

four hours of washing. The rapid declines in feather mineral concentrations after over two hours of exposure to seawater were probably due to the emptying of feather-matrix channels bearing seawater (Table 5). The more gradual loss of mineral content with the passage of time reflected diffusion from deeper sites. Levels of zinc, iron, and copper, the last showing unusual scatter, did not change in any consistent manner. The results for calcium, magnesium, and sodium can be compared with the results of the experiment described above in which aliquots of untreated feathers were washed for one-, four-, and eight-hour periods and no appreciable change was observed in feather mineral composition after the one-hour and longer washing periods. Presumably this bird had never been exposed to saline water and the complement of mobile ions in the feather structure was essentially removed after one hour of washing. Perhaps most of these feather mineral losses can be attributed to that complement of ions incorporated into the matrix during feather formation.

These results indicate that potassium is not taken up to any extent by the feather exposed to salt water; in fact, the ion seems to be excluded (Table 4). Iron, copper, and zinc were not changed either by exchange reactions with the cations in seawater or by leaching during washing with distilled water. During washing, sodium and magnesium were removed from the feather rapidly, within one or two hours, and small additional losses were experienced with longer washing. Losses of calcium were slower than losses of sodium or magnesium after the initial rapid flushing of ions from the feather. The four-hour washing period used in this study was judged sufficient to remove easily diffusible ions, leaving the feather with its profile of firmly bonded minerals that were incorporated in the protein matrix.

ILLUSTRATIONS

The evaluation of differences among samples, each characterized by an array of twelve variables, constituted a formidable undertaking and required time-consuming statistical procedures. We have been aided in our interpretations by diagrammatic expressions of the data and computer analysis. To gain an initial understanding of the data, the sophistication of modern computer analyses does not substitute for preliminary, subjective analyses of data based on a graphic method. To enable us to compare quickly and effectively one set of analyses of twelve minerals with another, a twelve-spoke diagram was devised (Figures 12–266, noninclusive). Somewhat similar graphs have been used in characterizing water and rock chemistries. The diagram first depicts the macroelements—calcium, magnesium, sodium, potassium, and phosphorus, next the minor or trace elements—iron, zinc, manganese, copper, silicon, and lastly the inert or nonessential elements—boron and aluminum. Four scales were required to accommodate about 95 percent of the variation found. An attempt was made to strike a balance between minimizing and exaggerating differences observed and, at the same time, achieving a scale that would be readily interpretable. The center circle was used to prevent small values from being obscured by drafted lines. It can be argued that the various mineral values, being separate entities, should not be connected. However, the purpose of the graph was to make the findings more or less instantly interpretable; therefore, connecting lines best provided the desired visual impact. The patterns resulting from the plotted data have enabled us to detect, almost at a glance, quantitative differences in the mineral patterns and to assemble similar patterns into subgroups for separate statistical analyses.

Sketch maps of the region of each colony area, showing principal geological formations to the extent that they are known, are included in this report. We have embellished the report with a considerable number of oblique aerial photographs taken at low altitudes. We believe that these photographs add immeasurably to an appreciation of arctic terrains, because relatively few people have either visited these remote areas or seen them from the air. The casual reader may make the judgment from superficial features evident in the photographs that most of the colony areas "look alike," but even impressions of this kind contribute to an understanding of the natural history of geese on their breeding grounds. The series of terrain photographs presented constitute a unique and nearly complete record of the northern breeding areas of two species of geese.

STATISTICAL TREATMENT

The characterization of the four major breeding populations by a series of variables from banded geese of known origin was admirably satisfied by the use of the discriminant function analysis; assignment of geese of unknown origin to these populations was accomplished by classification. The spectrographic method we used was less precise than we desired. However, we were confined to the optical-emission method because of financial considerations, and, of course, a portion of these considerations involved the employment of an analyst and the dedication of laboratory space and equipment over a period of seven years during which time the sample load varied widely. For nearly all batches of feather samples, which were analyzed in groups of 70 to 120, two finely ground and well-mixed plant samples were carried through the preparative and analytical steps. The means and variances for the mineral-content determinations of these plant samples represent the inherent uncertainty of the instrument, the variability associated with some dozen analysts, the sample variability, and, of course, that variation for which we cannot account. The averages of coefficients of variation for these plant samples were: calcium, 11.6; magnesium, 27.0; sodium, 25.8; potassium, 12.0; phosphorus, 33.9; iron, 22.7; zinc, 19.1; manganese, 35.8; copper, 27.5; boron, 33.6; silicon, 40.4; aluminum, 16.7; ash, 5.5. These results compare unfavorably with results of the replicate analyses carried out on subsamples of a single wing and presented in Table 1. There the variation was one-fourth to one-half of that noted for the plants over the entire study. Similarly, atomic absorption analyses of subsamples used in the experiment presented in Table 3 were lower in variances. These comparisons suggested that for any one analytical run the variability will be much lower than that associated with a body of analyses obtained over a considerable length of time. This conclusion is relevant to the feather data because feathers from any one colony were usually run as parts of several batches often separated by years, thus introducing between-batch variation. We recognized this limitation early in the development of the program, but faced with the realities discussed above in addition to the uncertainties of collecting banded birds and of making collections on the breeding grounds, we pressed the analyses to evaluate our hypotheses.

Hudson Bay Colonies

Feather composition of blue and lesser snow geese banded on Baffin and Southampton islands, at the mouth of the McConnell River, and on Cape Henrietta Maria were subjected to discriminant analysis (Figure *10*). Functions and their eigenval-

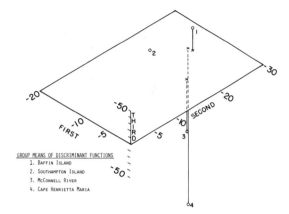

10. Three-dimensional representation of the relationship between feather mineral values for the major colonies of blue and lesser snow geese in the Hudson Bay area, as expressed by the discriminant function test.

ues are listed along with relevant composition data in Table 6. The thirty-eight variables surpass the number of samples for the Southampton Island and Cape Henrietta Maria colonies and equal the number from the McConnell River colony, a situation that should preclude the use of discriminant analysis. Previous trials with thirteen variables equivalent to twelve chemical elements analyzed for and ash weight gave barely significant functions, having less discriminatory power among the breeding populations than when ratios of the elements were included. For the functions calculated on thirty-eight variables (Table 6) the index of total discriminatory power for the three functions was 0.805, a value indicating that about 80 percent of the variability in discriminant space was attributable to observed group differences. The chi square for overall significance was 203.0 with 114 degrees of freedom ($P < .0005$). Inspection of the absolute magnitudes of the standardized vectors indicated that, for the first discriminant function, iron, sodium, aluminum, manganese, and magnesium and ratios of iron, manganese, and magnesium to zinc and of potassium to magnesium contributed most to differentiation. For the second function, iron (negative) and calcium and the ratios of iron to zinc and calcium to zinc (negative) contributed most. The third function did not add significant discriminatory power to the battery although it was used in the classification of birds of unknown origins discussed later.

6. Mean mineral composition in parts per million and mineral ratios found in vane portions of primary feathers of blue and lesser snow geese from colonies on Baffin Island, Southampton Island, at McConnell River, and on Cape Henrietta Maria, and discriminant functions calculated from colony parameters

	Factor	Baffin Island (n=59)	Southampton Island (n=23)	McConnell River (n=38)	Cape Henrietta Maria (n=24)	Normalized Vectors			Standardized Vectors		
						Y1	Y2	Y3	Y'1	Y'2	Y'3
1	Zn	142	135	164	133	−.00042	−.00058	0.00005	−0.1133	−0.1568	0.0137
2	Fe	224	274	246	166	0.00053	−.00049	0.00045	0.9547	−1.8794	0.8017
3	Mn	5.7	15	6.0	5.0	−.01020	0.00951	−.00523	−1.2069	0.0012	−1.6188
4	Mg	345	379	342	326	−.00047	−.00008	−.00005	−1.0283	−1.1764	−1.1064
5	Ca	1303	951	1720	1046	0.00007	0.00014	−.00010	0.4341	0.9316	−1.6694
6	Na	313	330	359	379	0.00046	0.00016	0.00039	0.7009	0.2518	0.6065
7	Al	100	114	113	64	0.00054	0.00036	−.00019	0.4700	0.3126	−1.1697
8	Cu	9.2	11	11	8.8	−.00307	0.00102	−.00117	−1.1250	0.0414	−1.0476
9	P	200	333	182	220	−.00003	0.00002	−.00004	−1.0414	0.0370	−1.0612
10	K	54	58	51	64	−.00049	0.00008	−.00028	−1.1457	0.0228	−1.0849
11	Fe/Zn	1.6	2.1	1.5	1.3	−.07215	0.05622	−.04489	−.9795	0.7632	−1.6095
12	Mn/Zn	0.042	0.12	0.035	0.040	0.92427	−.62844	0.24022	0.9804	−1.6666	0.2548
13	Mg/Zn	2.5	2.9	2.0	2.5	0.04077	0.00069	0.01667	0.6886	0.0116	0.2815
14	Ca/Zn	9.1	7.0	10	7.8	−.00419	−.02160	0.01081	−1.1925	−1.9932	0.4972
15	Na/Zn	2.2	2.5	2.1	2.9	−.03500	−.00423	−.05462	−1.3602	−1.0435	−1.5621
16	Al/Zn	0.71	0.88	0.66	0.51	−.00258	−.03448	0.02734	−1.0166	−1.2218	0.1759
17	K/Zn	38	32	32	18	−.00019	−.00020	−.00012	−1.0628	−1.0666	−1.0391
18	Mn/Fe	0.031	0.058	0.023	0.032	−.00056	−.42080	0.77469	−1.0002	−1.1853	0.3412
19	Mg/Fe	1.8	1.8	1.5	2.2	0.01665	0.00256	−.01875	0.1814	0.0279	−1.2043
20	Ca/Fe	6.5	5.0	8.1	7.2	−.00240	0.00312	0.00753	−1.0898	0.1168	0.2818
21	Na/Fe	1.8	1.7	1.8	2.8	0.00064	−.00808	−.01760	0.0091	−1.1156	−1.2518
22	Al/Fe	0.45	0.47	0.43	0.40	−.10849	−.05593	0.07711	−1.2357	−1.1215	0.1675
23	K/Fe	23	17	21	14	0.00074	0.00002	−.00069	0.0944	0.0021	−1.0876
24	Mg/Mn	90	54	73	90	−.00009	−.00134	−.00055	−1.0478	−1.7236	−1.2937
25	Ca/Mn	396	200	393	376	−.00006	0.00018	0.00004	−1.1746	0.5237	0.1215
26	Na/Mn	94	59	84	126	−.00018	0.00020	0.00031	−1.1482	0.1706	0.2628
27	Al/Mn	26	14	20	19	−.00137	−.00079	−.00038	−1.2225	−1.1289	−1.0610
28	K/Mn	1328	579	992	722	0.00004	0.00003	0.00001	0.3986	0.2574	0.0559
29	Ca/Mg	4.2	3.2	5.4	3.7	0.01252	−.00375	−.01107	0.2610	−1.0782	−1.2309
30	Na/Mg	1.0	1.0	1.1	1.4	−.04236	−.03721	−.03289	−1.2631	−1.2311	−1.2043
31	Al/Mg	0.30	0.33	0.32	0.21	0.21475	−.12421	0.26890	0.5247	−1.3035	0.6571
32	K/Mg	16	12	15	7.7	−.00646	0.00148	−.00441	−1.8090	0.1853	−1.5520
33	Na/Ca	0.30	0.44	0.21	0.43	0.00642	−.14333	0.09892	0.0158	−1.3525	0.2433
34	Al/Ca	0.085	0.14	0.060	0.074	−.27134	0.61830	−.49557	−1.2315	0.5275	−1.4228
35	K/Ca	4.2	5.0	2.9	2.4	0.00876	−.00651	0.00600	0.3360	−1.2495	0.2302
36	Al/Na	0.42	0.46	0.32	0.20	−.05989	−.01223	−.01918	−1.2680	−1.0547	−1.0858
37	K/Na	22	17	15	7.2	0.00054	−.00101	0.00080	0.1217	−1.2260	0.1792
38	K/Al	51	38	48	36	−.00011	−.00064	0.00105	−1.0290	−1.1739	0.2835

Function	Y1	Y2	Y3
Eigenvalue	1.3264	0.6681	0.3602
Contribution (%)	56.3	28.4	15.3
χ^2	103	62	38
df	40	38	36
P	<.001	<.01	<.5

14

The soil samples studied fall into two general categories based on their origins: 1) those traceable ultimately to carbonate terranes and subsequently derived from glacial till, marine sediments, or both; and 2) glacial tills or deltaic sediments derived from igneous and metamorphic terranes (Table 7).

As would be expected, soils on the carbonate terranes are significantly higher in calcium and magnesium than are those derived from Canadian Shield areas. Conversely, the significantly higher levels of sodium and potassium in soils derived from Precambrian rocks of the Canadian Shield as compared with those of soils on sedimentary limestone areas reflect the markedly higher levels of these elements in the noncarbonate rocks (Table 7). Levels of such important nutrients as phosphorus and iron did not differ between terranes in our limited sampling. The highest iron value was recorded for the sample from the McConnell River delta, a finding which relates to the deposit of silicious iron found by Lord (1953) thirty miles up the McConnell River (see Figure 71). The Cape Henrietta Maria feather series is also high in iron (Table 7); these high values can very likely be attributed to the transfer of iron-rich sediments from the Sutton Lake hills by the Sutton, Aquatuk, and Kinusheo rivers and an unnamed river (mouth at lat. 55°8′ N, long. 82°42′ W) all of which have their origins or headwaters in or near this outcrop of Precambrian rock (see Figures 11, 48). These hills are noted for their iron-rich beds of Superior-type sedimentary ores (Lang et al. 1970: 170–74) although they are not regarded currently as economical to mine.

Zinc and copper levels are lowest in carbonate rocks and soils derived from them, but the pattern that we found for these trace elements is not consistent between terranes, perhaps because of the limited sampling and the low concentrations derived from the parent rocks. Copper values for the two terranes do not differ significantly, but the higher values for iron, zinc, and copper in a soil sample taken along the Soper River are notable. Manganese is about twice as abundant in soils derived from igneous terranes as it is in soils traceable to carbonate terranes, an indication of its relatively greater concentration in igneous rocks. The pertinence of the few samples that were available for analyses make it all the more regrettable that all areas could not be sampled and characterized in detail.

CHAPTER THREE

Some Soil and Plant Relationships and Biogeochemical Parameters

PLANT RELATIONSHIPS

Although our limited series of plant samples do not permit comparison of mineral levels in the same species of plants from area to area, we believe the findings have considerable biological as well as statistical significance. Levels of all macroelements except magnesium were higher in the plants from the carbonate terranes than they were in the McConnell River series (Table 8). Samples recorded in Table 8 were taken in late July or August, 1966–68; and calculations of means and t tests were made neglecting censored aspects of data for sodium, aluminum, and silicon. We believe the higher levels of nitrogen and phosphorus found in the plants of the Paleozoic Basin area of Hudson Bay are particularly significant and may explain the ability of these areas to support dense populations of blue and lesser snow geese in summer. If species-related differences in nutrient composition do not explain the differences observed, more than likely the relatively high level of these elements in food plants represent recycling of elements from the goose feces as comparable differences were not found for phosphorus in the respective soils. However, nitrogen and phosphorus do not appear to be limiting for the McConnell River colony, which has been rapidly growing and may number a quarter of a million birds at the end of the breeding season. In relation to the size of their feeding range, the density of this colony is still probably well below that of the Cape Henrietta Maria and Baffin Island colonies.

The nitrogen levels of plants growing on carbonate terranes, particularly those from Cape Henrietta Maria and Cape Churchill, are note-

7. Concentrations of selected elements in soils derived from carbonate and igneous rock terranes of the breeding grounds of blue and lesser snow geese in the Hudson Bay region

Geologic Terranne and Locality	Sample Number	Elements (%)				Elements (parts per million)				
		Ca	Mg	Na	K	P	Fe	Zn	Mn	Cu
CARBONATE TERRANES										
Baffin Island, N.W.T.										
(Bluegoose Prairie)	135	9.13	0.80	0.34	1.04	450	7850	19	110	6.3
Cape Henrietta Maria,	118	4.75	1.24	1.02	1.27	558	9870	30	303	8.6
Ont. (Brant River area)	119	5.23	1.58	1.01	1.28	573	9870	35	321	8.2
	120	4.99	1.46	0.86	1.02	668	9870	39	285	7.6
	121	6.16	1.04	1.28	1.28	508	120	19	249	8.6
Cape Churchill, Man.	106	8.39	1.88	0.98	1.04	379	5250	15	127	6.9
(La Pérouse Bay area)	107	1.94	0.78	0.55	0.26	428	3660	7.6	43	3.1
	108	7.32	2.44	1.30	1.06	624	8170	27	144	9.2
	109	0.16	0.09	0.07	0.41	1460	360	12	196	3.7
Kendall Island, N.W.T.	133	4.47	1.71	0.52	1.13	894	360	68	249	15
(Mackenzie River delta)	134	4.85	1.71	0.52	1.12	879	320	59	249	16
Mean		5.22	1.33	0.77	0.99	675	5064	30	207	8.5
Standard error of the mean		0.79	0.19	0.12	0.10	93	1279	5.8	27	1.2
IGNEOUS/METAMORPHIC TERRANES										
Soper River, Baffin	111	1.72	1.01	2.24	2.04	1030	720	59	413	21
Island (inland areas)	112	1.87	0.81	2.04	1.82	902	760	27	515	12
	113	1.75	0.64	2.26	1.90	758	490	51	394	10
	114	2.17	1.04	2.00	1.84	1130	1070	72	624	13
	115	3.40	0.48	1.13	1.26	2110	9870	165	507	110
McConnell River delta, N.W.T.										
(2 mi. inland)	129	0.62	0.37	2.17	1.72	631	12180	23	214	6.9
Seal River delta,										
Man. (1 mi. inland)	127	1.06	0.44	1.66	1.84	713	450	35	431	18
Caribou River delta,										
Man. (1 mi. inland)	128	0.99	0.46	1.98	1.64	713	450	27	376	11
Mean		1.70	0.66	1.94	1.76	998	3249	57	434	25
Standard error of the mean		0.30	0.13	0.13	0.08	170	1712	16	43	12.2
t		.05*	.05*	.01	.01	ns†	na*	ns	.01	na*

* Variances dissimilar.   † ns = not significant.

worthy as they are of the magnitude reported for corn (*Zea mays*, whole plant and leaves) during early development and up to the time of flowering (N = 3.11–3.76, mean range of several thousand plants in Illinois over a three-year period). Similarities in values are notable because modern strains of corn are notable for the efficiency of their nitrogen utilization, having been selected for protein production.

Mean iron values were significantly higher for the McConnell River plants, but the second highest value—and other high values—recorded for plants from Cape Henrietta Maria were in keeping with the high iron content of soil samples from this area (Tables 7–8).

Manganese levels in the plants from the McConnell River delta are consistent with soil-manganese relationships for the two terranes. Both manganese and iron levels were high, as will be shown and discussed, in the primary feathers of some blue and lesser snow geese from the Cape Henrietta Maria and McConnell River areas.

Copper levels were significantly higher in plants from soils overlying carbonates than were those from soils of Canadian Shield origin. Doubtless this anomaly is a product of having plant samples representing only one river basin. The deltaic soils of the McConnell River are apparently low in copper, whereas the soils of the Seal River delta, only 125 miles to the south of the McConnell River, are relatively high in copper (Table 7).

Boron levels in the plants on the carbonate terranes were significantly higher than those on soils derived from the Canadian Shield. This relationship is in keeping with the generally recognized geochemical abundance of boron in different rock types.

Neither silicon nor aluminum varied significantly in plants of the two lithologic areas. Grasses and *Equisetum* typically accumulate silicon, and this is evident from the data in Table 8, the two

highest values being recorded for *Arctagrostris* and *Poa*. Conversely, attention is called to the low silicon values for the halophyte arrow-grass (*Triglochin maritima*), a soft, succulent plant much sought after by Canada geese.

BIOGEOCHEMICAL PARAMETERS OF GOOSE POPULATIONS

Four parameters reflecting the biogeochemistry of geese received attention in the course of the present study: 1) the weight of the ashed feather vane material expressed as a percentage of the oven-dried weight of the vane; 2) the color of the ash; 3) characteristics of the gizzard grit; and 4) the

11. An aerial view of a portion of Sutton Ridge in the vicinity of Aquatuk Lake, northern Ontario. Iron derived from Superior-type formations found in this ridge system are believed to account for high iron levels found in soils, plants, and primaries of some blue and lesser snow geese in the Cape Henrietta Maria colony.

mineral content of the feather ash. Ash weight and ash color proved to be inexact indicators of origins of geese, but these parameters, nevertheless, often reflected some aspect of the ecosystem of the breeding grounds.

Gizzard Grit

It was thought that gizzard grit might prove to be a helpful "biological tracer." The validity of the use of grit as a tracer would, of course, depend on the rapidity with which it was turned over and the number of days between an individual's leaving the breeding grounds and its being shot by a hunter. If only 5 to 10 percent of the grit from the breeding grounds still remained in the gizzard when the bird was shot and if the original grit were distinctive, the usefulness of this approach would be proved. Similarly, if a distinctive grit were

8. Mineral content of plants collected in or near blue or lesser snow goose colony sites in the eastern Canadian Arctic

Geologic Terrane, Locality, and Plant Species	Elements (%)						Elements (parts per million)					Elements (%)		Ash Weight (%
	Ca	Mg	Na	K	N	P	Fe	Zn	Mn	Cu	B	Si	Al	
CARBONATE TERRANES														
Bluegoose Prairie, Baffin Island														
Carex aquatilis var. stans (Drej) Boott	1.08	0.13	719	1.18	1.54	0.22	1260	64	236	14	21	1.96	104	5.
Unidentified sedge	0.90	0.20	1280	0.95	1.35	0.05	244	90	105	20	6	3.64	436	5.
Cape Henrietta Maria, Ont. (Brant River area)														
Scirpus cespitosus L.	0.38	0.11	458	1.76	2.74	0.06	188	44	144	13	14	1.25	66	4.
Scirpus cespitosus L.	0.87	0.16	1360	1.12	—[a]	0.08	809	48	232	18	18	1.65	125	5.
Carex sp.	0.44	0.18	563	2.32	2.29	0.30	634	43	204	12	12	2.70	155	7.
Arctagrostis latifolia (R. Br.) Griseb.	0.97	0.32	413	1.53	2.41	0.13	1486	52	263	10	15	>4.69	555	11.
Eriophorum (E. callitrix Cham.?)	0.24	0.14	413	1.44	2.40	0.40	84	60	177	14	8	<0.20	22	3.
Cape Churchill, Man. (La Pérouse Bay area)														
Scirpus cespitosus var. callosus Bigel	0.26	0.11	281	1.08	1.97	0.04	41	29	340	7	10	0.54	<10	2.
Eleocharis sp.	0.76	0.36	>1400	4.03	2.50	0.38	602	44	140	14	20	2.85	181	17.
Triglochin maritima L.[b]	0.96	0.35	>1400	4.54	3.93	0.23	182	20	58	12	20	<0.20	39	15.
Triglochin maritima L.[b]	0.77	0.28	>1400	2.91	3.79	0.17	47	20	47	10	16	<0.20	25	9.
Triglochin maritima L.[b]	0.56	0.24	>1400	3.34	2.20	0.06	52	17	10	10	30	<0.20	24	8.
Poa sp., Unidentified grass	0.94	0.55	>1400	2.53	2.16	0.36	829	40	176	14	24	4.29	382	13.
Unidentified sedge	0.54	0.26	1256	3.33	3.81	0.30	158	28	107	12	20	1.06	32	8
Mean	0.69[c]	0.24	981[c]	2.29[c]	2.54[c]	0.20[c]	472	43	159[c]	13	17[c]	1.82	154[c]	8.
IGNEOUS/METAMORPHIC TERRANES														
McConnell River delta, N.W.T.														
Carex aquatilis Wahl.	0.27	0.14	284	1.59	1.84	0.09	830	27	376	9	5	1.45	39	4
Carex rotunda Wahl.	0.39	0.11	62	0.34	1.15	0.03	286	16	217	5	5	2.73	24	4
Carex williamsii Britt.	0.52	0.28	518	1.29	1.56	0.06	1224	47	507	10	6	4.14	153	5
Carex sp.	0.39	0.18	581	1.10	1.39	0.08	1590	44	851	10	10	4.14	192	6
Carex sp.	0.35	0.14	203	1.23	1.56	0.06	942	35	365	7	5	3.49	84	4
Eriophorum sp.	0.32	0.13	306	1.13	1.44	0.04	751	36	455	6	5	1.92	78	5
Mean	0.37[d]	0.16	326[d]	1.11[e]	1.49[d]	0.06[d]	937	34	462[e]	8[d]	6[d]	2.98	95	5

[a] No data.   [b] From willow zone east of naval station, Churchill.   [c] Variances dissimilar.
[d] Means significantly different at 1% level.   [e] Means significantly different at 5% level.

picked up during migration, useful information would be gained. For example, if snow geese from Southampton Island or the McConnell River colony fed for a sufficient time at Cape Churchill or Cape Henrietta Maria to pick up new, identifiable grit, this fact should be evident from the presence of the distinctive grit from these latter areas in the gizzards of these geese when shot near waterfowl refuges in the Dakotas or northern Missouri.

A thorough, definitive study of the usefulness of grit as a tracer would be expensive. It would be dependent on large-scale grit collections from all of the known breeding grounds and northern stop-over areas and on laborious, quantitative identification of the rock and mineral species present in the grit. We carried out a preliminary visual inspection of washed and dried grit samples, and the results are suggestive of the potential of this identification method. The characteristics of a series of sixty-five grit samples collected from hunting camps near the mouths of the Attawapiskat, Kapiskau, and Albany rivers indicated that about 85 percent of the geese represented in the kill sample examined from this midsector of the west coast of James Bay originated from the Cape Henrietta Maria colony or had fed there for an appreciable period. If feather ash color and gizzard grit, either alone or together, proved to be useful biological tracers they would have a decided advantage over the use of more costly and complex elemental analyses of the feather vane to determine colony or population of origin.

Ash Weights of Vanes

The most recent review of the integument of birds is that of Stettenheim (1972), but, surprisingly, his thorough and scholarly treatise makes no mention of the mineral content of feathers in his discussion of their chemical composition. The ash weights of the primary-feather vanes of several populations of blue and snow geese were substantial, but colony means did not vary sufficiently to be significantly different for most comparisons among colonies (Table 9). Ash weight values for individual minerals and the diversity of mineral patterns are much more extreme for Canada goose populations than for blue, snow, or Ross' goose populations. For example, the mean percentage of ash for the vanes of primary feathers of Canada geese from south central Yukon was 4.0 (range, 0.99–7.77) and 3.20 (range, 0.67–7.54) for immatures and adults, respectively; for Belcher Islands geese, 2.8 percent (range, 0.82–5.85); and for molting geese taken off Akimiski Island in James Bay, 3.03 percent (range, 1.25–5.38), the Yukon Canada geese exceeding in this respect all other populations of North American geese studied to date.

Ash Color of Vanes

The color of the ash of the feather vane was recorded (Munsell system) after it was noted that there were marked color differences between some samples. These color differences are related to the relative amounts of the various oxides, iron oxides, particularly, contributing red hues. These color differences could be used to distinguish geese from the various colonies, but preliminary computer analyses of the relation of color to mineral values failed to be significant. Nevertheless, there were, for example, some marked differences in the frequency of color values between the various colonies nesting in the Hudson Bay region (Table 10). The darkest ash colors were recorded

9. Ash weights of vanes of primary feathers (percent of dry weight) of blue, lesser snow, and Ross' geese from major colonies in Canada and the U.S.S.R.

Colony	n	Mean (%)	Standard Deviation	Range	V
Baffin Island, N.W.T.	59	0.83 ± .08	0.63	0.27–3.25	76
Southampton Island, N.W.T.	23	0.62 ± .06	0.29	0.16–1.77	60
Cape Henrietta Maria, Ont.	24	0.85 ± .08	0.40	0.39–1.84	82
McConnell River, N.W.T.	38	0.75 ± .07	0.42	0.25–2.22	68
Cape Churchill, Man.	4	0.80 ± .12	0.25	0.51–1.02	125
Karrak Lake, N.W.T. (Ross' geese)	20	1.32 ± .05	0.24	0.99–1.75	54
Banks Island, N.W.T.	3	0.92 ± .09	0.16	0.73–1.04	92
Anderson River, N.W.T.	5	0.55 ± .08	0.19	0.32–0.79	85
Kendall Island, N.W.T.	7	0.87 ± .05	0.13	0.68–1.02	49
Wrangel Island, U.S.S.R.	17	0.80 ± .13	0.52	0.18–2.38	126

10. Prevalence of ash colors of primary feathers of blue and lesser snow geese from four colonies in the Hudson Bay area, Wrangel Island, U.S.S.R., and Ross' geese from Karrak Lake, N.W.T.

Ash Color			Baffin Island		Cape Henrietta Maria		Southampton Island		McConnell River		Wrangel Island		Karrak Lake	
Hue	Value	Chroma	n	%	n	%	n	%	n	%	n	%	n	%
2.5	5	4							1	2.6				
2.5	6	6							1	2.6				
Subtotal or average									2	5.2				
5	5	2			1	4.2								
5	6	4							2	5.3				
5	8	4							2	5.3				
Subtotal or average					1	4.2			4	10.6				
7.5	4	2	2	3.4			1	4.4					2	10.0
7.5	5	2	6	10.2	2	8.3							9	45.0
7.5	5	4			1	4.2			2	5.3	1	5.9		
7.5	6	2	2	3.4	2	8.3	1	4.4					4	20.0
7.5	6	4	5	8.5	2	8.3			2	5.3			3	15.0
7.5	7	2			1	4.2							1	5.0
7.5	7	4	13	22.0	1	4.2	4	17.4	11	28.9	5	29.4		
7.5	7	6			1	4.2	1	4.4	1	2.6				
7.5	8	2	1	1.7	1	4.2	2	8.7			2	11.7		
7.5	8	4	24	40.6	5	20.6	7	30.1	10	26.3	5	29.4	1	5.0
7.5	8	6					1	4.4	1	2.6	1	5.9		
Subtotal or average			53	89.8	16	66.5	17	73.8	27	71.0	14	82.3	20	100.0
10	3	2									1	5.9		
10	5	1	1	1.7										
10	6	3			1	4.2								
10	7	4									2	11.8		
10	8	2			1	4.2								
10	8	3			1	4.2	1	4.4						
10	8	4	1	1.7	1	4.2	1	4.4	2	5.3				
10	8.5	4	4	6.8	1	4.2	4	17.4	3	7.9				
10	8	6			2	8.3								
Subtotal or average			6	10.2	7	29.3	6	26.2	5	13.2	3	17.7		
Total			59	100.0	24	100.0	23	100.0	38	100.0	17	100.0	20	100.0

for geese banded in the McConnell River area, presumably reflecting the relatively high amounts of iron or manganese found in the primary vanes of some of these geese. The feather ash of the Cape Churchill geese was concentrated notably in the 10 YR or yellow-red hue. Thus, ash color of the primaries was at least suggestive of the origins of some geese. The distribution of ash colors for the Baffin Island and Southampton Island geese can be considered identical, which is consistent with aforementioned difficulties in attempts to separate these colonies on the basis of the mineral content of their primaries. In contrast, values of ash colors for geese from the Cape Henrietta Maria colony were scattered throughout the recorded ash-color spectrum.

FEATHER MINERAL PATTERNS—INTRACOLONY VARIATION

As numerous intracolony variations in feather mineral patterns were found for most colonies, it was most fortunate that significant intercolony differences were also found. Variations in mineral patterns were found that could be related to the age and sex of the individual and to subdivisions of the range in which the bird matured or molted (i.e., mineral differences in the ecosystem that in turn relate to distinct differences in geology and drainage patterns). In addition, it was found that the partially developed primaries of geese collected on the breeding grounds might not provide the same mineral patterns as would the completely grown feathers of geese that had been banded in the same area and shot elsewhere in the fall as free-flying individuals.

Differences Associated with Age

Our analyses of differences in feather minerals related to age were restricted to geese banded on Bluegoose Prairie, Baffin Island, and on Cape Henrietta Maria, Ontario, and later shot by hunters. It was hoped that variations in mineral profiles related to intercolony variations in the ecosystem would not wholly mask differences related to age. In adult geese, differences associated with sex which result from the differential effects of the sex hormones on mineral metabolism could further obscure the significance of any comparison between age classes (see Hanson and Jones 1974 and the discussion of copper metabolism in chapter 7).

Values for potassium, iron, silicon, and aluminum and ash weight were significantly higher for adults banded on the Bluegoose Prairie than they were for immatures (Figure 12). The relationship was the reverse for the Cape Henrietta Maria

geese; for these samples, values for the immatures were significantly higher in respect to iron, aluminum, and silicon, and ash weight (Figure 13). The small size (four) of the adult sample casts doubt on the biological validity of this comparison despite statistical significance.

The data used to represent the Churchill colony in our preliminary study (Hanson and Jones 1968) were based on partially grown feathers from immatures and adults. The series of five goslings and eleven adults differed significantly only in respect to phosphorus, but both series were significantly higher in respect to sodium, potassium, iron, zinc, copper, and aluminum from the four adults banded on Cape Churchill and shot there as flying birds in September (Figures 14, 93). Fairly similar differences between the mineral content of primaries in growth stages and mineral values for reference feathers from banded immatures and adults shot away from the breeding grounds were also exhibited by blue lesser snow geese from the McConnell River (Figure 15) and the Anderson River colonies (Figure 16). No significant differences

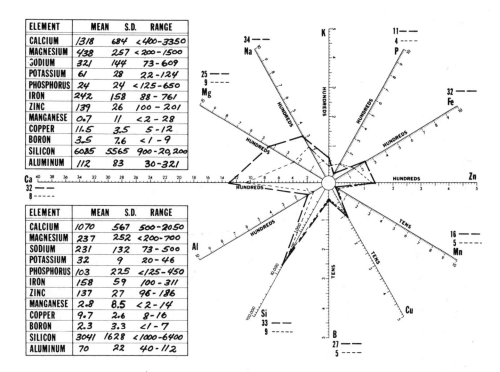

ELEMENT	MEAN	S.D.	RANGE
CALCIUM	1318	684	<400-3350
MAGNESIUM	438	257	<200-1500
SODIUM	321	144	73-609
POTASSIUM	61	28	22-124
PHOSPHORUS	24	24	<125-650
IRON	242	158	88-761
ZINC	139	26	100-201
MANGANESE	0.7	11	<2-28
COPPER	11.5	3.5	5-12
BORON	3.5	7.6	<1-9
SILICON	6025	5565	900-20,200
ALUMINUM	112	83	30-321

ELEMENT	MEAN	S.D.	RANGE
CALCIUM	1070	567	500-2050
MAGNESIUM	237	252	<200-700
SODIUM	231	132	73-500
POTASSIUM	32	9	20-46
PHOSPHORUS	103	225	<125-450
IRON	158	59	100-311
ZINC	137	27	96-186
MANGANESE	2.8	8.5	<2-14
COPPER	9.7	2.6	8-16
BORON	2.3	3.3	<1-7
SILICON	3041	1628	<1000-6400
ALUMINUM	70	22	40-112

12. Feather mineral patterns of immature and adult blue and lesser snow geese banded on the Bluegoose Prairie, Baffin Island, and shot elsewhere. Potassium, iron, silicon, and aluminum values are significantly higher in adults (heavy dashed line) than in immatures (n = 35 and 10 respectively).

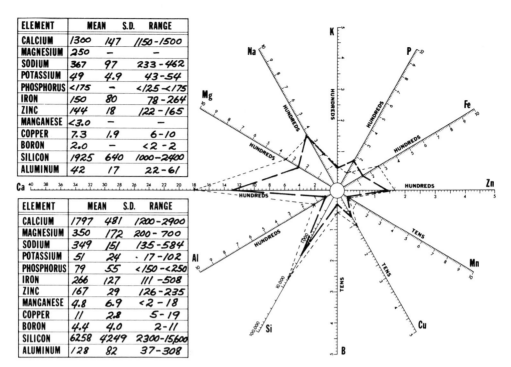

13. Feather mineral patterns for immature and adult blue and lesser snow geese banded on Cape Henrietta Maria, Ontario (n = 19 and 4 respectively). Immatures had significantly higher values for iron, aluminum, and silicon.

ELEMENT	MEAN	S.D.	RANGE
CALCIUM	1300	147	1150-1500
MAGNESIUM	250	-	-
SODIUM	367	97	233-462
POTASSIUM	49	4.9	43-54
PHOSPHORUS	<175	-	<125-<175
IRON	150	80	78-264
ZINC	144	18	122-165
MANGANESE	<3.0	-	-
COPPER	7.3	1.9	6-10
BORON	2.0	-	<2-2
SILICON	1925	640	1000-2400
ALUMINUM	42	17	22-61

ELEMENT	MEAN	S.D.	RANGE
CALCIUM	1797	481	1200-2900
MAGNESIUM	350	172	200-700
SODIUM	349	151	135-584
POTASSIUM	51	24	17-102
PHOSPHORUS	79	55	<150-<250
IRON	266	127	111-508
ZINC	167	29	126-235
MANGANESE	4.8	6.9	<2-18
COPPER	11	2.8	5-19
BORON	4.4	4.0	2-11
SILICON	6258	4249	2300-15,600
ALUMINUM	128	82	37-308

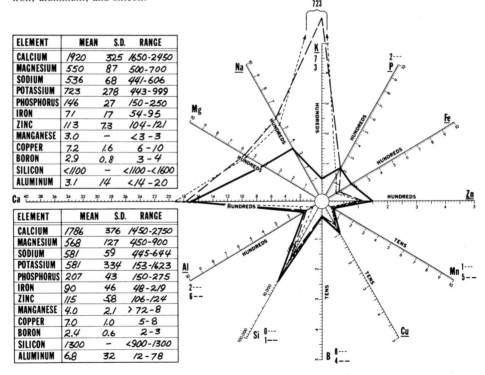

14. Feather mineral patterns of partially developed primaries of gosling and adult blue and lesser snow geese from Cape Churchill, Manitoba (n = 5 and 11 respectively). Solid line pattern depicts mineral values for four adults banded on Cape Churchill and shot in September.

ELEMENT	MEAN	S.D.	RANGE
CALCIUM	1920	325	1650-2450
MAGNESIUM	550	87	500-700
SODIUM	536	68	441-606
POTASSIUM	723	278	443-999
PHOSPHORUS	146	27	150-250
IRON	71	17	54-95
ZINC	113	7.3	104-121
MANGANESE	3.0	-	<3-3
COPPER	7.2	1.6	6-10
BORON	2.9	0.8	3-4
SILICON	<1100	-	<1100-<1600
ALUMINUM	3.1	14	<14-20

ELEMENT	MEAN	S.D.	RANGE
CALCIUM	1786	376	1450-2750
MAGNESIUM	568	127	450-900
SODIUM	581	59	445-644
POTASSIUM	581	334	153-1623
PHOSPHORUS	207	43	150-275
IRON	90	46	48-219
ZINC	115	58	106-124
MANGANESE	4.0	2.1	>72-8
COPPER	7.0	1.0	5-8
BORON	2.4	0.6	2-3
SILICON	1300	-	<900-1300
ALUMINUM	6.8	32	12-78

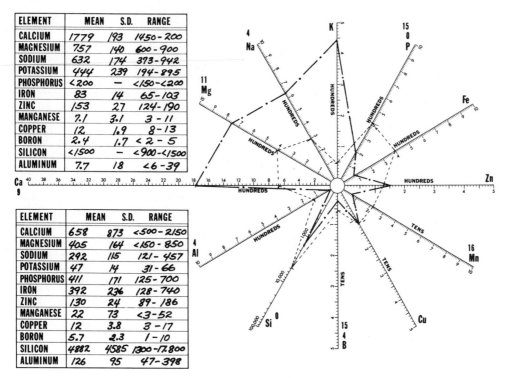

ELEMENT	MEAN	S.D.	RANGE
CALCIUM	1779	193	1450-200
MAGNESIUM	757	140	600-900
SODIUM	632	174	373-942
POTASSIUM	444	239	194-895
PHOSPHORUS	<200	–	<150-<200
IRON	83	14	65-103
ZINC	153	27	124-190
MANGANESE	7.1	3.1	3-11
COPPER	12	1.9	8-13
BORON	2.4	1.7	<2-5
SILICON	<1500	–	<900-<1500
ALUMINUM	7.7	18	<6-39

ELEMENT	MEAN	S.D.	RANGE
CALCIUM	658	873	<500-2150
MAGNESIUM	405	164	<150-850
SODIUM	292	115	121-457
POTASSIUM	47	14	31-66
PHOSPHORUS	411	171	125-700
IRON	392	236	128-740
ZINC	130	24	89-186
MANGANESE	22	73	<3-52
COPPER	12	3.8	3-17
BORON	5.7	2.3	1-10
SILICON	4882	4585	1300-17.800
ALUMINUM	126	95	47-398

15. Feather mineral patterns of partially developed primaries of goslings from the McConnell River colony and from immatures banded there and shot elsewhere (n = 5 and 17 respectively).

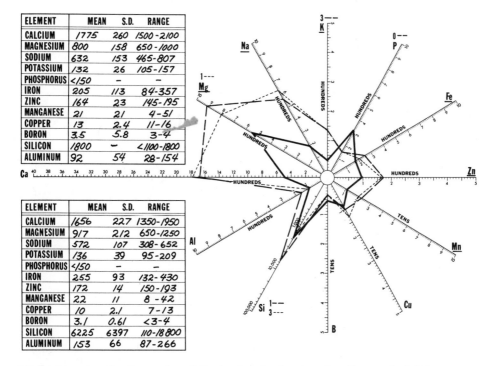

ELEMENT	MEAN	S.D.	RANGE
CALCIUM	1775	260	1500-2100
MAGNESIUM	800	158	650-1000
SODIUM	632	153	465-807
POTASSIUM	132	26	105-157
PHOSPHORUS	<150	–	–
IRON	205	113	84-357
ZINC	164	23	145-195
MANGANESE	21	21	4-51
COPPER	13	2.4	11-16
BORON	3.5	5.8	3-4
SILICON	1800	–	<1100-1800
ALUMINUM	92	54	28-154

ELEMENT	MEAN	S.D.	RANGE
CALCIUM	1656	227	1350-1950
MAGNESIUM	917	212	650-1250
SODIUM	572	107	308-652
POTASSIUM	136	39	95-209
PHOSPHORUS	<150	–	–
IRON	255	93	132-430
ZINC	172	14	150-193
MANGANESE	22	11	8-42
COPPER	10	2.1	7-13
BORON	3.1	0.61	<3-4
SILICON	6225	6397	110-18800
ALUMINUM	153	66	87-266

16. Feather mineral patterns of partially developed primaries of goslings and adult lesser snow geese from the Anderson River delta, District of Mackenzie (n = 4 and 9 respectively). Solid line pattern depicts pattern of five adults banded at the Anderson River and shot elsewhere.

23

appeared between the feather mineral contents of the goslings and molting adults in the latter colony, but one or both of the latter series differed significantly from the reference feathers of geese banded there and shot in autumn in respect to sodium, potassium, iron, zinc, and manganese, and ash weight.

Differences Associated with Sex

The only series of flight feathers available to us from geese of known age and sex that were relatively homogeneous in their nutritional experience during the molt period of the previous year were from a collection of Ross' geese made by John P. Ryder at Karrak Lake, N.W.T., between mid-June and 9 July 1968. The adult sex classes of these geese as well as those of a small series from California differed significantly in respect to copper and sodium, values for the females being higher for both elements than those for males (Figure 17). The significance of these differences is treated under the respective elements discussed in Chapter 7.

Use of Molted Primaries

The mineral content of freshly molted primaries would seemingly be representative of the feather mineral content of all geese of a colony, but we advise that such samples be treated with caution. Perhaps freshly molted primaries should not be used in cases where there is not substantial evidence that the history of the geese that dropped the primaries was similar in the previous year to the history of geese that undergo the molt in any given sector of the colony area. Data for primaries found molted near the Brant River, Cape Henrietta Maria, Ontario, are given in Figure 18 and compared with growing primaries sampled in two years. The molted feathers differed significantly in mineral content from samples of growing primaries for one or both years in respect to calcium, magnesium, sodium, phosphorus, iron, copper, boron, and aluminum. Also, little resemblance occurred between the pattern for molted primaries and that of feathers from banded geese shot in the fall (Figures 18, 52–59).

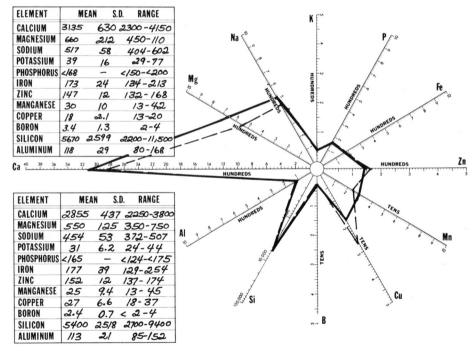

ELEMENT	MEAN	S.D.	RANGE
CALCIUM	3135	630	2300-4150
MAGNESIUM	660	212	450-110
SODIUM	517	58	404-602
POTASSIUM	39	16	29-77
PHOSPHORUS	<168	–	<150-<200
IRON	173	24	134-213
ZINC	147	12	132-168
MANGANESE	30	10	13-42
COPPER	18	2.1	13-20
BORON	3.4	1.3	2-4
SILICON	5670	2599	2200-11,500
ALUMINUM	118	29	80-168

ELEMENT	MEAN	S.D.	RANGE
CALCIUM	2855	437	2250-3800
MAGNESIUM	550	125	350-750
SODIUM	454	53	372-507
POTASSIUM	31	6.2	24-44
PHOSPHORUS	<165	–	<124-<175
IRON	177	39	129-254
ZINC	152	12	137-174
MANGANESE	25	9.4	13-45
COPPER	27	6.6	18-37
BORON	2.4	0.7	< 2-4
SILICON	5400	2518	2700-9400
ALUMINUM	113	21	85-152

17. Feather mineral patterns of adult males and adult females collected at Karrak Lake, N.W.T. (n = 10 for both samples). Adult females were significantly higher than males in respect to copper and sodium.

Differences Associated with Feather Growth

We initially believed that the sheath-free portions of growing primaries from geese collected on their breeding grounds would represent a colony as fully as fledged birds shot in the hunting season. However, we subsequently learned that the sheath material around the opening feathers could contaminate the vane portion of the primaries and bias the analytical results. To gain insight into the relative usefulness of partially developed feathers as reference material, the following samples from geese at Cape Henrietta Maria, Ontario, were compared: sheath material, partially grown primaries collected from birds sacrificed on 13 July 1968 (only the vane portions that had emerged from the sheath material and expanded fully were used), and primary feathers from fully feathered birds shot on the Cape on 20 September 1968 (Table 11). Significant differences were found between the feathers of birds in the molt and those of flying birds in respect to calcium, sodium, zinc, and copper. Sodium, potassium, and phosphorus were found to be high in sheath material, but of these, only sodium was higher in the growing feathers that had recently emerged, suggesting a naturally higher level in the newly formed vane.

Year-to-Year Differences

We have shown that partially grown primary feathers should not be used as colony reference material; further, despite the fact that geese may be collected year after year in banding drives in the same general area within the feeding range of a colony, mineral patterns may not be duplicated in all respects from year to year. Data for geese collected in banding drives near the Brant River, Cape Henrietta Maria, in 1968 and 1969 are shown in Figure 18. These patterns differ significantly in respect to sodium, potassium, and copper. We suspect that annual variations are associated with differences in the use by the birds of local feeding areas rather than with important year-to-year changes in the mineral content of the plants ingested.

11. Concentrations of minerals (ppm as weight of dry feathers) in sheath material, composite of primary feathers picked up from ground, primary feathers from molting lesser snow geese, and primary feathers from birds shot on Cape Henrietta Maria shortly before migration

Sample	Ca	Mg	Na	K	P	Fe	Zn	Mn	Cu	B	Si	Al	Ash (%)
Sheath	750	450	>1400	8600	2550	48	47	<5	9	<3	<2000	<10	5.59
Composite	950	200	342	33	<100	74	133	<2	12	<1	<600	<16	0.40
Molting													
2597	1050	200	356	19	<200	84	176	<3	14	<2	<1400	36	0.37
2598	1350	200	330	18	<200	80	160	<3	14	<2	<1300	10	0.49
2599	1400	<250[a]	415	57	<300	115	179	<5	14	<3	<2000	39	0.47
2600	1250	200	318	13	<200	71	170	<4	16	<2	<1400	8	0.31
2601	1250	200	394	12	<200	69	158	<3	16	<2	<1300	9	0.31
Mean	1260	210	363	24		83	169		14.8			20	0.39
Standard error of the mean	60	10	18	8		8	4		0.5			7	0.04
Flying													
3109	1000	200	84	42	<150	79	137	<2	11	2	900	42	0.37
3110	1100	200	232	62	<150	82	127	<2	13	<1	<900	31	0.40
3111	1050	175	164	32	<150	98	136	<2	10	<2	<1000	37	0.34
3112	950	150	91	37	<150	82	125	<2	7	<1	<900	34	0.30
3113	1150	200	256	46	<150	92	132	<2	8	2	<900	37	0.38
Mean	1050	185	165	44		87	131		9.8			36	0.36
Standard error of the mean	35	10	35	5		4	2		1.1			2	0.02
t	<.05	ns[b]	<.01	ns		ns	<.01		<.01			ns[c]	ns

Note: Differences between means for molting and flying birds were tested for significant separation.
[a] Considered 250 ppm in t calculation.   [b] ns = not significant.   [c] Variances dissimilar.

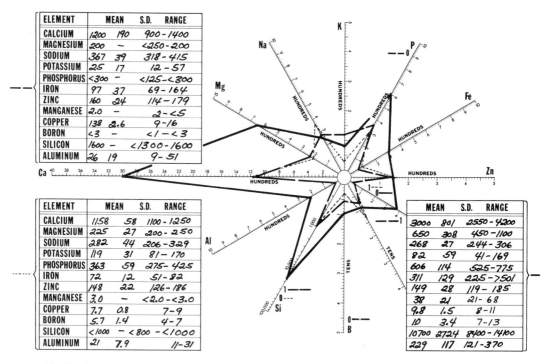

18. Feather mineral patterns for partially developed primaries of blue and lesser snow geese collected in 1968 and 1969 near the Kinapuwao (Brant) River, Cape Henrietta Maria, Ontario (n = 6 for both series). Outlying pattern represents mineral values for four random collections of molted feathers.

ELEMENT	MEAN	S.D.	RANGE
CALCIUM	1200	190	900-1400
MAGNESIUM	200	~	<250-200
SODIUM	367	39	318-415
POTASSIUM	25	17	12-57
PHOSPHORUS	<300	-	<125-<300
IRON	97	37	69-164
ZINC	160	24	114-179
MANGANESE	2.0	-	2-<5
COPPER	138	2.6	9-16
BORON	<3	-	<1-<3
SILICON	1600	-	<1300-1600
ALUMINUM	26	19	9-51

ELEMENT	MEAN	S.D.	RANGE
CALCIUM	1158	58	1100-1250
MAGNESIUM	225	27	200-250
SODIUM	282	44	206-329
POTASSIUM	119	31	81-170
PHOSPHORUS	363	59	275-425
IRON	72	12	51-82
ZINC	148	22	126-186
MANGANESE	3.0	-	<2.0-<3.0
COPPER	7.7	0.8	7-9
BORON	5.7	1.4	4-7
SILICON	<1000	-	<800-<1000
ALUMINUM	21	7.9	11-31

MEAN	S.D.	RANGE
3000	801	2550-4200
650	308	450-1100
268	27	244-306
82	59	41-169
606	114	525-775
311	129	225->501
149	28	119-185
38	21	21-68
9.8	1.5	8-11
10	3.4	7-13
10700	2724	8400-14100
229	117	121-370

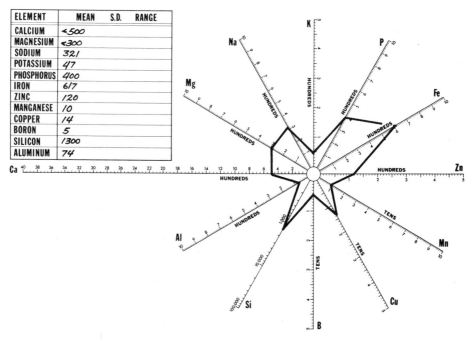

ELEMENT	MEAN	S.D.	RANGE
CALCIUM	<500		
MAGNESIUM	<300		
SODIUM	321		
POTASSIUM	47		
PHOSPHORUS	400		
IRON	617		
ZINC	120		
MANGANESE	10		
COPPER	14		
BORON	5		
SILICON	1300		
ALUMINUM	74		

19. Feather mineral pattern of lesser snow goose banded near the McConnell River, N.W.T., and shot the same year near the mouth of the Severn River, Ontario.

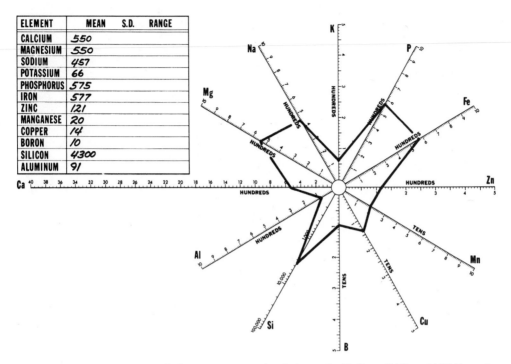

ELEMENT	MEAN	S.D.	RANGE
CALCIUM	550		
MAGNESIUM	550		
SODIUM	457		
POTASSIUM	66		
PHOSPHORUS	575		
IRON	577		
ZINC	121		
MANGANESE	20		
COPPER	14		
BORON	10		
SILICON	4300		
ALUMINUM	91		

20. Feather mineral pattern of a lesser snow goose banded near the McConnell River, N.W.T., and shot the same year near the mouth of the Pipawatin River, Ontario. The mouth of the Pipawatin River is four miles west of the Severn River delta.

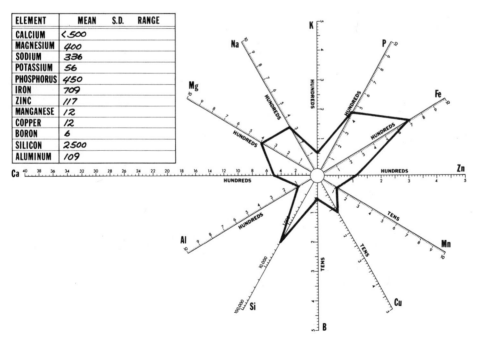

ELEMENT	MEAN	S.D.	RANGE
CALCIUM	<500		
MAGNESIUM	400		
SODIUM	336		
POTASSIUM	56		
PHOSPHORUS	450		
IRON	709		
ZINC	117		
MANGANESE	12		
COPPER	12		
BORON	6		
SILICON	2500		
ALUMINUM	109		

21. Feather mineral pattern of a lesser snow goose banded near the McConnell River, N.W.T., and shot the same year near the mouth of the Severn River, Ontario.

27

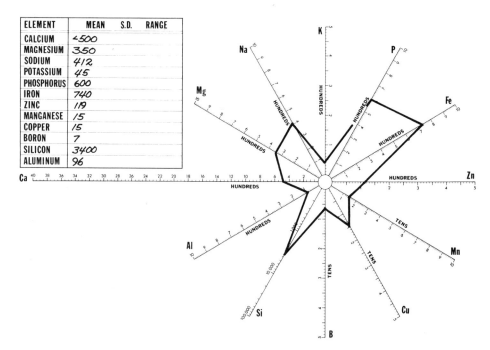

ELEMENT	MEAN	S.D.	RANGE
CALCIUM	<500		
MAGNESIUM	350		
SODIUM	412		
POTASSIUM	45		
PHOSPHORUS	600		
IRON	740		
ZINC	119		
MANGANESE	15		
COPPER	15		
BORON	7		
SILICON	3400		
ALUMINUM	96		

22. Feather mineral pattern of a lesser snow goose banded near the McConnell River, N.W.T., and shot the same year near the mouth of the Pipawatin River, N.W.T.

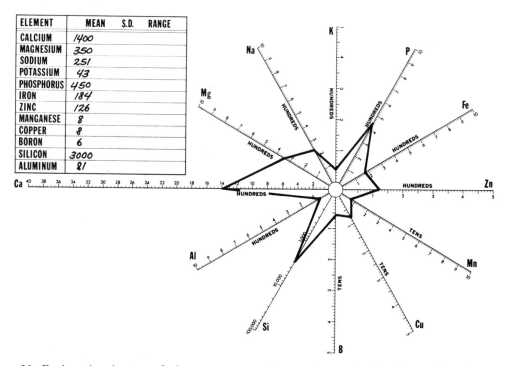

ELEMENT	MEAN	S.D.	RANGE
CALCIUM	1400		
MAGNESIUM	350		
SODIUM	251		
POTASSIUM	43		
PHOSPHORUS	450		
IRON	184		
ZINC	126		
MANGANESE	8		
COPPER	8		
BORON	6		
SILICON	3000		
ALUMINUM	81		

23. Feather mineral pattern of a lesser snow goose shot near the mouth of the Severn River, Ontario.

28

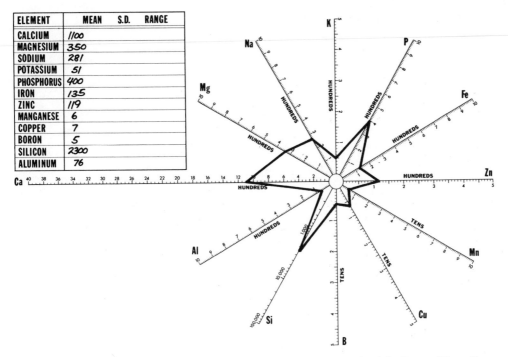

ELEMENT	MEAN	S.D.	RANGE
CALCIUM	1100		
MAGNESIUM	350		
SODIUM	281		
POTASSIUM	51		
PHOSPHORUS	400		
IRON	135		
ZINC	119		
MANGANESE	6		
COPPER	7		
BORON	5		
SILICON	2300		
ALUMINUM	76		

24. Feather mineral pattern of a lesser snow goose shot near the mouth of the Severn River, Ontario.

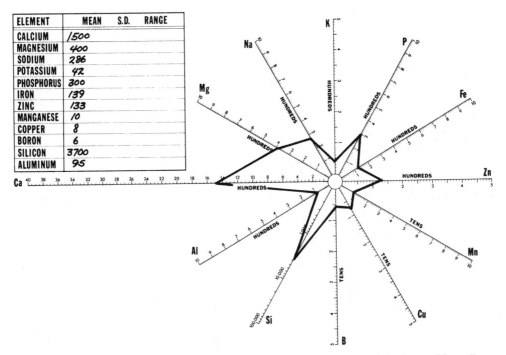

ELEMENT	MEAN	S.D.	RANGE
CALCIUM	1500		
MAGNESIUM	400		
SODIUM	286		
POTASSIUM	42		
PHOSPHORUS	300		
IRON	139		
ZINC	133		
MANGANESE	10		
COPPER	8		
BORON	6		
SILICON	3700		
ALUMINUM	95		

25. Feather mineral pattern of a lesser snow goose shot near the mouth of the Severn River, Ontario.

29

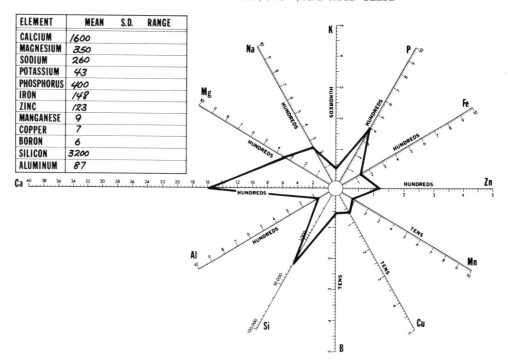

ELEMENT	MEAN	S.D.	RANGE
CALCIUM	1600		
MAGNESIUM	350		
SODIUM	260		
POTASSIUM	43		
PHOSPHORUS	400		
IRON	148		
ZINC	123		
MANGANESE	9		
COPPER	7		
BORON	6		
SILICON	3200		
ALUMINUM	87		

26. Feather mineral pattern of a lesser snow goose shot near the mouth of the Severn River, Ontario.

Differences Within Colony Areas

Four of the six colonies of blue and lesser snow geese in the Hudson Bay area are located on Paleozoic limestones. Yet, because these colonies, with the exception of the Cape Churchill colony, are adjacent to areas of Precambrian rocks, primaries having diverse mineral patterns are produced. These subcolony differences are discussed under each colony.

FEATHER MINERAL PATTERNS—SIMILARITIES

Within Species

If undue emphasis appears to have been given to intracolony pattern variations, it should be stressed that the similarity of patterns within subcolony groupings has demonstrated repeatedly the strength of the feather mineral technique for indicating the origins of geese. Two series of examples can be given: 1) groups of similar feather mineral patterns from geese of known origin shot in the same area the same year (Figures 19–22), and 2) a series of similar patterns from geese banded on unknown breeding grounds, but also shot in the same area the same year (Figures 23–26). Both series, from geese shot on the coast of Hud-

son Bay in the vicinity of Ft. Severn, are believed to represent groups of geese that associated together in the same flock on the breeding grounds (McConnell River, in the case of the first series), had similar nutritional experiences, and subsequently migrated together. (If space permitted, examples of extended series of nearly identical feather mineral patterns could be given. Notable in this respect are a series of patterns for black brant, *Branta bernicla nigricans,* from Padulla Bay, Washington. Clearly, this population originates from a single locality.)

Between Species

The question that commonly arises in discussions about the utility of the feather mineral technique is whether, provided the same experimental diet, two species of waterfowl would show the same feather mineral pattern. Part of the answer and the results of such a test would depend on whether all components of the diet were actually ingested. Given dry foods, ducks commonly swill each beakful in water. Which particles, representing different grains from different soils, are actually eaten probably differs from species to species because of differences in bill structure and food preference (Goodman and Fisher 1962). Differences in their digestive tracts would further mitigate against diverse species of ducks absorbing the same components of an identical diet.

In the Arctic it would not be surprising to find that Ross' geese tend to feed on slightly more weakly structured or younger growth stages of grasses and sedges than do the more robust snow geese, but in areas where both feed together it is likely that the food plants used, growing in the same soil, are similar in mineral content. The close relationship of these geese and their similar digestive tracts suggest a priori that if they feed on the same area in the wild, they would absorb similar mineral fractions in their diets.

This conclusion can be inferred from feather mineral patterns of lesser snow and Ross' geese collected in the Rio Grande Valley (Bosque del Apache N.W.R.) although, admittedly, circular reasoning is involved. (National Wildlife Refuge is abbreviated as N.W.R. throughout.) These facts or premises comprise the links in this circular chain of reasoning: 1) the Rio Grande Valley constitutes a unique flyway and wintering ground, highly isolated from all others; 2) the Queen Maud Gulf lowlands in the central Canadian Arctic is the only region in the Arctic where both lesser snow and Ross' geese commonly occur together during the breeding season; 3) the finding that both the lesser snow and Ross' geese from the Rio Grande Valley had similar feather mineral patterns (see Figures 153–54, 215) suggests that both species originated from the same sector of a common breeding ground; and 4) the association of these two species on an isolated wintering ground is, in itself, indicative of a common geographical origin and a probable association, if only loosely, in migration although neither would be dependent on the other for guidance. These premises, in turn, rest on the premise that, if they consumed the same portions of the same plant foods grown on the same soil, two closely related species of waterfowl would produce similar feather mineral patterns. Although the evidence is largely circumstantial, we believe that these inferred relationships will eventually be well established by a banding and marking program.

Inferences can also be made regarding an adult male greater snow goose shot in fall at Ft. Severn, Ontario. This locality is roughly 700 miles to the west of the normal fall migration of most greater snow geese that migrate southward through the midsection of the Ungava-Labrador peninsula. The feather mineral pattern of this goose (Figure 136) is, however, strikingly similar to that of four blue and lesser snow geese banded in the Foxe Plain of Baffin Island. Because Ft. Severn lies within the fall dispersal range of southward-bound blue and lesser snow geese from the Baffin Island colony, it is reasonable to suspect that the greater snow in question underwent the molt in the Foxe Plain of Baffin Island and then joined a segment of this population, more likely a flock of white-phase birds, in its southward migration. Being an adult male, and possibly unmated, these findings and conclusions could more readily explain the occurrence of this greater snow goose far to the west of the normal range of the race than a supposition that it had summered in the area of its birthplace and had merely "strayed" westward in migration.

In the case of two diverse, distantly related species of geese with somewhat different feeding habits, this reasoning would be less valid. For example, Canada geese, which are chiefly grazers, may not ingest as much iron and manganese as do snow and Ross' geese, noted for their root grubbing. These elements, relative to other cations, tend to be more concentrated in roots than in shoots of plants (Sutcliffe 1962:121).

Between Areas

A series of diverse feather mineral patterns from geese originating from an extensive, continuous breeding ground indicate diverse mineral inputs (hence, diverse geologic terranes) into the nutrient chain; conversely, similar feather patterns from two geographically separated breeding areas are indicative of similar geologic terranes. Striking examples of the role that feather mineral patterns can play as biogeochemical indicators are the patterns of lesser snow geese from Jenny Lind Island and Sherman Inlet (Figures 1, 95–96, 137). The bedrock of Jenny Lind Island consists of limestones of uncertain age (Figure 94, and Geological map of Canada, Map 1250A1); the carbonate rocks that occur on the mainland around Sherman Inlet are Ordovician and Silurian in age (Heywood 1961). However, the highly distinctive and similar feather mineral patterns from these two areas, particularly in respect to calcium-magnesium ratios, strongly indicate that the sedimentary rocks of the areas where these geese underwent the molt are of similar type.

CHAPTER FOUR

Geology, Soils, and Feather Mineral Patterns

EASTERN ARCTIC COLONIES

The six major colonies of blue and lesser snow geese in the Hudson Bay area are: 1) the Baffin Island colony bordering Foxe Basin; 2) the Cape Henrietta Maria colony; 3) the East Bay colony located around the apex of East Bay, Southampton Island; 4) the Boas River colony at the apex of the Bay of Gods Mercy, Southampton Island; 5) the McConnell River colony (Figure *1*); and 6) the Cape Churchill colony. The history, color-phase composition, and productivity of the three mainland colonies have recently been reported upon by Hanson et al. (1972). Miscellaneous records of blue and lesser snow geese along the coast of Hudson Bay and on Akimiski Island are also shown in Figures *2* and *3*. The Baffin Island and McConnell River colonies are each comprised of several groups that may be regarded as subcolonies.

Baffin Island Colony

The Baffin Island colony is by far the largest of the six Hudson Bay colonies both in area and number of birds it contains. It has been most recently studied by Kerbes (1969). The geese that leave this colony in autumn may exceed 750,000. The colony occupies the coastal section of the Great Plain of the Koukdjuak and extends from the apex of Bowman Bay to as far north as Taverner Bay, a distance along the coastline of 160 miles, the exact northern limits of the range depending on the phenology of the spring season (Figure *27*). Three subdivisions of the colony may be recognized (Cooch 1961): Bowman Bay, Cape Dominion, and the Koukdjuak River. These distinctions relate

more to physiography than to the distribution of the nesting geese, which is more or less continuous (Figure *27*). In 1966 the senior author was privileged to accompany Richard H. Kerbes on an aerial survey of this colony, and seen from the air, it must surely rank as one of the great sights of the wildlife world. During the entire 160-mile flight, a constant procession of flocks flushed ahead of the plane (Figure *28*).

Geology and Soils. Physiographically, the coastal plain occupied by this colony is part of the Foxe Plain. Ordovician limestones underlie the Great Plain of the Koukdjuak (Figures *29–30*). In the area immediately east of Bowman Bay, an east-facing scarp of crystalline rocks outcrops as a low ridge parallel to the east shore of Bowman Bay (Figures *29–30*). As a result, this area is better drained than the featureless, waterlogged Great Plain of the Koukdjuak (Figure *31*) (Blackadar 1966). There are no descriptions of the soils of the region.

Feather Mineral Patterns. Our study fortunately was carried out during the two seasons of banding in this colony by Richard H. Kerbes of the Canadian Wildlife Service. As a result of banding by Kerbes on the Great Plain of the Koukdjuak and an earlier banding program by Louis Lemieux along the Koukdjuak, wings of fifty-nine banded geese were obtained as basic reference material. The first question that arose in evaluating the feather mineral analyses was whether significant differences existed between samples from geese banded in the three sectors of the breeding grounds. None was found. Hence, values for the colony could be expressed for a single geographical entity (Figure *32*). Nevertheless, the nutritional experiences of the banded individuals must have varied considerably, as it was possible to recognize fifteen feather mineral pattern variations (Figures *33–47*). No pattern represented more than eight geese. Basic similarities between several of the patterns can be noted, the result of averaging what at first appeared to be a group of distinct individual patterns. Nevertheless, we believed that there would be a net gain in understanding (albeit distorted) the nature of variations in these reflections of the mineral ecosystem by our publishing these pattern variations and those for the other colonies. A portion of these Baffin Island variations must be attributed in large measure to sediments transported onto the Foxe Plain from the Precambrian sectors of the watersheds. Some flocks that have nested on the Bluegoose Prairie wander and feed along the more geologically varied south coast of Bowman Bay

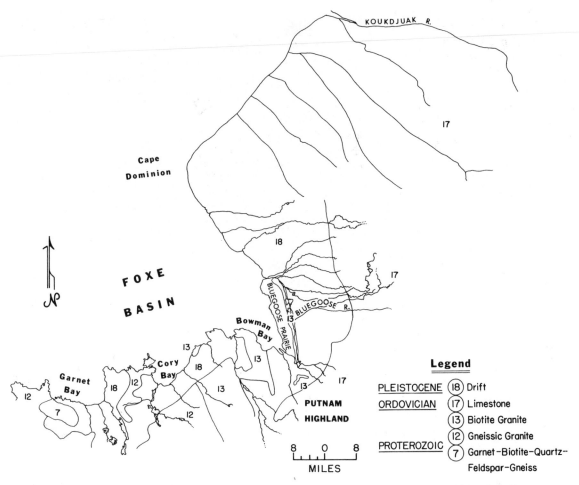

KOUKDJUAK R.

Cape
Dominion

17

FOXE

BASIN

18

17

BLUEGOOSE PRAIRIE
BLUEGOOSE R.
13

Bowman
Bay

13

Cory
Bay
18

13

13

18
12

Garnet
Bay

18

12

12

7

13

12

17

PUTNAM
HIGHLAND

Legend

PLEISTOCENE	(18)	Drift
ORDOVICIAN	(17)	Limestone
	(13)	Biotite Granite
	(12)	Gneissic Granite
PROTEROZOIC	(7)	Garnet–Biotite–Quartz–Feldspar–Gneiss

8   0   8
MILES

27. Geologic setting of the blue and snow goose breeding grounds in the Foxe Plain of Baffin Island.

and up the Aukpar River. In these areas they would be likely to develop diverse mineral patterns.

Cape Henrietta Maria Colony

The geese of the Cape Henrietta Maria colony nest mainly in the vicinity of Kawanabiskak Lake (Figures *48–50*), but minor nesting areas and the feeding range of the flightless adults and goslings extend along Hudson Bay from the Kinapuwao (Brant) River to the base of the Cape (Figure 2). This colony, or at least its dramatic expansion in range and numbers, may date back only as far as the 1950s. In 1973 this colony contained 79,000 adult-plumaged birds (R. H. Kerbes, in press). The history of this colony and those of the Cape Churchill and McConnell River colonies are detailed in Hanson et al. (1972).

Geology and Soils. Cape Henrietta Maria, part of the Hudson Bay Lowlands, is underlain by Silurian limestones, but the rivers that dissect the Cape either traverse or have their headwaters in or near isolated outcrops of Precambrian rocks in the Sutton Ridge (Figures *11, 48*). The isostatic rebound of the area has proceeded at the rate of about 1.2 m per 100 years (Webber et al. 1970). At this rate of uplift the present range of the colony has only been available to nesting geese in the last 250 years (Hanson et al. 1973). Consequently, the soils are derived from recent marine clays. Nearshore sediments are mainly ice-rafted sand and sandy gravel which become progressively finer offshore. The fine sediments are light olive gray in color and consist chiefly of carbonate rock and shale fragments. Calcium carbonate constitutes about 30 percent of the finer sediments as compared with 10

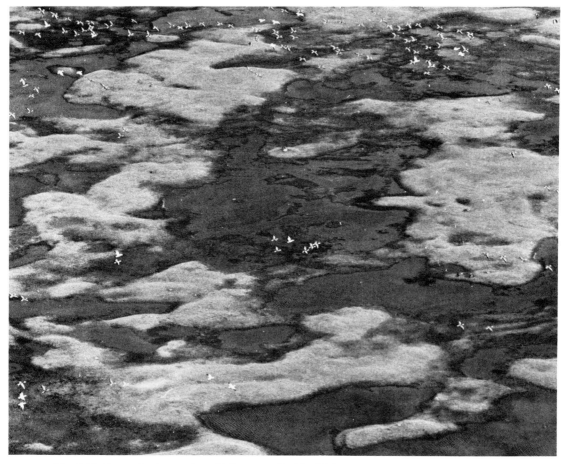

28. A highly productive section of Bluegoose Prairie. Meltwater occupies the areas interspersed among the grassland patches.

percent offshore of the McConnell River colony, 10 to 20 percent offshore of the Cape Churchill colony, and 60 percent off southern Baffin Island (Pelletier et al. 1968:600).

Feather Mineral Patterns. Mean values for the twenty-four samples of reference feathers are given in Figure *51.* The individual feather mineral patterns were sorted into eight groups of like patterns; five of these accounted for twenty of the geese represented (Figures *52–59,* 83 percent of the total sample). Each of the remaining four feather mineral patterns were classified as being distinctive (Figures *56–59*). Surprisingly, all patterns are relatively low in calcium. A basic similarity of most of the patterns can be noted. Several, as often is the case, are merely expanded and contracted versions of the same basic ratios of minerals.

Southampton Island Colonies

Two widely separated colonies of blue and lesser snow geese exist on Southampton Island, one at the apex of East Bay and the other at the mouth of the Boas River (Figures *1, 60*). In 1973, these Southampton colonies contained 208,000 adult-plumaged birds (R. H. Kerbes, in press). The Boas River colony is easily the larger of the two (Figures *61–63*). According to Manning (1942), nesting is confined almost entirely to the grassy islands of the braided, coastal plain estuary. The nesting habitat of the East Bay colony is basically similar to the Boas River colony although not physiographically related to the estuary of a single river. A week after the goslings have hatched, the families at Boas River begin a rapid movement inland (Bray 1943); as a result the minerals in their foods relate entirely

29. The Koukdjuak River, Baffin Island. The northern boundary of the Baffin Island colony lies in the vicinity of this river in most years, its exact location in a given year being related to the northward extent of the largely snow-free area at time of nest initiation.

to a freshwater environment. We assume that the post-nesting movements of the East Bay colony are essentially similar. According to Richard H. Kerbes (personal communication, 1973) the Boas River flocks move as far north as the headwaters of that river. In this area the mineral content of their feeding grounds is more immediately affected by drainage from the adjacent Precambrian rocks (Figure 60).

Geology and Soils. Both colony sites are similar geologically, being underlain by Ordovician and Silurian limestone (Armstrong 1947). The drainage basin of Boas River and the range of the East Bay colony lie mainly within the area of Paleozoic formations; consequently, there is little reason to expect the minerals of the nutrient chain in either area to differ significantly from those in the other

area. The soils of Southampton Island do not appear to have been studied for their mineral content. A rock sample collected by one of us from an area of felsenmeer at Coral Harbour was essentially pure calcium carbonate. Marine sediments off the south coast of Southampton Island reflect the calcareous nature of the region, having a calcium carbonate content of greater than 50 percent (Leslie 1964).

Feather Mineral Patterns. Twenty-three samples of reference feathers from Southampton Island were analyzed, eight from East Bay and fifteen from Boas River. No significant differences were found between averages of these two series. Data for the combined samples are given in Figure 64. Six sub-patterns were distinguished (Figures 65–70). Three are based on single birds and one

30. One of the numerous shallow rivers of the Great Plain of the Koukdjuak. This area was sufficiently snow-free on June 24, 1966, to attract nesting pairs.

(Figure *65*) is representative of thirteen individuals (57 percent of the total sample).

McConnell River Colony

The principal concentration of nesting geese in the McConnell River colony is in the braided estuary of the McConnell River (Figures *1, 3, 71, 72*), but the colony may, in effect, include several subcolonies. Probably scattered nestings also occur along most of the seventy miles of coastal marsh from the Tha-anne River, where blue and lesser snow geese were reported in 1950 by Hawkins et al. (1950), to within less than a mile from the settlement of Eskimo Point, where nests were found in 1968 (Hanson et al. 1972). The principal feeding grounds lie parallel to and probably within five miles of the coast (Figures *72–73*). Additional small colonies are now found to the north at the mouth of the Maguse River and on Austin Island, where approximately three thousand geese were observed by H. G. Lumsden and George W. Swenson in 1964 and where large flocks were photographed by H. C. Hanson and Eugene H. Bossenmaier in 1969 (Figure *71*). Because of the presence of considerable numbers of nesting geese north of Eskimo Point, there is no assurance that blue and lesser snow geese that pass Eskimo Point in fall migration necessarily originate from colonies on Southampton Island. Currently there are at least 595,000 adult-plumaged geese in the McConnell River colony (R. H. Kerbes, in press; Dzubin et al. 1973). This colony may have been initiated as late as the 1930s, but in earlier years probably few people visited this coastal area during summer; hence, the inception of this colony cannot be

31. A section of Bluegoose Prairie adjacent to Bowman Bay. View looking eastward.

known with any exactness (see Hanson et al. 1972 for more detailed information on this colony).

Wings of geese banded at McConnell River were obtained for our study from a number of points on the Ontario coasts of Hudson and James bays, suggesting that many of the geese in the Mc-Connell River colony become widely scattered along the south and west coasts of the bays in the fall. A more complete series of band recoveries, evaluated in relation to kill, indicated that as many as 25 percent of this colony may feed at the south end of James Bay before continuing their southward migration (Table 23). The relative paucity of band recoveries from the Attawapiskat-Kapiskau-Albany area suggests an overflight of this middle sector of the James Bay coast.

Geology and Soils. Acidic, igneous rocks of the Canadian Shield ranging from granite to dacite in composition underlie the west coast of Hudson Bay from the Churchill River northward. South of the Maguse River, however, there are few or no outcrops near the coast (Lord 1953); rather, most of the coastal plain between the Churchill and the Thlewiaza rivers is distinguished by a largely silt-buried boulder field, the northern limits of which mark the southern limits of the coastal feeding range of this colony. The geology and mineralogy of the southern part of the District of Keewatin are still little known, due in part to the scarcity of rock outcrops, although an economic mineral potential is postulated for the area (Lord 1953). For example, a silicious iron formation is known to lie between the McConnell River and one of its tributaries at a distance of thirty miles from the coast (Lord 1953; Gross 1965) (Figure *71*). This occurrence of iron in the drainage area of the McConnell River has

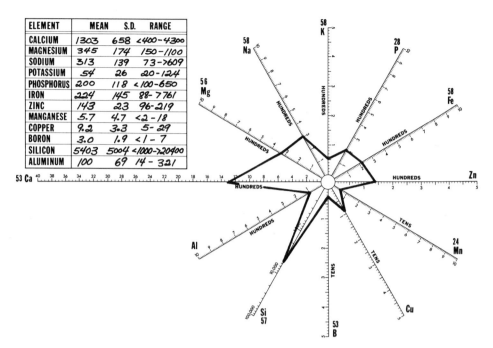

ELEMENT	MEAN	S.D.	RANGE
CALCIUM	1303	658	<400-4300
MAGNESIUM	345	174	150-1100
SODIUM	313	139	73->609
POTASSIUM	54	26	20-124
PHOSPHORUS	200	118	<100-650
IRON	224	145	88-7761
ZINC	143	23	96-219
MANGANESE	5.7	4.7	<2-18
COPPER	9.2	3.3	5-29
BORON	3.0	1.9	<1-7
SILICON	5403	5004	<1000->20400
ALUMINUM	100	69	14-321

32. Feather mineral pattern representing 100 percent of the fifty-nine blue and lesser snow geese banded on the Foxe Plain of Baffin Island and shot elsewhere.

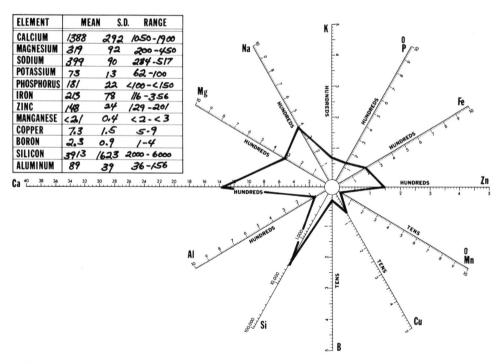

ELEMENT	MEAN	S.D.	RANGE
CALCIUM	1388	292	1050-1900
MAGNESIUM	319	92	200-450
SODIUM	399	90	284-517
POTASSIUM	73	13	62-100
PHOSPHORUS	181	22	<100-<150
IRON	213	78	116-356
ZINC	148	24	129-201
MANGANESE	<2.1	0.4	<2-<3
COPPER	7.3	1.5	5-9
BORON	2.3	0.9	1-4
SILICON	3913	1623	2000-6000
ALUMINUM	89	39	36-156

33. Feather mineral pattern A representing four (7 percent) of fifty-nine blue and lesser snow geese banded on the Foxe Plain of Baffin Island and shot elsewhere.

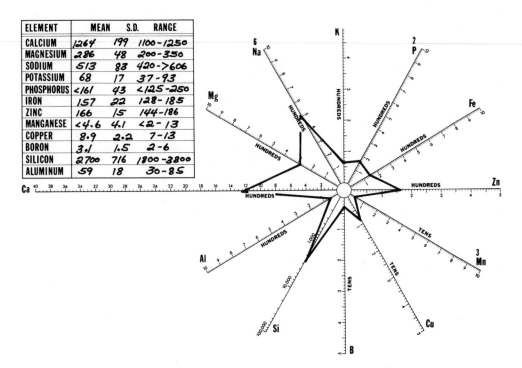

ELEMENT	MEAN	S.D.	RANGE
CALCIUM	1264	199	1100-1250
MAGNESIUM	286	48	200-350
SODIUM	513	83	420->606
POTASSIUM	68	17	37-93
PHOSPHORUS	<161	43	<125-250
IRON	157	22	128-185
ZINC	166	15	144-186
MANGANESE	<4.6	4.1	<2-13
COPPER	8.9	2.2	7-13
BORON	3.1	1.5	2-6
SILICON	2700	716	1800-3800
ALUMINUM	59	18	30-85

34. Feather mineral pattern B representing four (7 percent) of fifty-nine blue and lesser snow geese banded on the Foxe Plain of Baffin Island and shot elsewhere.

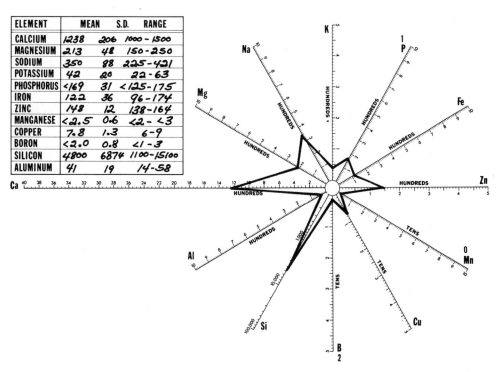

ELEMENT	MEAN	S.D.	RANGE
CALCIUM	1238	206	1000-1500
MAGNESIUM	213	48	150-250
SODIUM	350	88	225-421
POTASSIUM	42	20	22-63
PHOSPHORUS	<169	31	<125-175
IRON	122	36	96-174
ZINC	148	12	138-164
MANGANESE	<2.5	0.6	<2-<3
COPPER	7.8	1.3	6-9
BORON	<2.0	0.8	<1-3
SILICON	4800	6874	1100-15100
ALUMINUM	41	19	14-58

35. Feather mineral pattern C representing six (10 percent) of fifty-nine blue and lesser snow geese banded on the Foxe Plain of Baffin Island and shot elsewhere.

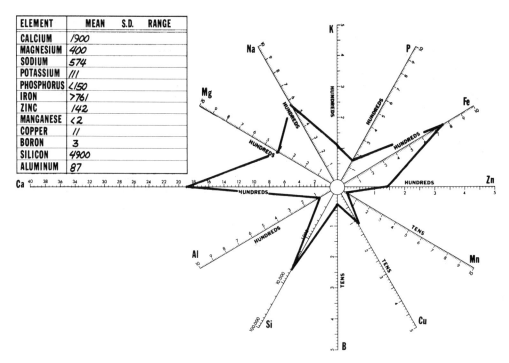

ELEMENT	MEAN	S.D.	RANGE
CALCIUM	1900		
MAGNESIUM	400		
SODIUM	574		
POTASSIUM	111		
PHOSPHORUS	<150		
IRON	>761		
ZINC	142		
MANGANESE	<2		
COPPER	11		
BORON	3		
SILICON	4900		
ALUMINUM	87		

36. Feather mineral pattern D representing four (7 percent) of fifty-nine blue and lesser snow geese banded on the Foxe Plain of Baffin Island and shot elsewhere.

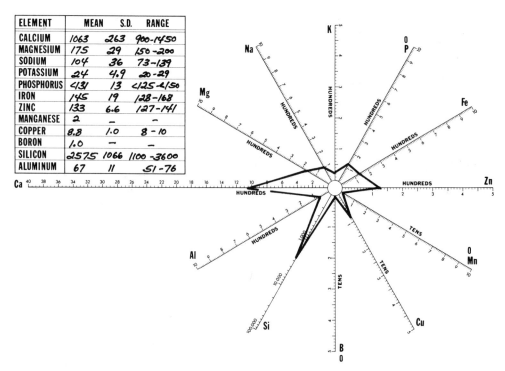

ELEMENT	MEAN	S.D.	RANGE
CALCIUM	1063	263	900-1450
MAGNESIUM	175	29	150-200
SODIUM	104	36	73-139
POTASSIUM	24	4.9	20-29
PHOSPHORUS	<131	13	<125-<150
IRON	145	19	128-168
ZINC	133	6.6	127-141
MANGANESE	2	—	—
COPPER	8.8	1.0	8-10
BORON	1.0	—	—
SILICON	2575	1066	1100-3600
ALUMINUM	67	11	51-76

37. Feather mineral pattern E representing seven (12 percent) of fifty-nine blue and lesser snow geese banded on the Foxe Plain of Baffin Island and shot elsewhere.

40

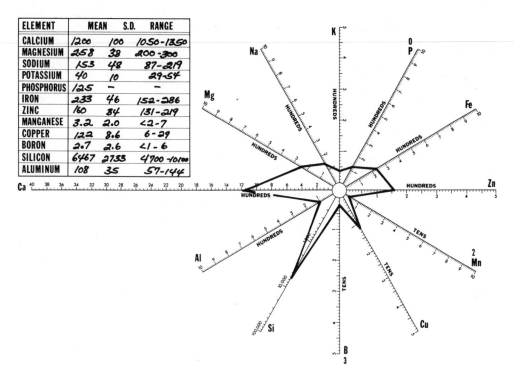

ELEMENT	MEAN	S.D.	RANGE
CALCIUM	1200	100	1050-1350
MAGNESIUM	258	38	200-300
SODIUM	153	48	87-219
POTASSIUM	40	10	29-54
PHOSPHORUS	125	–	
IRON	233	46	152-286
ZINC	160	84	131-219
MANGANESE	3.2	2.0	<2-7
COPPER	122	8.6	6-29
BORON	2.7	2.6	<1-6
SILICON	6467	2733	4900-10100
ALUMINUM	108	35	57-144

38. Feather mineral pattern F representing eight (13 percent) of fifty-nine blue and lesser snow geese banded on the Foxe Plain of Baffin Island and shot elsewhere.

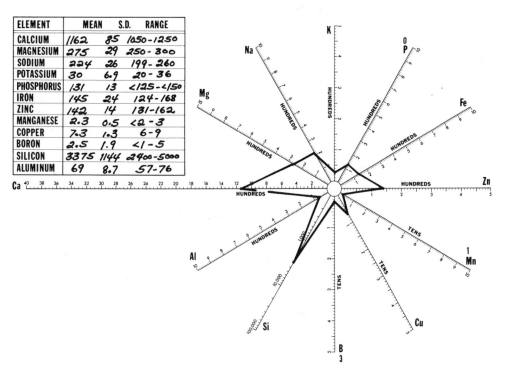

ELEMENT	MEAN	S.D.	RANGE
CALCIUM	1162	85	1050-1250
MAGNESIUM	275	29	250-300
SODIUM	224	26	199-260
POTASSIUM	30	6.9	20-36
PHOSPHORUS	131	13	<125-<150
IRON	145	24	124-168
ZINC	142	14	131-162
MANGANESE	2.3	0.5	<2-3
COPPER	7.3	1.3	6-9
BORON	2.5	1.9	<1-5
SILICON	3375	1144	2400-5000
ALUMINUM	69	8.7	57-76

39. Feather mineral pattern G representing two (3 percent) of fifty-nine blue and lesser snow geese banded on the Foxe Plain of Baffin Island and shot elsewhere.

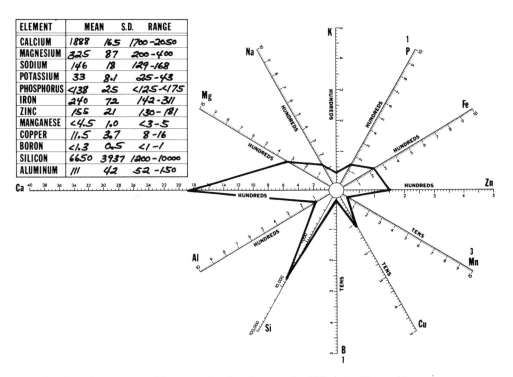

ELEMENT	MEAN	S.D.	RANGE
CALCIUM	1888	16.5	1700-2050
MAGNESIUM	325	87	200-400
SODIUM	146	18	129-168
POTASSIUM	33	8.1	25-43
PHOSPHORUS	<138	25	<125-<175
IRON	240	72	142-311
ZINC	155	21	130-181
MANGANESE	<4.5	1.0	<3-5
COPPER	11.5	3.7	8-16
BORON	<1.3	0.5	<1-1
SILICON	6650	3937	1200-10000
ALUMINUM	111	42	52-150

40. Feather mineral pattern H representing five (8 percent) of fifty-nine blue and lesser snow geese banded on the Foxe Plain of Baffin Island and shot elsewhere.

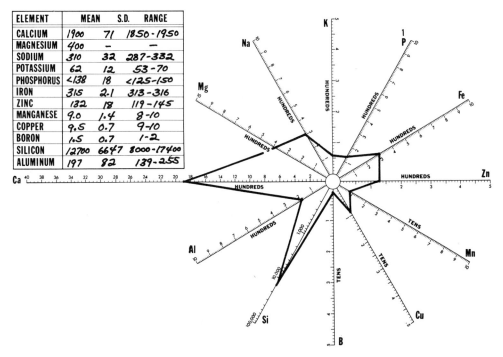

ELEMENT	MEAN	S.D.	RANGE
CALCIUM	1900	71	1850-1950
MAGNESIUM	400	—	—
SODIUM	310	32	287-332
POTASSIUM	62	12	53-70
PHOSPHORUS	<138	18	<125-150
IRON	315	2.1	313-316
ZINC	132	18	119-145
MANGANESE	9.0	1.4	8-10
COPPER	9.5	0.7	9-10
BORON	1.5	0.7	1-2
SILICON	12700	6647	8000-17400
ALUMINUM	197	82	139-255

41. Feather mineral pattern I representing five (8 percent) of fifty-nine blue and lesser snow geese banded on the Foxe Plain of Baffin Island and shot elsewhere.

42

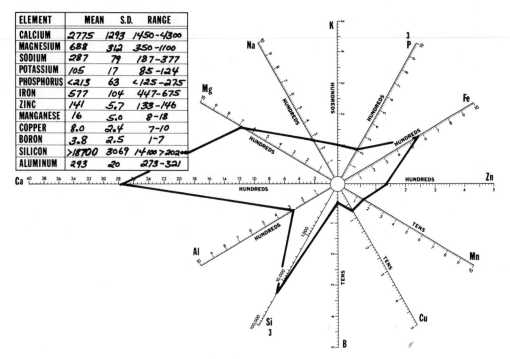

ELEMENT	MEAN	S.D.	RANGE
CALCIUM	2775	1293	1450-4300
MAGNESIUM	688	312	350-1100
SODIUM	287	79	187-377
POTASSIUM	105	17	85-124
PHOSPHORUS	<213	63	<125-275
IRON	577	104	447-675
ZINC	141	5.7	133-146
MANGANESE	16	5.0	8-18
COPPER	8.0	2.4	7-10
BORON	3.8	2.5	1-7
SILICON	>18700	3069	14100 >20200
ALUMINUM	293	20	273-321

42. Feather mineral pattern J representing three (5 percent) of fifty-nine blue and lesser snow geese banded on the Foxe Plain of Baffin Island and shot elsewhere.

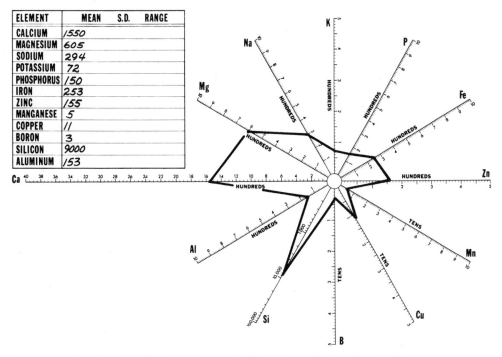

ELEMENT	MEAN	S.D.	RANGE
CALCIUM	1550		
MAGNESIUM	605		
SODIUM	294		
POTASSIUM	72		
PHOSPHORUS	150		
IRON	253		
ZINC	155		
MANGANESE	5		
COPPER	11		
BORON	3		
SILICON	9000		
ALUMINUM	153		

43. Feather mineral pattern K representing four (7 percent) of fifty-nine blue and lesser snow geese banded on the Foxe Plain of Baffin Island and shot elsewhere.

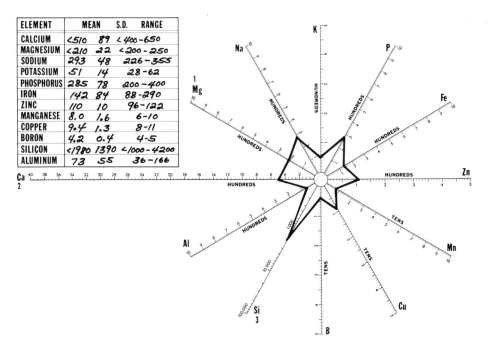

ELEMENT	MEAN	S.D.	RANGE
CALCIUM	<510	89	<400-650
MAGNESIUM	<210	22	<200-250
SODIUM	293	48	226-355
POTASSIUM	51	14	28-62
PHOSPHORUS	285	78	200-400
IRON	142	84	88-290
ZINC	110	10	96-122
MANGANESE	8.0	1.6	6-10
COPPER	9.4	1.3	8-11
BORON	4.2	0.4	4-5
SILICON	<1980	1390	<1000-4200
ALUMINUM	73	55	36-166

44. Feather mineral pattern L representing four (7 percent) of fifty-nine blue and lesser snow geese banded on the Foxe Plain of Baffin Island and shot elsewhere.

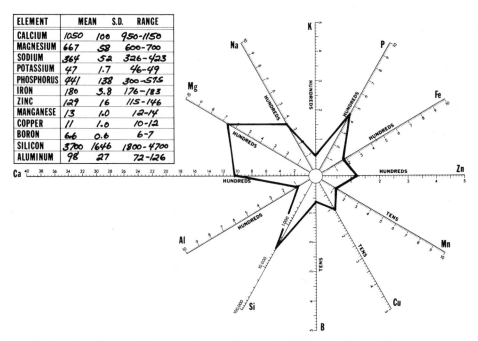

ELEMENT	MEAN	S.D.	RANGE
CALCIUM	1050	100	950-1150
MAGNESIUM	667	58	600-700
SODIUM	364	52	326-423
POTASSIUM	47	1.7	46-49
PHOSPHORUS	441	138	300-575
IRON	180	3.8	176-183
ZINC	129	16	115-146
MANGANESE	13	1.0	12-14
COPPER	11	1.0	10-12
BORON	6.6	0.6	6-7
SILICON	3700	1646	1800-4700
ALUMINUM	98	27	72-126

45. Feather mineral pattern M representing one (2 percent) of fifty-nine blue and lesser snow geese banded on the Foxe Plain of Baffin Island and shot elsewhere.

44

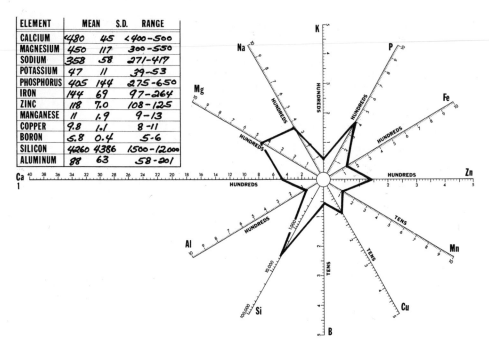

ELEMENT	MEAN	S.D.	RANGE
CALCIUM	<480	45	<400-500
MAGNESIUM	450	117	300-550
SODIUM	358	58	271-417
POTASSIUM	47	11	39-53
PHOSPHORUS	405	144	275-650
IRON	144	69	97-264
ZINC	118	7.0	108-125
MANGANESE	11	1.9	9-13
COPPER	9.8	1.1	8-11
BORON	5.8	0.4	5-6
SILICON	4260	4386	1500-12,000
ALUMINUM	88	63	58-201

46. Feather mineral pattern N representing one (2 percent) of fifty-nine blue and lesser snow geese banded on the Foxe Plain of Baffin Island and shot elsewhere.

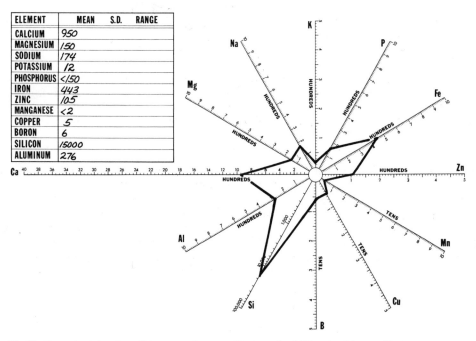

ELEMENT	MEAN	S.D.	RANGE
CALCIUM	950		
MAGNESIUM	150		
SODIUM	174		
POTASSIUM	12		
PHOSPHORUS	<150		
IRON	443		
ZINC	105		
MANGANESE	<2		
COPPER	5		
BORON	6		
SILICON	15000		
ALUMINUM	276		

47. Feather mineral pattern O representing one (2 percent) of fifty-nine blue and lesser snow geese banded on the Foxe Plain of Baffin Island and shot elsewhere.

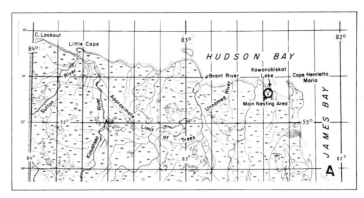

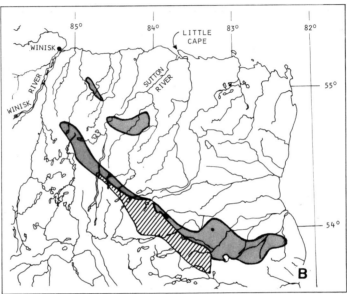

48. A) Geographic setting of the Cape Henrietta Maria colony, Ontario. B) Relation of Precambrian rock outcrops to rivers draining the Cape Henrietta Maria area. Crosshatched area denotes granitic rocks; stippled areas denote volcanic and sedimentary rocks. The latter contain iron-bearing strata. Unshaded areas are underlain by limestones of Silurian age.

special significance in relation to our findings on the feather minerals of a portion of the geese in this colony (see chapter 6).

The coastal marine sediments—which underlie the coastal plain—are sandy in nature and may contain 10 to 20 percent calcium carbonate (Leslie 1964). Analyses of soil samples from the braided estuary of the McConnell River collected for us by Paul Prevett are given in Table 7.

Feather Mineral Patterns. Reference feathers were analyzed for thirty-eight banded geese. The pattern and mean values are give in Figure *74.* Fourteen pattern variations were recognized (Figures *75–88*), but it must be admitted that at least

one (Pattern E) includes more variations than is desirable; the alternate course was more "splitting" than seemed advisable. Five patterns included 71 percent of the samples. Two patterns included two birds each, and seven patterns one bird each. Considering the extensive range of these geese along a coastal plain receiving drainage from a diversity of Precambrian rocks, it is not surprising that many diverse patterns were found. For example, two patterns—D and L (Figures *78, 86*)—are notably high in iron; the geese from which these samples came possibly ingested relatively high amounts of iron by feeding in areas of the coastal plain that received sediments from the iron ore deposit mentioned above. Relationships of these kinds will be amplified in Chapter 6.

Cape Churchill Colony

The nesting of geese in the Cape Churchill colony is largely confined to areas peripheral to LaPérouse Bay, but its feeding range is extended to a roughly triangular area that flanks Hudson Bay for about thirteen miles west of the Cape and extends seven miles south (Figures *1, 89, 90*). This area, like the other colony areas, is a low, flat, coastal plain and is dissected by only a few small shallow streams (Figures *91–92*). The history of this colony, the beginning of which may not date much earlier than the 1940s, has been summarized by Hanson et al. (1972). This population numbered eight thousand adult-plumaged geese in 1966 (R. H. Kerbes, in press).

Geology and Soils. Sedimentary rocks of Ordovician age underlie the Cape and the soils of the area, and the silt and clay sediments in which they developed are highly calcareous. The sediments offshore are 10 to 20 percent calcium carbonate. The Churchill River drains an enormous area of Precambrian shield, but any enrichment that the river might presumably bring to the nearby coastal waters is not reflected in the mineral content of the reference feathers.

The sediments of Hudson Bay have been studied by Leslie (1964). He found that sediments off the southwest coast of Cape Henrietta Maria ranged from 30 to 50 percent in calcium carbonate content, whereas markedly lower values were found off the south coast of Cape Churchill. The calcium carbonates were attributed to three sources: 1) shell material, 2) detrital limestone and dolomite fragments, and 3) redeposited glacial rock flour.

Feather Mineral Patterns. Average feather mineral values for this colony are based on only four samples of reference feathers (Figure *93*). Of the six colonies located in the Hudson Bay area, reference feathers from the Cape Churchill and Cape

49. A portion of the Kawanabiskak Lake area at low tide. The primary nesting grounds of the Cape Henrietta Maria colony lie adjacent to the western sectors of this lake.

Henrietta Maria colonies are similar to having the highest calcium values.

CENTRAL ARCTIC POPULATIONS

No sizable colonies of blue and lesser snow geese exist in the northern sector of the central Canadian Arctic; rather, they occur as sparsely scattered populations in the region of marine submergence south of Queen Maud Gulf, and along the southeast coase of Victoria Island (Barry 1960) (Figure 4). Many of the snow geese breeding south of Queen Maud Gulf nest in the vicinity of the Ross' goose colonies (Hanson et al. 1956; Ryder 1967: 40; Ryder 1971) (see Figure 137). A small population of about two hundred birds is also found on Jenny Lind Island (Parmalee et al. 1967) (Figures 4, 94). In the more southern sector of the mainland Arctic, a colony of five to six hundred lesser snow geese is found on an island at the east end of Beverly Lake, one of a chain of lakes in the course of the Thelon River.

Geology and Soils. The southern half of Victoria Island is underlain by Ordovician and Silurian limestones, Jenny Lind Island by Paleozoic sedimentary rocks (Figure 94), and the mainland south of Queen Maud Gulf mainly by Precambrian rock (see Figure 137). The latter area is discussed in detail in relation to our data on Ross' geese. We do not have reference material for either the small Beverly Lake colony or the Victoria Island population, but feather mineral patterns for the latter might show resemblances to those for Jenny Lind Island and the Sherman Inlet area because of lithological similarities.

Feather Mineral Patterns. Clippings of primaries from the left wing of the single lesser snow goose collected on Jenny Lind Island were kindly made available to us by David F. Parmalee. While a single sample cannot define the flock (Figure 95), the extreme characteristics of this pattern are such that additional samples are likely to be recognizably similar. The pattern of a single museum specimen from Sherman Inlet (see Figure 137) is re-

50. Mouth of the Mukataship (Black Duck) River, six miles west of the base of the Cape Henrietta Maria peninsula.

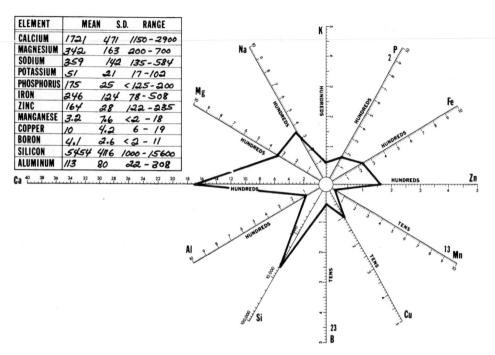

ELEMENT	MEAN	S.D.	RANGE
CALCIUM	1721	471	1150-2900
MAGNESIUM	342	163	200-700
SODIUM	359	142	135-584
POTASSIUM	51	21	17-102
PHOSPHORUS	175	25	<125-200
IRON	246	124	78-508
ZINC	164	28	122-235
MANGANESE	3.2	7.6	<2 - 18
COPPER	10	4.2	6 - 19
BORON	4.1	2.6	<2 - 11
SILICON	5454	4116	1000-15600
ALUMINUM	113	80	22-308

51. Feather mineral pattern representing twenty-four or 100 percent of the blue and lesser snow geese banded on Cape Henrietta Maria, Ontario, and shot elsewhere.

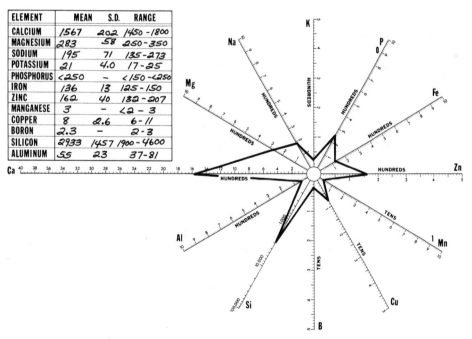

ELEMENT	MEAN	S.D.	RANGE
CALCIUM	1567	202	1450-1800
MAGNESIUM	283	58	250-350
SODIUM	195	71	135-273
POTASSIUM	21	4.0	17-25
PHOSPHORUS	<250	-	<150-<250
IRON	136	13	125-150
ZINC	162	40	132-207
MANGANESE	3	-	<2 - 3
COPPER	8	2.6	6-11
BORON	2.3	-	2-3
SILICON	2933	1457	1900-4600
ALUMINUM	55	23	37-81

52. Feather mineral pattern A representing three (13 percent) of twenty-four blue and lesser snow geese banded on Cape Henrietta Maria, Ontario, and shot elsewhere.

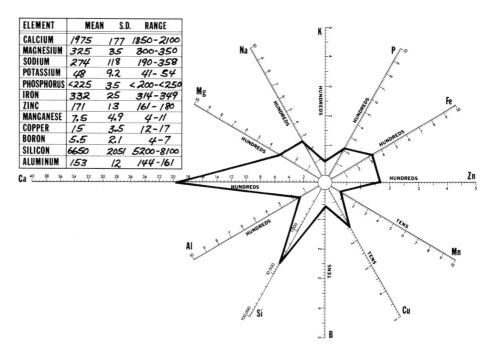

ELEMENT	MEAN	S.D.	RANGE
CALCIUM	1975	177	1850-2100
MAGNESIUM	325	35	300-350
SODIUM	274	118	190-358
POTASSIUM	48	9.2	41-54
PHOSPHORUS	<225	35	<200-<250
IRON	332	25	314-349
ZINC	171	13	161-180
MANGANESE	7.5	4.9	4-11
COPPER	15	3.5	12-17
BORON	5.5	2.1	4-7
SILICON	6650	2051	5200-8100
ALUMINUM	153	12	144-161

53. Feather mineral pattern B representing two (8 percent) of twenty-four blue and lesser snow geese banded on Cape Henrietta Maria, Ontario, and shot elsewhere.

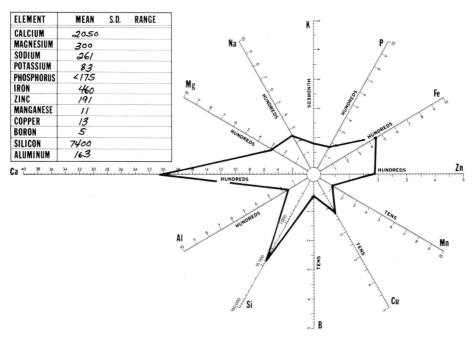

ELEMENT	MEAN	S.D.	RANGE
CALCIUM	2050		
MAGNESIUM	300		
SODIUM	261		
POTASSIUM	83		
PHOSPHORUS	<175		
IRON	460		
ZINC	191		
MANGANESE	11		
COPPER	13		
BORON	5		
SILICON	7400		
ALUMINUM	163		

54. Feather mineral pattern C representing one (4 percent) of twenty-four blue and lesser snow geese banded on Cape Henrietta Maria, Ontario, and shot elsewhere.

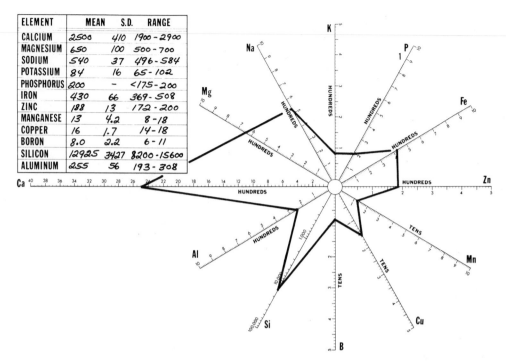

ELEMENT	MEAN	S.D.	RANGE
CALCIUM	2500	410	1900-2900
MAGNESIUM	650	100	500-700
SODIUM	540	37	496-584
POTASSIUM	84	16	65-102
PHOSPHORUS	200	–	<175-200
IRON	430	66	369-508
ZINC	188	13	172-200
MANGANESE	13	4.2	8-18
COPPER	16	1.7	14-18
BORON	8.0	2.2	6-11
SILICON	12925	3427	8200-15600
ALUMINUM	255	56	193-308

55. Feather mineral pattern D representing four (16 percent) of twenty-four blue and lesser snow geese banded on Cape Henrietta Maria, Ontario, and shot elsewhere.

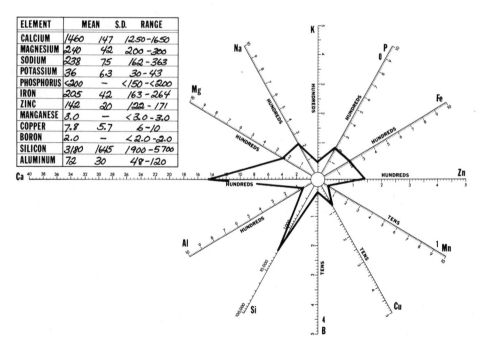

ELEMENT	MEAN	S.D.	RANGE
CALCIUM	1460	147	1250-1650
MAGNESIUM	240	42	200-300
SODIUM	238	75	162-363
POTASSIUM	36	6.3	30-43
PHOSPHORUS	<200	–	<150-<200
IRON	205	42	163-264
ZINC	142	20	122-171
MANGANESE	3.0	–	<3.0-3.0
COPPER	7.8	5.7	6-10
BORON	2.0	–	<2.0-2.0
SILICON	3180	1645	1900-5700
ALUMINUM	72	30	48-120

56. Feather mineral pattern E representing five (21 percent) of twenty-four blue and lesser snow geese banded on Cape Henrietta Maria, Ontario, and shot elsewhere.

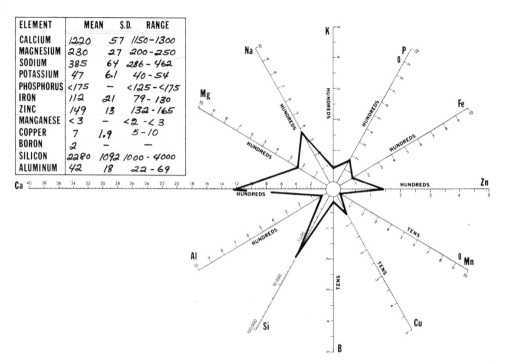

ELEMENT	MEAN	S.D.	RANGE
CALCIUM	1220	57	1150-1300
MAGNESIUM	230	27	200-250
SODIUM	385	64	286-462
POTASSIUM	47	6.1	40-54
PHOSPHORUS	<175	—	<125-<175
IRON	112	21	79-130
ZINC	149	13	132-165
MANGANESE	<3	—	<2-<3
COPPER	7	1.9	5-10
BORON	2	—	—
SILICON	2280	1092	1000-4000
ALUMINUM	42	18	22-69

57. Feather mineral pattern F representing five (21 percent) of twenty-four blue and lesser snow geese banded on Cape Henrietta Maria, Ontario, and shot elsewhere.

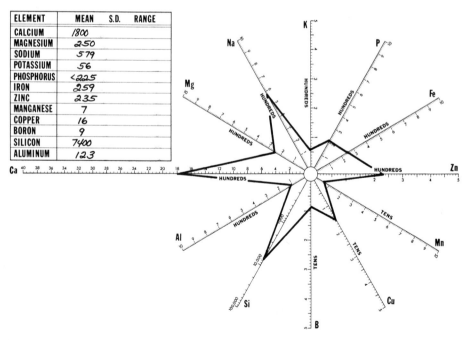

ELEMENT	MEAN	S.D.	RANGE
CALCIUM	1800		
MAGNESIUM	250		
SODIUM	579		
POTASSIUM	56		
PHOSPHORUS	<225		
IRON	259		
ZINC	235		
MANGANESE	7		
COPPER	16		
BORON	9		
SILICON	7400		
ALUMINUM	123		

58. Feather mineral pattern G representing one (4 percent) of twenty-four blue and lesser snow geese banded on Cape Henrietta Maria, Ontario, and shot elsewhere.

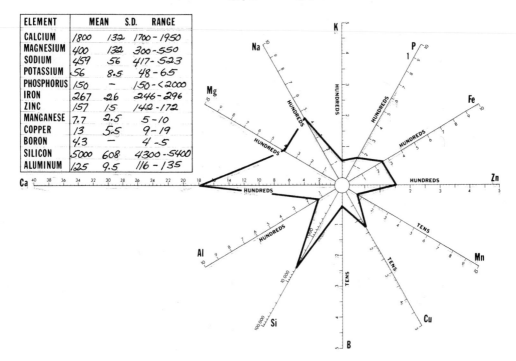

ELEMENT	MEAN	S.D.	RANGE
CALCIUM	1800	132	1700 - 1950
MAGNESIUM	400	132	300 - 550
SODIUM	459	56	417 - 523
POTASSIUM	56	8.5	48 - 65
PHOSPHORUS	150	—	150 - <2000
IRON	267	26	246 - 296
ZINC	157	15	142 - 172
MANGANESE	7.7	2.5	5 - 10
COPPER	13	5.5	9 - 19
BORON	4.3	—	4 - 5
SILICON	5000	608	4300 - 5400
ALUMINUM	125	9.5	116 - 135

59. Feather mineral pattern H representing three (13 percent) of twenty-four blue and lesser snow geese banded on Cape Henrietta Maria, Ontario, and shot elsewhere.

markably similar to that of the Jenny Lind Island bird except in respect to manganese (Figure 96). The similarity of these two patterns indicates that the carbonate rocks of the two areas are of similar age and chemical composition.

WESTERN ARCTIC COLONIES

Three colonies of blue and lesser snow geese occur in the western sector of the Canadian Arctic—the Banks Island (Egg River) colony, the Anderson River colony, and the Kendall Island colony in the mouth of the Mackenzie River. The colonies on Wrangel Island off the northeast tip of Siberia, U.S.S.R., are included in this category because in winter geese from these colonies intermingle in California and other western states with wintering populations from the central and western Canadian Arctic.

Banks Island Colony

The Banks Island colony is distributed along both sides of the lower Egg River for six miles, terminating at the confluence of the Egg with the Big River sixteen miles from the coast (Figures 4, 97). The area of the nesting colony is estimated at be-

tween ten and twelve square miles. As in the case of many of the other rivers associated with nesting colonies, the lower portion of the Egg River is a shallow, braided stream (Figures 98–99). After hatching, the geese disperse toward the coast, the main feeding grounds being the Big River flats. McEwen (1958) estimated that at the end of the nesting season the colony may have totaled about 100,000 birds; in 1971 the colony was estimated to have increased to approximately 200,000 birds (Dzubin et al. 1973).

Physiography, Geology, and Soils. Thorsteinsson and Tozer (1962:12–13) have given an excellent description of the lowlands of central and western Banks Island, the area which includes the Big River drainage basin: "A low plain of gently rolling hills, shallow valleys, and alluvial flats and benches occupies most of central and western Banks Island. The plain is largely within 500 feet of sea-level but its southeastern, eastern, and northeastern parts rise gradually to elevations of 800 to 1,000 feet. Local relief ranges from a few tens of feet to 300 feet. The plain is underlain by weakly consolidated sandstone and shale of the Eureka Sound Formation. This formation is overlain by gravel and sand of the Beaufort Formation,

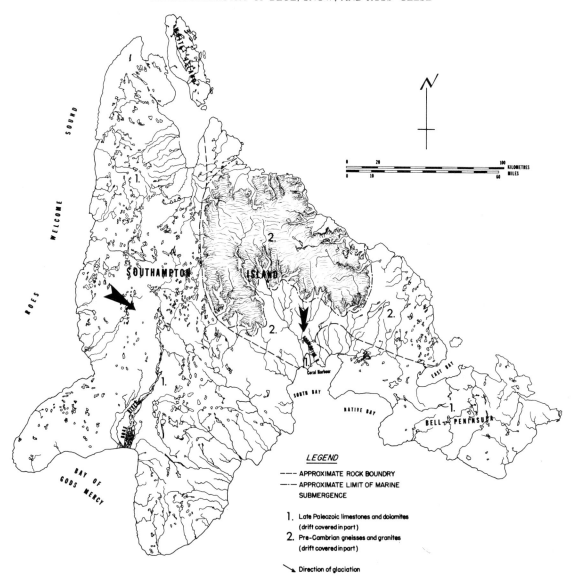

60. Geologic map of Southampton Island showing primary geologic features and location of Boas River colony at the head of Bay of Gods Mercy and the East Bay colony at the head of East Bay.

various deposits of glacial origin that probably relate to more than one pre-Wisconsin (?) glaciation, interglacial and postglacial silt and peat, and alluvium of both present and former rivers. These surficial deposits range up to about 200 feet in thickness, but in many parts of the region they are represented by only a few inches or feet of gravelly material formed by mass wasting of the above deposits.

''The region is drained by a series of long, westflowing rivers that rise in the moraine only a few miles from the east coast. The main rivers occupy exceedingly broad shallow valleys floored by gravelly and sandy alluvium and swampy tundra with large polygons and rounded shallow ponds. In places gravelly terraces stand a few feet to a few tens of feet above the valley floors. The lower reaches of the rivers are strikingly braided but the upstream portions are partly braided and partly meandering. The large rivers enter the Beaufort Sea through broad swampy deltas which barely fill the mouths of the valleys. Estuaries at the mouths of small rivers carrying less sediment have clearly been drowned by the sea. Headlands along the west coast bear wave-cut cliffs up to 150 feet high, whereas the lower intervening coast is fringed by sandy to gravelly spits and bars.

"For a few miles inland from the coast, the interstream parts of the lowland plain are characterized by gently rolling, degraded, but almost undissected ground-moraine topography. Small local streams flow on the surface of the plain or in shallow draws and are disorganized to subdendritic in pattern. The fairly numerous ground-moraine ponds are smoothly rounded in outline as though substantially modified by solifluction. The surface material is typically boulder-strewn, gravelly colluvium apparently formed from the intermingling of glacial deposits with the underlying Beaufort gravel and/or sandy to shaly bedrock.

"In contrast, the interior, major part of the lowland plain is dissected by a dendritic network of river valleys and gullies that has reduced the region to a stage of late youth and locally even to early maturity. Typically the valleys and gullies are only a few feet to tens of feet deep. Most of the surface remaining between the valleys is similar to the undissected parts of the lowland adjacent to the coast and probably has had a similar history. However, parts of the dissected surface are high-level fluvial benches on which evidence of glaciation has not been found. The valleys and gullies are mostly broad and shallow and only locally V-shaped. Although some are being actively eroded by modern streams, many are floored by peat, contain lakes, or are occupied by underfit streams. The time of erosion of most of the abandoned and underfit valleys is not known. Some however have been cut by meltwater emanating from glacial ice along the east Banks Island moraine and others by meltwater from an (earlier?) ice-sheet that extended across the interior of Banks Island; alluvial plains are associated with the meltwater channels of both generations. Some of the dissection, however, is much

61. A portion of the Boas River delta, site of the largest colony of nesting blue and lesser snow geese on Southampton Island.

62. A sector of the breeding grounds of the East Bay colony of blue and lesser snow geese on Southampton Island.

older and appears to have preceded glaciation, but whether it is interglacial or preglacial is not known.''

Fyles (1962:8) has stated: "Clearly defined raised strand lines have not been found in western Banks Island, but doubtful emergent shore features occur in a few places. Drowned estuaries at the mouths of rivers point to recent encroachment of the sea over the land.''

"On the basis of microfloras, Beaufort beds are dated as varying from late Tertiary to early Pleistocene'' (Thorsteinsson and Tozer 1962). No significant ore-bearing formations are known.

Tedrow and Douglas (1964) described and reported data for four different soil series developed in Beaufort Formation materials in north-central Banks Island. Certain soils on this part of the Island reflect the antiquity of the landscape, being relatively highly developed for Arctic soils. For example, the Beaufort soil series had a yellowish-red B horizon that has 4.0 percent extractable iron (Fe_2O_3). Tedrow and Douglas summarize their observations thus: "Soils of Banks . . . , are predominantly well-drained with many high arctic affinities. Because of the dry desert-like appearance of most of the soils, they are designated collectively as Polar Desert rather than Tundra. Many of the soils have salts accumulated at the surface, and within the well-drained soils pedogenic carbonate accumulation within the solum is an unusual occurrence.''

Feather Mineral Patterns. Our representations of the feather mineral patterns for the Banks Island colony are based on clippings of primaries from three museum specimens (Cape Kellett and Sachs Harbor, Banks Island) and a series of six birds shot in the fall on the mainland tundra west of the mouth of the Mackenzie River (Figures *100–101*).

According to Tom Berry, who furnished the latter collection, this area is the primary feeding and resting area of the Banks Island geese when making their mainland landfall on their southward migration. Although both collections are representative of the Banks Island population, they bear little resemblance to each other. These diverse patterns can probably be readily explained. Most of the nonbreeding yearling geese—and presumably many of the failed breeders as well—make a molt migration to the valley of the Thomsen River at the north end of the island (Wayne Speller, personal communication, 1974). Geese constituting this group would likely have feather mineral patterns different from those of the Egg River colony birds, as the surficial geologies of the two areas differ (Figure 97). We suspect that the feather mineral pattern for the geese shot at Cape Kellett and Sachs Harbor is most nearly representative of the adults

and immature birds from the Egg River colony area and that the high iron pattern (Figure *101*) for the birds shot at the mouth of the Mackenzie is more nearly representative of the birds that summer in the Thomsen River valley. This conclusion is consonant with the findings of Tedrow and Douglas (1964) that the subsurface horizons of soils of the Beaufort series in the north-central part of the island contain 4.0 percent extractable iron (Fe_2O_3).

Feather mineral patterns of lesser snow geese collected in the Imperial Valley, California, probably the major wintering grounds of the Banks Island geese, show basic similarities with the pattern representing the three birds collected on Banks Island (Figure *100*), whereas feather mineral patterns D and E from birds wintering at the Gray Lodge Refuge, California (see Figures *237–38*) more closely resemble the average pattern for the

63. A low altitude oblique photograph of a portion of the East Bay colony. Note nesting pairs.

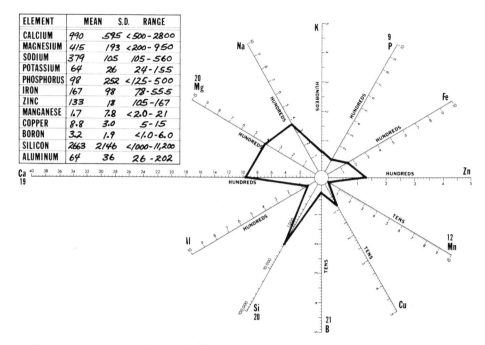

ELEMENT	MEAN	S.D.	RANGE
CALCIUM	990	595	<500-2800
MAGNESIUM	415	193	<200-950
SODIUM	379	105	105-560
POTASSIUM	64	26	24-155
PHOSPHORUS	98	252	<125-500
IRON	167	98	78-555
ZINC	133	18	105-167
MANGANESE	4.7	7.8	<2.0-21
COPPER	8.8	3.0	5-15
BORON	3.2	1.9	<1.0-6.0
SILICON	2663	2146	<1000-11,200
ALUMINUM	64	36	26-202

64. Feather mineral pattern for combined samples consisting of twenty-three geese banded at Boas River and East Bay, Southampton Island, and shot elsewhere.

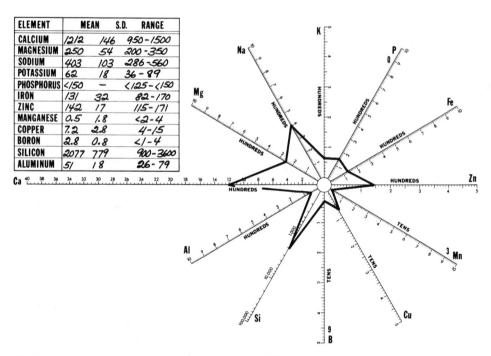

ELEMENT	MEAN	S.D.	RANGE
CALCIUM	1212	146	950-1500
MAGNESIUM	250	54	200-350
SODIUM	403	103	286-560
POTASSIUM	62	18	36-89
PHOSPHORUS	<150	—	<125-<150
IRON	131	32	82-170
ZINC	142	17	115-171
MANGANESE	0.5	1.8	<2-4
COPPER	7.2	2.8	4-15
BORON	2.8	0.8	<1-4
SILICON	2077	779	900-3600
ALUMINUM	51	18	26-79

65. Feather mineral pattern A representing thirteen (57 percent) of twenty-three blue and lesser snow geese banded on Southampton Island and shot elsewhere.

58

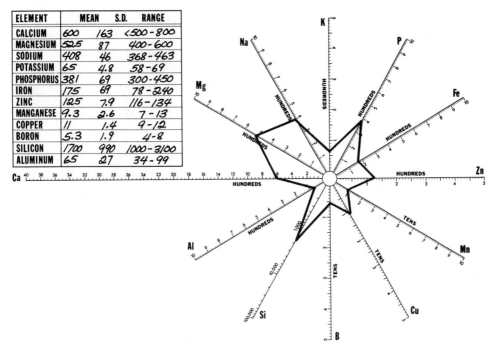

ELEMENT	MEAN	S.D.	RANGE
CALCIUM	600	163	<500-800
MAGNESIUM	525	87	400-600
SODIUM	408	46	368-463
POTASSIUM	65	4.8	58-69
PHOSPHORUS	381	69	300-450
IRON	175	69	78-240
ZINC	125	7.9	116-134
MANGANESE	9.3	2.6	7-13
COPPER	11	1.4	9-12
BORON	5.3	1.9	4-8
SILICON	1700	990	1000-3100
ALUMINUM	65	27	34-99

66. Feather mineral pattern B representing four (18 percent) of twenty-three blue and lesser snow geese banded on Southampton Island and shot elsewhere.

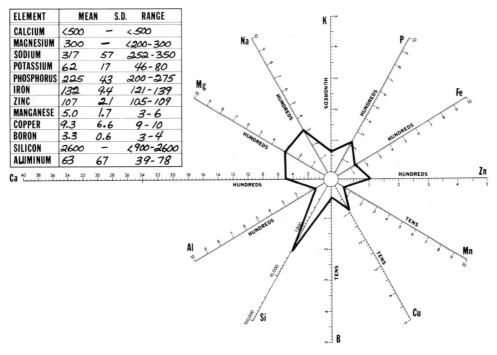

ELEMENT	MEAN	S.D.	RANGE
CALCIUM	<500	—	<500
MAGNESIUM	300	—	<200-300
SODIUM	317	57	252-350
POTASSIUM	62	17	46-80
PHOSPHORUS	225	43	200-275
IRON	132	9.4	121-139
ZINC	107	2.1	105-109
MANGANESE	5.0	1.7	3-6
COPPER	9.3	6.6	9-10
BORON	3.3	0.6	3-4
SILICON	2600	—	<900-2600
ALUMINUM	63	67	39-78

67. Feather mineral pattern C representing three (13 percent) of twenty-three blue and lesser snow geese banded on Southampton Island and shot elsewhere.

59

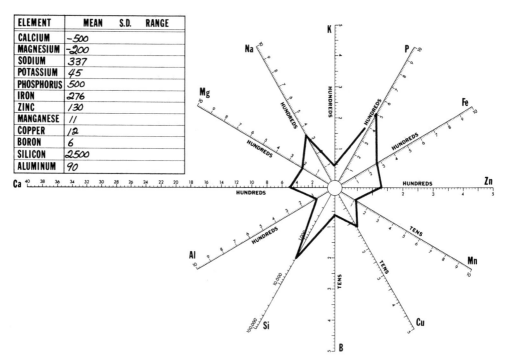

ELEMENT	MEAN	S.D.	RANGE
CALCIUM	-500		
MAGNESIUM	-200		
SODIUM	337		
POTASSIUM	45		
PHOSPHORUS	500		
IRON	276		
ZINC	130		
MANGANESE	11		
COPPER	12		
BORON	6		
SILICON	2500		
ALUMINUM	90		

68. Feather mineral pattern D representing one (4 percent) of twenty-three blue and lesser snow geese banded on Southampton Island.

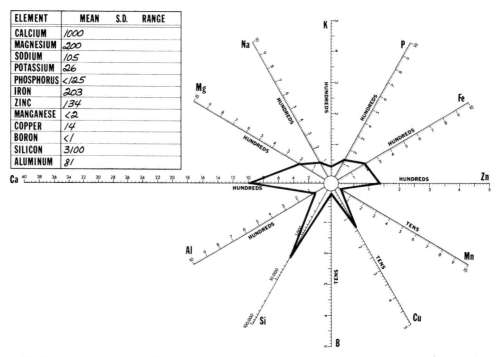

ELEMENT	MEAN	S.D.	RANGE
CALCIUM	1000		
MAGNESIUM	200		
SODIUM	105		
POTASSIUM	26		
PHOSPHORUS	<125		
IRON	203		
ZINC	134		
MANGANESE	<2		
COPPER	14		
BORON	<1		
SILICON	3100		
ALUMINUM	81		

69. Feather mineral pattern E representing one (4 percent) of twenty-three blue and lesser snow geese banded on Southampton Island and shot elsewhere.

60

ELEMENT	MEAN	S.D.	RANGE
CALCIUM	2800		
MAGNESIUM	950		
SODIUM	450		
POTASSIUM	155		
PHOSPHORUS	475		
IRON	555		
ZINC	120		
MANGANESE	21		
COPPER	11		
BORON	6		
SILICON	11200		
ALUMINUM	202		

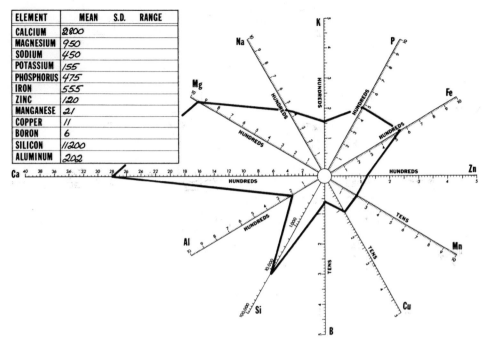

70. Feather mineral pattern F representing one (4 percent) of twenty-three blue and lesser snow geese banded on Southampton Island and shot elsewhere.

six lesser snow geese collected at the northwest corner of the mouth of the Mackenzie River (Figure *101*).

Anderson River Colony

The Anderson River colony is located on the low deltaic islands at the mouth of Anderson River (Figures *4, 102–3*). It has numbered about fifteen thousand birds in recent years. After the eggs hatch, the young and the adults feed on the adjacent mainland coastal areas. Nagel (1969) believes that the Anderson River geese make up the principal component of the flight through Utah.

Geology and Soil. The delta of the Anderson River lies at the eastern edge of the low Arctic Coastal Plain that extends inland in this area for as much as thirty miles (Figure *102*). The geology of the region, glaciated between 10,000 and 13,000 BP, is known from a preliminary survey by Yorath and Balkwill (1970). The basin of the Anderson River contains rocks ranging in age from the Proterozoic to Recent, and over much of its lower course the Anderson River is deeply incised. Lithologies are mostly silt and clay-rich sedimentary rocks of Cretaceous, Ordovician, and Devonian periods. Near the coast the river traverses rock units rich in sulphides, sulphates (see appendix 2),

and the magnesium-bearing clay, bentonite (Figure *104*). Certain shales in the area are quite bituminous and burn spontaneously. Rock units in the lower sectors of the basin (Beaufort Formation) are believed to have close affinities with those on Banks Island.

Feather Mineral Patterns. Primary feathers were obtained for analyses from five lesser snow geese banded at the delta of the Anderson River and shot by hunters in California (Figure *105*). The composite feather mineral pattern for these five geese is unique among those of geese breeding in the western colonies in respect to calcium and magnesium values and the ratio of calcium to magnesium (see Table 25). The explanation for the relatively high magnesium levels in the feathers of these geese was apparent when the geology of the basin of the Anderson River was reviewed (Yorath and Balkwill 1970). Adjacent to the lower portions of the Anderson River are extensive deposits of bentonite clays [$Mg_{3-x} Fe_x Si_{4-y} Al_y O_{10} (OH)_2$], which are notably rich in magnesia (4–6 percent MgO). These areas are indicated as Kb, the bentonitic zone, in Figure *102*. According to C. J. Yorath (personal communication, 1974), the composition of the soils of the region shown in Figure *104*, as determined from X-ray diffraction analy-

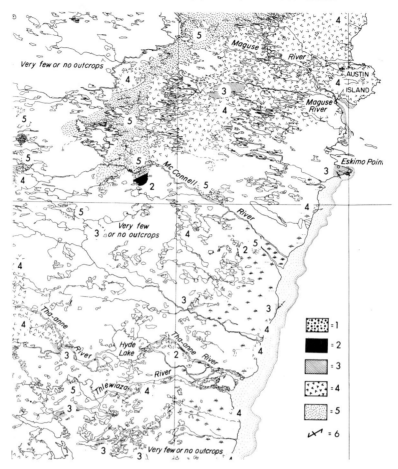

71. Reconnaissance geology of the coastal range of the McConnell River colony and the more immediate areas of the inland drainage basins. 1) Undifferentiated Early Proterozoic rocks, quartzite; dolomite and/or limestone. 2) Mainly quartzite; minor intercalated conglomerate, pebbly quartzite, and siliceous iron formations. 3) Volcanic and/or sedimentary rocks of Archaean and/or Early Proterozoic age; derived schist and gneiss, commonly with some granitic material. 4) Granodiorite, granite, quartz diorite, pegmatite, and allied rocks; in part gneissic, with many bands of, probably, partly assimilated sedimentary and volcanic rocks. 5) Andesitic and dacitic greenstones; rhyolite, tuff, agglomerate; derived amphibole schist and gneiss; minor quartz-mica schist and magnetite iron formation. 6) Schistosity, gneissosity (inclined, vertical, dip unknown). (After Lord 1953)

ses, is chlorite and montmorillonite, 22–30 percent; illite, 12–14 percent; kaolinite, 8–0 percent; quartz, 40–52 percent; and feldspar, under 6–7 percent. In addition, deposits tentatively identified by Yorath and Balkwill (1970) as Beaufort Formation (Tb in Figure *104*) contain pebbles of dolomite. Food plants growing directly on deposits of bentonitic soils or, much less likely, dolomitic grit, or a combination of both transported downriver to the delta readily account for the relatively high magnesium concentrations found in the primaries of the Anderson River geese.

Kendall Island Colony

The Kendall Island colony is the smallest of the western colonies and probably has not exceeded several thousand birds. The colony is located on a small deltaic island at the mouth of the Mackenzie River (Figures *4, 106*). The food resources of the island are inadequate both in kind and quantity for adults and their broods (Figures *107–8*). Hence, as soon as the goslings are hatched, the family parties head for Richards Island and a closely associated smaller island to feed. Richards Island can be regarded as a part of the mainland that was cut off by a channel of the Mackenzie River.

Geology and Soils. Rampton and Mackay (1971:1) have described the geology of the general area east of the mouth of Mackenzie River: "Exposures throughout Tuktoyaktuk Peninsula, Richards Island and nearby areas indicate that much of this area is underlain by fluvial sands and silts and fine-grained deltaic sands. Some sections expose marine clay at depth. The deposits described above are capped by wind-blown sands on the gravels and clay tills and till-like deposits of varying thickness on Richards Island and the southern and central parts of the Tuktoyaktuk Peninsula."

Nearby Mesozoic shales and sandstones outcrop in the Richardson Mountains at the west side of the delta, and poorly consolidated Tertiary sediments (gravels, sands, silts, and cretaceous shales in the south) constitute the lower Caribou Hills along the east side north of Inuvik (Fyles et al. 1972:4). The Mackenzie River, of course, drains extensive terranes of diverse rock types to the south (see appendix 2 and Table 45).

Feather Mineral Patterns. Average feather mineral values for the Kendall Island colony are based on seven adults collected on the summer range (Figure *109*). As these geese had incompletely grown primaries, sample values may be somewhat high for sodium and potassium as compared with those of fully fledged birds that have been on the wing for several weeks. If not, birds from this colony will be distinguishable on the basis of these elements. Although we believe only one bird of unknown origin taken on the wintering grounds could have come from the Kendall Island colony, even the occurrence of this "Kendall Island designate" was surprising in view of sample sizes and the small size of the colony. Panarctic Oil Company has drilled for oil on Kendall Island, and the future of this colony is in doubt.

Wrangel Island Colony

The main Wrangel Island (U.S.S.R.) colony, reported in 1960 to contain about 400,000 nesting snow geese, is located around the upper reaches of the Tundrovaya River near the base of Tundrovoi

72. Delta of the McConnell River, N.W.T. The nesting of the majority of the blue and lesser snow geese in this colony takes place on islands in this delta.

Peak (Figure *110*) (Upenski 1965). This colony, at least in past years, may have been exceeded in size only by the Baffin Island population. The colony area, largely surrounded by low mountains, has a more equitable climate than if it were located near the coast and exposed to winds coming off the ice pack that surrounds the island most of the summer. Two small satellite colonies are found on the south coast, presumably associated with the braided river deltas of the Khishnikov and Mamontovaya rivers (Figure *110*). Possibly in earlier times the delta at the mouth of the Gusinaya (Goose) River supported a colony, judging from the name. The geese of the Wrangel Island colony winter on the Fraser River delta of British Columbia and in the central valleys of the Pacific Coast states.

According to Dement'ev and Gladkov (1967), the Wrangel Island snow geese (presumably after nesting) concentrate at the mouths of the rivers along the coasts. The diet of the Russian populations of lesser snow geese is reported to include small fresh- and saltwater invertebrates.

Geology and Soils. Wrangel Island (Figure *4*) lies ninety miles off the coast of northeast Siberia and is about ninety miles long and forty-eight miles wide. Portions of it are mountainous, some peaks reaching elevations of 2,000 to 4,000 meters. We are indebted to Neely H. Bostick of the Illinois State Geological Survey for his preparation of the geological map of Wrangel Island (Figure *110*). Basing his summary on studies by Tilman et al. (1964), Gnibidenko (1968), and Ivanov (1973), Bostick provided these annotations on the various geological units:

RECENT SEDIMENTS. Sand and shingle beaches and bars; sandy and clay loams near river mouths; peat bogs; rock debris in valleys.
UPPER TERTIARY. Alluvial conglomerates and sandstones, lake or marsh clays, and sands with minor peat. Older river and coastal terrace deposits. Tundra Akademii is mainly clay and sandy loams. The Recent and Tertiary sediments are no doubt derived from the older rocks on the island. The main difference between these sediments and the older rocks would be the much higher content of organic matter in the Recent and Tertiary sediments, their probable lack of carbonates, and their high content of clay products from weathering of the older rocks.

73. Typical feeding area adjacent to the Hudson Bay coast for family parties from the McConnell River colony.

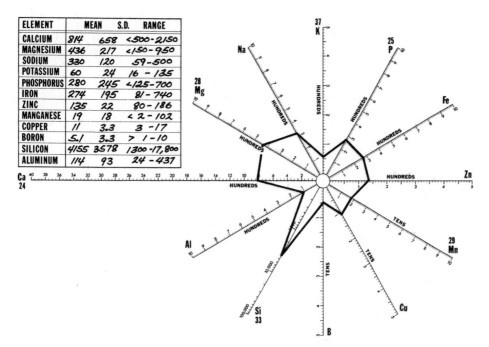

ELEMENT	MEAN	S.D.	RANGE
CALCIUM	814	658	<500-2150
MAGNESIUM	436	217	<150-950
SODIUM	330	120	59-500
POTASSIUM	60	24	16-135
PHOSPHORUS	280	245	<125-700
IRON	274	195	81-740
ZINC	135	22	80-186
MANGANESE	19	18	<2-102
COPPER	11	3.3	3-17
BORON	5.1	3.3	>1-10
SILICON	4155	3578	1300-17,800
ALUMINUM	114	93	24-437

74. Feather mineral pattern representing thirty-eight or 100 percent of the blue and lesser snow geese banded in the vicinity of the McConnell River, N.W.T., and shot elsewhere.

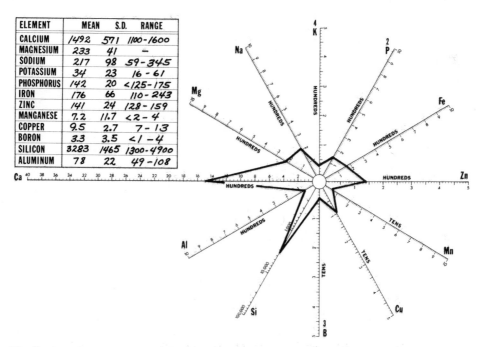

ELEMENT	MEAN	S.D.	RANGE
CALCIUM	1492	571	1100-1600
MAGNESIUM	233	41	–
SODIUM	217	98	59-345
POTASSIUM	34	23	16-61
PHOSPHORUS	142	20	<125-175
IRON	176	66	110-243
ZINC	141	24	128-159
MANGANESE	7.2	11.7	<2-4
COPPER	9.5	2.7	7-13
BORON	3.3	3.5	<1-4
SILICON	3283	1465	1300-4900
ALUMINUM	78	22	49-108

75. Feather mineral pattern A representing six (15 percent) of thirty-eight blue or lesser snow geese banded in the vicinity of the McConnell River delta, N.W.T., and shot elsewhere.

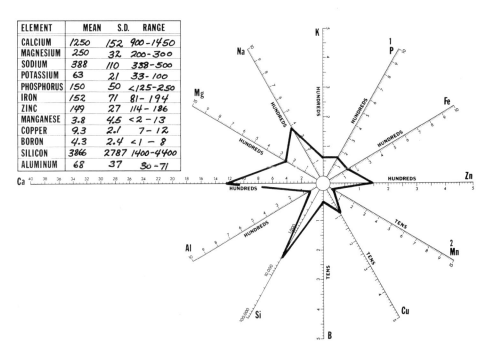

ELEMENT	MEAN	S.D.	RANGE
CALCIUM	1250	152	900-1450
MAGNESIUM	250	32	200-300
SODIUM	388	110	358-500
POTASSIUM	63	21	33-100
PHOSPHORUS	150	50	<125-250
IRON	152	71	81-194
ZINC	149	27	114-186
MANGANESE	3.8	4.5	<2-13
COPPER	9.3	2.1	7-12
BORON	4.3	2.4	<1-8
SILICON	3866	2787	1400-4400
ALUMINUM	68	37	30-71

76. Feather mineral pattern B representing six (15 percent) of thirty-eight blue and lesser snow geese banded in the vicinity of the McConnell River delta, N.W.T., and shot elsewhere.

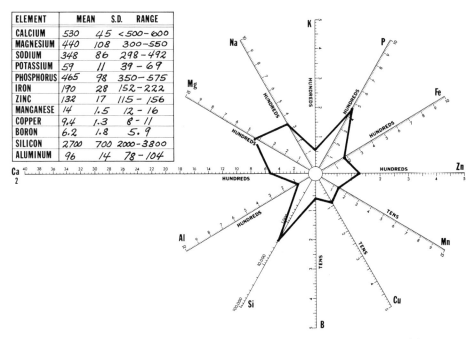

ELEMENT	MEAN	S.D.	RANGE
CALCIUM	530	45	<500-600
MAGNESIUM	440	108	300-550
SODIUM	348	86	298-492
POTASSIUM	59	11	39-69
PHOSPHORUS	465	98	350-575
IRON	190	28	152-222
ZINC	132	17	115-156
MANGANESE	14	1.5	12-16
COPPER	9.4	1.3	8-11
BORON	6.2	1.8	5.9
SILICON	2700	700	2000-3800
ALUMINUM	96	14	78-104

77. Feather mineral pattern C representing five (13 percent) of thirty-eight blue and lesser snow geese banded in the vicinity of the McConnell River delta, N.W.T., and shot elsewhere.

66

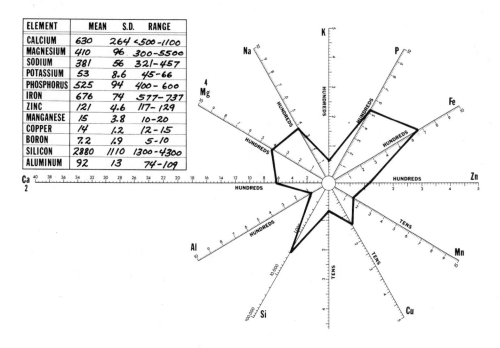

ELEMENT	MEAN	S.D.	RANGE
CALCIUM	630	264	<500-1100
MAGNESIUM	410	96	300-5500
SODIUM	381	56	321-457
POTASSIUM	53	8.6	45-66
PHOSPHORUS	525	94	400-600
IRON	676	74	577-737
ZINC	121	4.6	117-129
MANGANESE	15	3.8	10-20
COPPER	14	1.2	12-15
BORON	7.2	1.9	5-10
SILICON	2880	1110	1300-4300
ALUMINUM	92	13	74-109

78. Feather mineral pattern D representing five (13 percent) of thirty-eight blue and lesser snow geese banded in the vicinity of the McConnell River delta, N.W.T., and shot elsewhere.

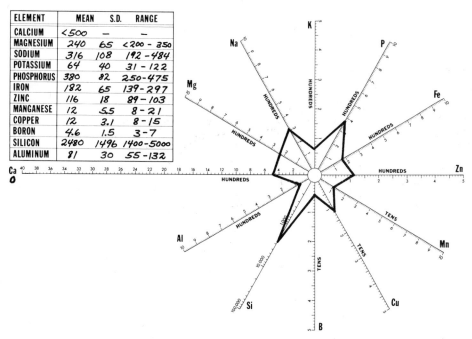

ELEMENT	MEAN	S.D.	RANGE
CALCIUM	<500	—	—
MAGNESIUM	240	65	<200 - 350
SODIUM	316	108	192 - 484
POTASSIUM	64	40	31 - 122
PHOSPHORUS	380	82	250-475
IRON	182	65	139-297
ZINC	116	18	89 - 103
MANGANESE	12	5.5	8 - 21
COPPER	12	3.1	8 - 15
BORON	4.6	1.5	3 - 7
SILICON	2480	1496	1400-5000
ALUMINUM	81	30	55 -132

79. Feather mineral pattern E representing five (13 percent) of thirty-eight blue and lesser snow geese banded in the vicinity of the McConnell River delta, N.W.T., and shot elsewhere.

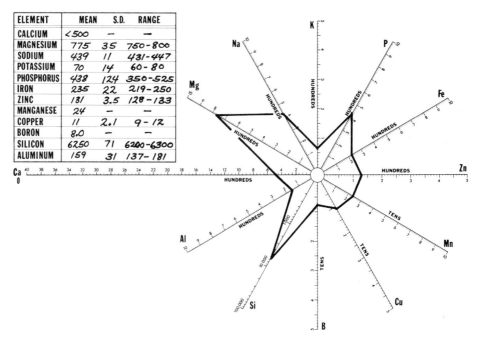

ELEMENT	MEAN	S.D.	RANGE
CALCIUM	<500	—	—
MAGNESIUM	775	35	750-800
SODIUM	439	11	431-447
POTASSIUM	70	14	60-80
PHOSPHORUS	438	124	350-525
IRON	235	22	219-250
ZINC	131	3.5	128-133
MANGANESE	24	—	—
COPPER	11	2.1	9-12
BORON	8.0	—	—
SILICON	6250	71	6200-6300
ALUMINUM	159	31	137-181

80. Feather mineral pattern F representing two (53 percent) of thirty-eight blue and lesser snow geese banded in the vicinity of the McConnell River, N.W.T., and shot elsewhere.

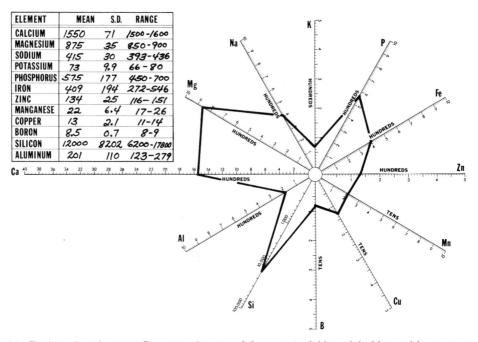

ELEMENT	MEAN	S.D.	RANGE
CALCIUM	1550	71	1500-1600
MAGNESIUM	875	35	850-900
SODIUM	415	30	393-436
POTASSIUM	73	9.9	66-80
PHOSPHORUS	575	177	450-700
IRON	409	194	272-546
ZINC	134	25	116-151
MANGANESE	22	6.4	17-26
COPPER	13	2.1	11-14
BORON	8.5	0.7	8-9
SILICON	12000	8202	6200-17800
ALUMINUM	201	110	123-279

81. Feather mineral pattern G representing two (5.3 percent) of thirty-eight blue and lesser snow geese banded in the vicinity of the McConnell River delta, N.W.T., and shot elsewhere.

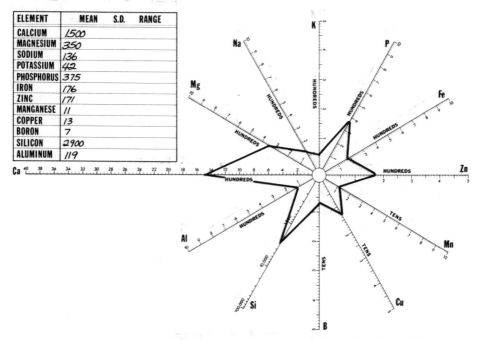

ELEMENT	MEAN	S.D.	RANGE
CALCIUM	1500		
MAGNESIUM	350		
SODIUM	136		
POTASSIUM	42		
PHOSPHORUS	375		
IRON	176		
ZINC	171		
MANGANESE	11		
COPPER	13		
BORON	7		
SILICON	2900		
ALUMINUM	119		

82. Feather mineral pattern H representing one (3 percent) of thirty-eight blue and lesser snow geese banded in the vicinity of the McConnell River delta, N.W.T., and shot elsewhere.

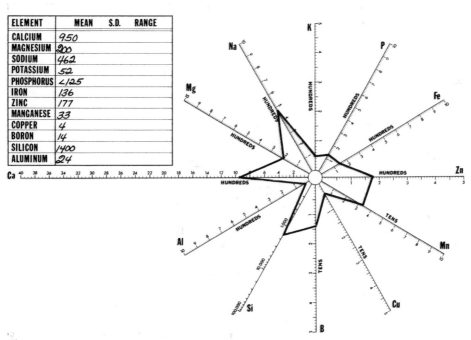

ELEMENT	MEAN	S.D.	RANGE
CALCIUM	950		
MAGNESIUM	200		
SODIUM	462		
POTASSIUM	52		
PHOSPHORUS	<125		
IRON	136		
ZINC	177		
MANGANESE	33		
COPPER	4		
BORON	14		
SILICON	1400		
ALUMINUM	24		

83. Feather mineral pattern I representing one (3 percent) of thirty-eight blue and lesser snow geese banded in the vicinity of the McConnell River delta, N.W.T., and shot elsewhere.

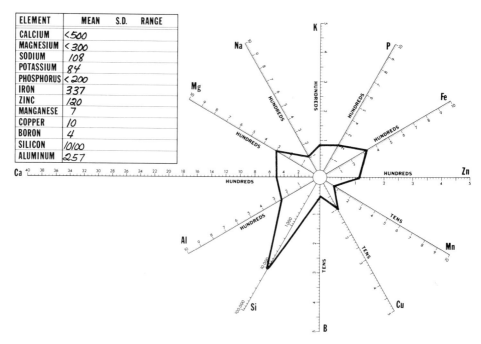

ELEMENT	MEAN	S.D.	RANGE
CALCIUM	<500		
MAGNESIUM	<300		
SODIUM	108		
POTASSIUM	84		
PHOSPHORUS	<200		
IRON	337		
ZINC	120		
MANGANESE	7		
COPPER	10		
BORON	4		
SILICON	10100		
ALUMINUM	257		

84. Feather mineral pattern J representing one (3 percent) of thirty-eight blue and lesser snow geese banded in the vicinity of the McConnell River delta, N.W.T., and shot elsewhere.

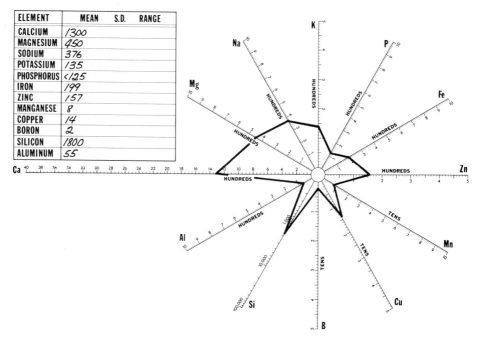

ELEMENT	MEAN	S.D.	RANGE
CALCIUM	1300		
MAGNESIUM	450		
SODIUM	376		
POTASSIUM	135		
PHOSPHORUS	<125		
IRON	199		
ZINC	157		
MANGANESE	8		
COPPER	14		
BORON	2		
SILICON	1800		
ALUMINUM	55		

85. Feather mineral pattern K representing one (3 percent) of thirty-eight blue and lesser snow geese banded in the vicinity of the McConnell River delta, N.W.T., and shot elsewhere.

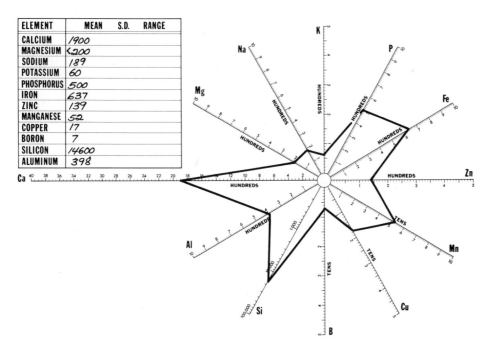

ELEMENT	MEAN	S.D.	RANGE
CALCIUM	1900		
MAGNESIUM	<200		
SODIUM	189		
POTASSIUM	60		
PHOSPHORUS	500		
IRON	637		
ZINC	139		
MANGANESE	52		
COPPER	17		
BORON	7		
SILICON	14600		
ALUMINUM	398		

86. Feather mineral pattern L representing one (3 percent) of thirty-eight blue and lesser snow geese banded in the vicinity of the McConnell River delta, N.W.T., and shot elsewhere.

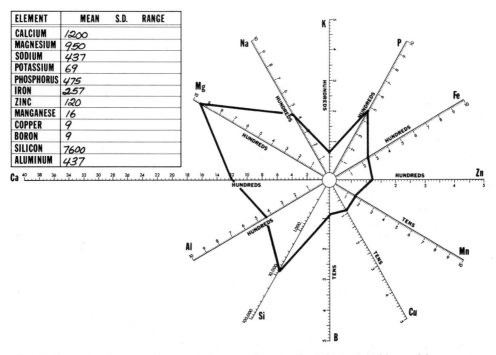

ELEMENT	MEAN	S.D.	RANGE
CALCIUM	1200		
MAGNESIUM	950		
SODIUM	437		
POTASSIUM	69		
PHOSPHORUS	475		
IRON	257		
ZINC	120		
MANGANESE	16		
COPPER	9		
BORON	9		
SILICON	7600		
ALUMINUM	437		

87. Feather mineral pattern M representing one (3 percent) of thirty-eight blue and lesser snow geese banded in the vicinity of the McConnell River delta, N.W.T., and shot elsewhere.

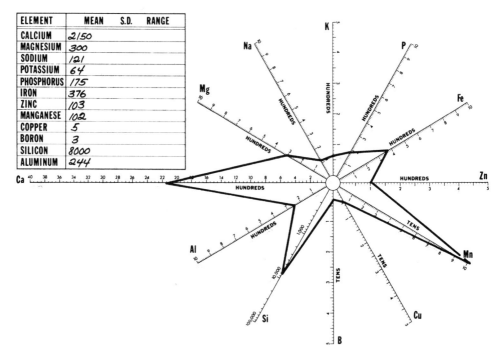

ELEMENT	MEAN	S.D.	RANGE
CALCIUM	2150		
MAGNESIUM	300		
SODIUM	121		
POTASSIUM	64		
PHOSPHORUS	175		
IRON	376		
ZINC	103		
MANGANESE	102		
COPPER	5		
BORON	3		
SILICON	8000		
ALUMINUM	244		

88. Feather mineral pattern N representing one (3 percent) of thirty-eight blue and lesser snow geese banded in the vicinity of the McConnell River delta, N.W.T., and shot elsewhere.

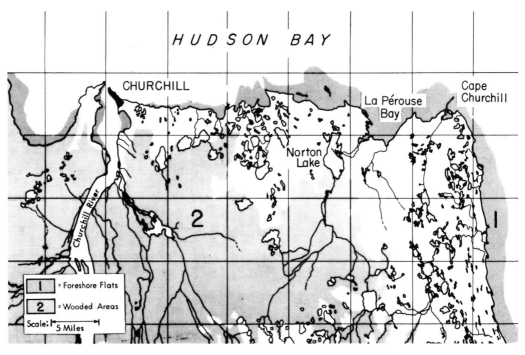

89. Geographic setting of the Cape Churchill colony.

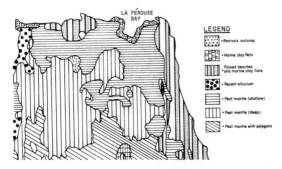

90. Surficial geology of the Cape Churchill, Manitoba, area. The map area is underlain by limestones of Ordovician age. (Redrawn from Ritchie 1962)

UPPER TRIASSIC. Lower part—shales; middle part—sandstones and siltstones; upper part—mainly sandstone.

MIDDLE (AND UPPER?) CARBONIFEROUS. Greenish calcareous siltstones interbedded with black pyritic shales and siltstones.

LOWER CARBONIFEROUS. Pillar Formation—Mainly siltstones, some shales, very subordinate interbedded sandstones. Uering Formation—Carbonates with minor shales and sandstones, mainly limestone, with some dolomitized limestone and siliceous limestone.

UPPER DEVONIAN? Metasedimentary rocks—slates, phyllites, and quartzites. Lower part—dark gray and black metasiltstones and sandy slates; upper part—quartzites and quartzose conglomerates. Age uncertain; similar to Devonian strata in the Brooks Range.

WRANGEL COMPLEX. Metamorphic rocks—quartz, albite, and muscovite—chlorite subfacies of the greenschists. Originally consisted of coarse quartzitic sandstones with subordinate arkoses and graywackes.

IGNEOUS INTRUSIONS. Saturate the Wrangel Complex and occur in a few sites in the Carboniferous strata. Amphibolites, gabbros, granosyenites and sodium-rich granites.

Bostick concludes: "There appear to be no metalliferous deposits on the island that might be expected to yield unusual concentrations of ions in the soil. Also, there are no serpentinites. If a map with more identification of streams and ridges were available, it would be possible from the literature to pinpoint areas of carbonate rocks of the Uering Formation and some black pyritic shales in the younger part of the Carboniferous section. For now, much of the area must be described as undifferentiated Carboniferous strata."

Svatkov (1958) has described in a general way the properties of soils on Wrangel Island. On the island, soil properties are closely related to topographic position and the effects of percolating water during short thaw periods. The pervasive influence of only 20 cm of precipitation per year is also evident. Arctic polygonal soils, arctic sod soils, and sod illuvial-humic soils are probably the most extensive soils in the breeding areas. Minor areas of gley and polygonal solonchak soils may also occur. Grasses and sedges are major components of the flora occupying sod-invested soils, and their roots—important goose foods—are largely confined to the uppermost 2–4 cm of soil, where they form mats and are readily available to the grazing birds. Svatkov gives virtually no compositional data for these soils, but one infers from those few available that the soils are not particularly fertile; phosphorus especially is low—on the order of 100 ppm (total). The solonchak soils possess carbonate efflorescences and are an expression of evaporative processes operating in the region. (The concept of a solonchack soil infers a substantial content of chlorides, sulfates, and bicarbonate salts—in amounts of at least one-half of one percent; see discussion of sulfur, appendix 2.) Although these soils are not important in their areal extent, one may conclude that salts might be abundant in areas used by snow geese.

"The fairly wide distribution of carbonate rocks in the mountainous portion of the island favors the accumulation of humus in the surface horizons of the soil in broad river valleys and in piedmont portions of the littoral plains to which carbonates may be transported by snowmelt. However, not much carbonate is evidently transported and the solid grainy structure which is usual for the dark-colored sod-gley-podzolic and sod-gley soils of the forest zone does not occur in the arctic gley-sod soils of Wrangel Island" (Svatkov 1958:84).

Feather Mineral Patterns. The feather mineral pattern for Wrangel Island is based on seventeen geese banded there and shot in Oregon and California (Figure *111*). Four sub-patterns were distinguished. Two of these are basically variations on a theme (Figures *112–13*); a third is notable for its high manganese value (Figure *114*). The fourth pattern (Figure *115*) is sufficiently similar to patterns of four Ross' geese and lesser snow geese from New Mexico to suggest that it could represent two geese, originally banded on Wrangel Island, that had spent the summer in the Queen Maud Gulf lowlands prior to being bagged by hunters. On numerical grounds alone, however, the odds make this latter conjecture very unlikely. More likely, pattern D represents a segment of the population on Wrangel Island, and this pattern resembles a basic pattern type that occurs in other colonies

91. La Pérouse Bay near Cape Churchill, Manitoba. The main nesting area of the Cape Churchill colony lies adjacent to this bay. Note the width of clay flats exposed at low tide.

92. View looking southeastward of La Pérouse Bay. The clumped vegetation consists of several species of willows.

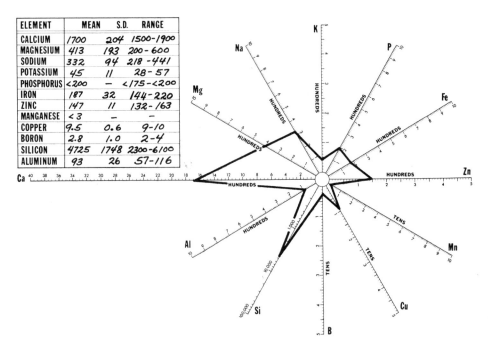

ELEMENT	MEAN	S.D.	RANGE
CALCIUM	1700	204	1500-1900
MAGNESIUM	413	193	200-600
SODIUM	332	94	218-441
POTASSIUM	45	11	28-57
PHOSPHORUS	<200	—	<175-<200
IRON	187	32	144-220
ZINC	147	11	132-163
MANGANESE	<3	—	—
COPPER	9.5	0.6	9-10
BORON	2.8	1.0	2-4
SILICON	4725	1748	2300-6100
ALUMINUM	93	26	57-116

93. Feather mineral pattern representing four (100 percent) of the blue and lesser snow geese banded on Cape Churchill, Manitoba, and shot elsewhere.

94. Jenny Lind Island, Queen Maud Gulf, N.W.T. View looking seaward from an area of ice-heaved limestone blocks and limestone detritus (*felsenmeer*). (Photograph courtesy of David F. Parmalee)

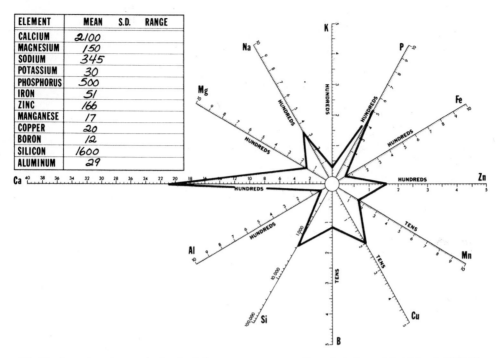

ELEMENT	MEAN	S.D.	RANGE
CALCIUM	2100		
MAGNESIUM	150		
SODIUM	345		
POTASSIUM	30		
PHOSPHORUS	500		
IRON	51		
ZINC	166		
MANGANESE	17		
COPPER	20		
BORON	12		
SILICON	1600		
ALUMINUM	29		

95. Feather mineral pattern of a single lesser snow goose collected on Jenny Lind Island, N.W.T.

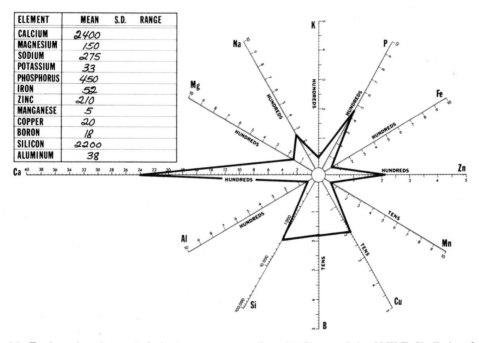

ELEMENT	MEAN	S.D.	RANGE
CALCIUM	2400		
MAGNESIUM	150		
SODIUM	275		
POTASSIUM	33		
PHOSPHORUS	450		
IRON	52		
ZINC	210		
MANGANESE	5		
COPPER	20		
BORON	18		
SILICON	2200		
ALUMINUM	38		

96. Feather mineral pattern of a lesser snow goose collected at Sherman Inlet, N.W.T. Similarity of the pattern of this goose with the pattern of the goose collected on Jenny Lind Island is indicative that the limestones of the two breeding areas are of the same age.

77

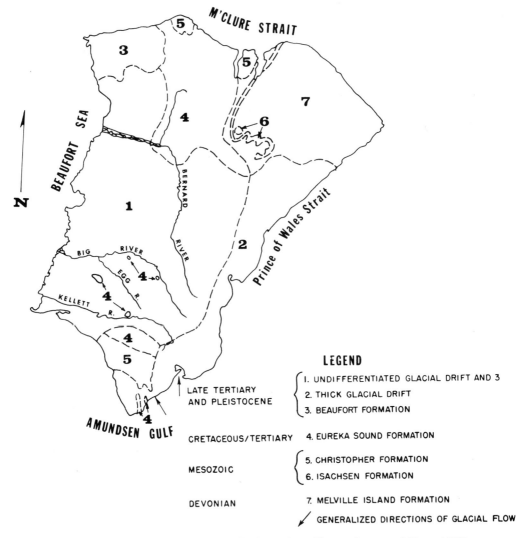

LEGEND

LATE TERTIARY
AND PLEISTOCENE

{ 1. UNDIFFERENTIATED GLACIAL DRIFT AND 3
{ 2. THICK GLACIAL DRIFT
{ 3. BEAUFORT FORMATION

CRETACEOUS/TERTIARY   4. EUREKA SOUND FORMATION

MESOZOIC

{ 5. CHRISTOPHER FORMATION
{ 6. ISACHSEN FORMATION

DEVONIAN

7. MELVILLE ISLAND FORMATION

GENERALIZED DIRECTIONS OF GLACIAL FLOW

97. Geologic map of Banks Island. (Redrawn from Thornsteinsson and Tozer 1962)

(Southampton Island, Figure 68; McConnell River, Figure 79). Some of the nonbreeding birds of the Wrangel Island colony (yearlings and failed adults) undergo the molt on the mainland coast at Cape Schmidt and at the mouths of the Indigirka and Kolyma rivers (F. G. Cooch, personal communication, 1969). A few of these birds could have been included in the feather sample of geese banded on Wrangel Island and thus would account for one of the disparate patterns found.

HIGH ARCTIC COLONIES—GREATER SNOW GEESE

The high Arctic region is the breeding range of greater snow geese. According to a brief summary of the range prepared for us by Richard H. Kerbes, "it includes Baffin Island (north of Cumberland Sound and Foxe Basin), Bylot Island, the Queen Elizabeth Islands, and northwest Greenland. The majority (at least three-fourths) of the population nests near river deltas, at the heads of fiords on northern Baffin Island, and on the coastal plateau of southern Bylot Island. These areas support fairly large concentrations or colonies of up to several hundred nests each. In less favoured parts of the range, greater snow goose nests are found singly or in widely scattered concentrations of usually less that 25 nests." The total population of this race in 1969 was approximately 66,000, of which less than 1,000 were believed to have originated in northwest Greenland (Heyland and Boyd 1970).

98. View of a portion of the Banks Island colony along the Egg River. Note blue-lesser snow pair in foreground. Blue phase birds are very rare in the western Arctic of Canada. (Photograph courtesy of E. H. McEwen)

99. Valley of the Egg River, Banks Island, as seen from an adjacent ridge. Fine white specks scattered over valley floor are nesting pairs of lesser snow geese. (Photograph courtesy of E. H. McEwen)

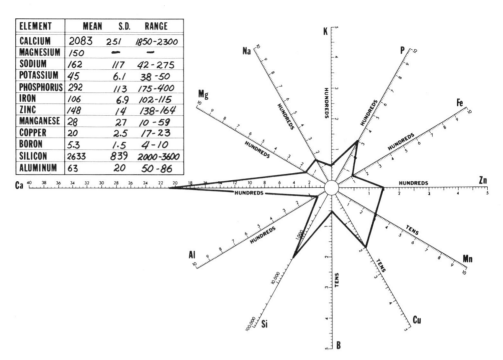

ELEMENT	MEAN	S.D.	RANGE
CALCIUM	2083	251	1850-2300
MAGNESIUM	150	—	—
SODIUM	162	117	42-275
POTASSIUM	45	6.1	38-50
PHOSPHORUS	292	113	175-400
IRON	106	6.9	102-115
ZINC	148	14	138-164
MANGANESE	28	27	10-59
COPPER	20	2.5	17-23
BORON	5.3	1.5	4-10
SILICON	2633	839	2000-3600
ALUMINUM	63	20	50-86

100. Feather mineral pattern for the lesser snow geese collected on Banks Island (Cape Kellett and Sachs Harbor).

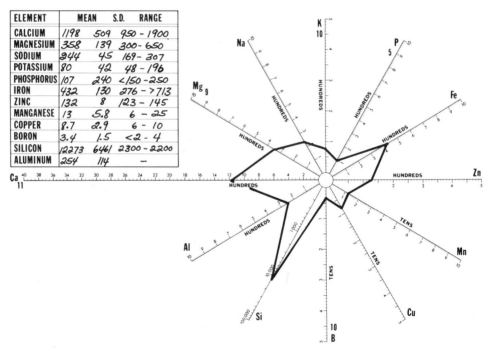

ELEMENT	MEAN	S.D.	RANGE
CALCIUM	1198	509	950-1900
MAGNESIUM	358	139	300-650
SODIUM	244	45	169-307
POTASSIUM	80	42	48-196
PHOSPHORUS	107	240	<150-250
IRON	432	130	276->713
ZINC	132	8	123-145
MANGANESE	13	5.8	6-25
COPPER	8.7	2.9	6-10
BORON	3.4	1.5	<2-4
SILICON	12273	6461	2300-2200
ALUMINUM	254	114	—

101. Feather mineral pattern of eleven lesser snow geese shot in September on the west side of the mouth of the Mackenzie River. The high iron content of these samples suggest that they are representative of the geese from the Egg River colony. The tundra area to the west of the mouth is a feeding area for flocks migrating south from Banks Island.

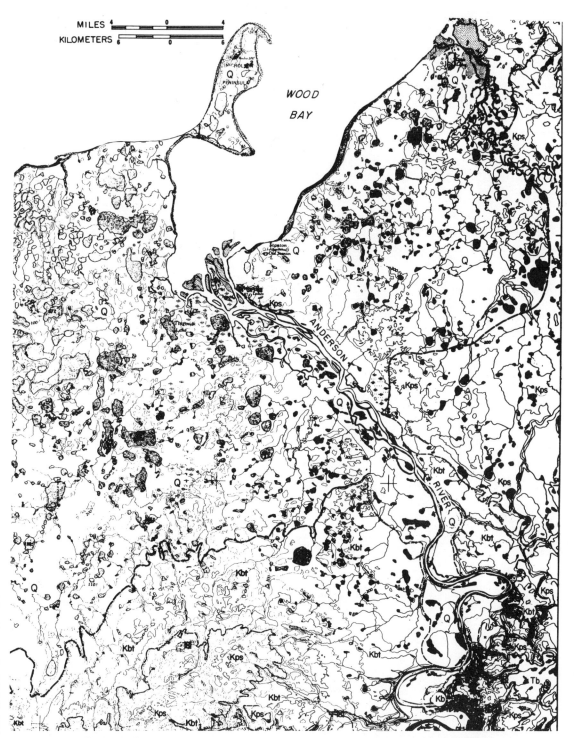

102. Geology of part of the lower portions of the Anderson River basin, District of Mackenzie, N.W.T. Geological units shown are: Q) Quaternary—glacial till, outwash, and alluvium; Tb) Tertiary and/or Quaternary—Beaufort formation (?), unconsolidated gravel and sand, quartzite, dolomite, and black chert pebbles; Kps) (Upper) Cretaceous—pale shale zone; Kbt) (Upper) Cretaceous—bituminous zone; Kb) (Lower) Cretaceous—bentonitic zone. Formation contact boundaries where not definitely identified are indicated by dashed lines. Topographic contour interval is one hundred feet. (Reprinted from Yorath and Balkwill 1970)

It will be a number of years before most of the smaller concentrations have been banded and adequate samples of primary feathers of banded birds have been collected for analyses. Because of the diverse geologies on which these geese nest, it can be anticipated that a diversity of feather mineral patterns will characterize populations of this race. This diversity is evident in our sample from the Cape Tourmente area of Quebec and Back Bay N.W.R., Virginia. Feathers from the latter area were taken from birds found dead.

Bylot Island Colony

Bylot Island is located off the northeast tip of Baffin Island, lying at 73° N and 80° W (Figure 4).

Most of its forty-two hundred square miles are mountainous and glacier covered; the southwest corner of the island consists of a two-hundred-foot-high plateau of gravel outwash that is deeply dissected along its outer edges by streams emanating from the glaciers (Figures 116–17). Nesting is confined wholly to this dissected plateau (Figure 118), nests being found both on ravine slopes and on the intervening ridges (Lemieux 1959).

During the molt, the nonbreeding geese that are at first distributed along the coastal polygon marsh around the edge of the plateau (Figure 119) gradually move inland, where they are soon joined by the newly hatched broods and the breeding adults. About a week after the hatching, family parties are

103. Islands in the outer portions of the delta of the Anderson River, N.W.T. Nesting of this colony is confined to these low grass-sedge covered islands.

104. A portion of the bentonitic zone along the lower Anderson River, N.W.T. Note dramatic steep-sloped erosion pattern that develops on these water-laden, bentonitic-clay soils. The montmorillite fraction of this soil has a water holding capacity 2.6–3.2 times its weight.

scattered over the entire plateau, some reaching the foot of the mountains. In August, when the young are three to four weeks old, the families move down to the coastal marsh (Figure *118*) where, presumably, the major portion of the wing primaries are grown. This population accounts for the greater portion of the spectacular concentration of greater snow geese that feed each fall in the Cape Tourmente area of the St. Lawrence River before continuing their migration to wintering quarters in the Chesapeake Bay area (Lemieux 1959).

Geology and Soils. The geology of Bylot Island is poorly known. The most recent geological map of Canada (Geological Survey of Canada: Map 1045A) shows the southwest corner of the island, the site of the colony, to be underlain by rocks classified only as sedimentary (Figure *116*).

Feather Mineral Patterns. Primary feathers of greater snow geese from three breeding areas were analyzed (Figures *120–22*). It is unfortunate that an adequate series of feather samples from the Bylot Island colony was not available for study, as probably the world's largest concentration of greater snow geese breed on the glacier-free, southern triangle of this island (Figure *116*). Per-

haps the mineral pattern of the primaries from a single museum skin from Pond Inlet, Baffin Island (Figures *116, 120*) depicts the salient characteristics of this colony. Similar hopes of representativeness must be held for the primaries of single specimens collected at Croker Bay, Devon Island, and Slidre Fiord, Ellesmere Island (Figures *121–22*). All three patterns are highly distinctive, reflecting the uniqueness of the geologies of the areas from which they came. Unfortunately, with the exception of pattern P (Figure *135*), which resembles pattern B (Figure *121*) in some respects, the patterns of other greater snow geese shot away from their breeding grounds (Figures *123–36*) show no resemblance to the above patterns.

A major stopover area for greater snow geese in their fall migration is the Cape Tourmente area in the St. Lawrence River. Two feather mineral patterns from geese shot there, E and F (Figures *124–25*), are basically variations on the same theme. Pattern D (Figure *123*) is fairly similar to patterns J and K from two greater snow geese collected near the Squaw Creek N.W.R., Missouri (Figures *128–29*) and to patterns K and L of geese from the Back Bay N.W.R., Virginia (Figures *130–31*).

Patterns M, N, and possibly O, also from the Back Bay N.W.R., may be variations representing one breeding area (Figures *132–34*). Pattern Q (Figure *136*) of a greater snow goose shot on the south coast of Hudson Bay at Ft. Severn represents a goose that may have summered on the Foxe Plain as discussed earlier. It also could represent another breeding segment of greater snow geese, and its partial similarity to pattern E (Figure *126*) should also be noted.

Pattern H (Figure *127*) is highly distinctive and must be presumed to represent still another small isolated breeding population sequestered somewhere in the high Arctic of North America, although one of several breeding areas in northern Greenland cannot be ruled out as the origin of this bird.

Three of the patterns are characterized by high copper values, two by high iron values, and one by a high level of manganese. These findings are indicative of the relatively high level of mineralization of many high Arctic areas.

ROSS' GOOSE COLONIES

Probably over 95 percent of the world's stock of Ross' geese nest in a 185-mile-wide (long. 96° to 104°) and 85-mile-deep (lat. 66° 6′ to 68°) region

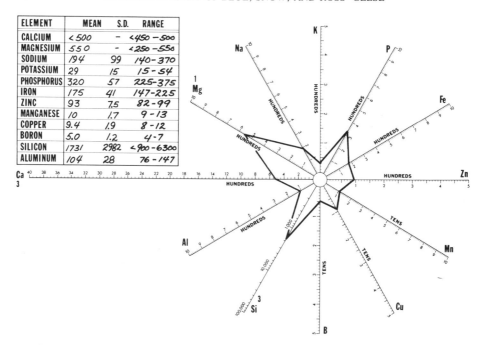

ELEMENT	MEAN	S.D.	RANGE
CALCIUM	<500	–	<450-500
MAGNESIUM	550	–	<250-550
SODIUM	194	99	140-370
POTASSIUM	29	15	15-54
PHOSPHORUS	320	57	225-375
IRON	175	41	147-225
ZINC	93	7.5	82-99
MANGANESE	10	1.7	9-13
COPPER	9.4	1.9	8-12
BORON	5.0	1.2	4-7
SILICON	1731	2982	<900-6300
ALUMINUM	104	28	76-147

105. Feather mineral pattern of five lesser snow geese banded in the Anderson River delta and shot elsewhere.

lying in the area of former marine submergence south of Queen Maud Gulf (Figures *4, 137–42*). Physiographically, the range is limited on the west by the higher and better-drained lands that lie to the east of Bathurst Inlet and on the east by Chantrey Inlet (Figure *137*).

Geology and Soils. The geology of the Queen Maud Gulf area has been described briefly by Queneau (Hanson et al. 1956), Heywood (1961), and Fraser (1964). The latter two reports are reconnaissance reports of the Canadian Geological Survey that are accompanied by geologic maps based on observations made along transects at six-mile intervals. The surficial geology of the northern portion of the District of Keewatin has been described by Craig (1961).

The bedrock of the area to the south of Queen Maud Gulf is composed chiefly of massive crystalline rocks of the Canadian Shield, particularly granite and gneissic granite (Heywood 1961; Figure *140*). The bedrock also includes scattered pods and zones of calcium-rich metamorphic rocks ranging from plagioclase-hornblende gneisses and schists to amphibolites (Figure *137*). Outcrops of limestones and dolomites of Ordovician and Silurian ages and perhaps Cambrian age occur to the east on Adelaide Peninsula. Copper-bearing rock was first reported from the region by Queneau

(Hanson et al. 1956), and at the present time (1974) the general region is being actively explored for minerals.

The zone of lakes used by nesting Ross' geese falls within the area of marine submergence which extends about eighty-five miles inland from the

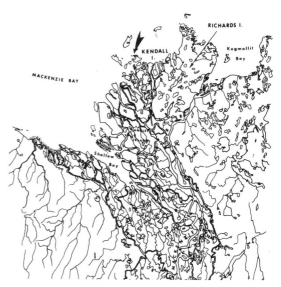

106. Geographic setting of the Kendall Island colony of lesser snow geese.

107. Mated pairs of lesser snow geese on Kendall Island, N.W.T. Note coarseness of vegetation. Probably because of this factor, broods are taken to the adjacent mainland for rearing.

108. Mated pairs of lesser snow geese on Kendall Island, N.W.T. Note that large areas of the island have been denuded of vegetation by the feeding activities of these geese. Photograph taken June 1967.

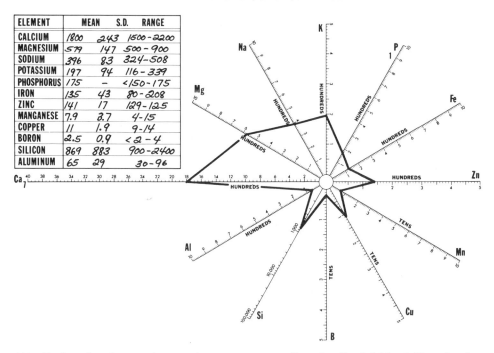

ELEMENT	MEAN	S.D.	RANGE
CALCIUM	1800	243	1500-2200
MAGNESIUM	579	147	500-900
SODIUM	396	83	324-508
POTASSIUM	197	94	116-339
PHOSPHORUS	175	–	<150-175
IRON	135	43	80-208
ZINC	141	17	129-125
MANGANESE	7.9	3.7	4-15
COPPER	11	1.9	9-14
BORON	2.5	0.9	<2-4
SILICON	869	883	900-2400
ALUMINUM	65	29	30-96

109. Feather mineral pattern for seven lesser snow geese collected on Kendall Island. The primaries of these geese were not fully grown.

coast (Figure *141*). Old beach ridges are prominent on the slopes of coastal hills. The sediments of the region are silty or clay rich and contain pebbles, and locally, range up to 300 feet thick along the Arrowsmith River south of Pelly Bay.

The soils fall within the general and rather non-descript classification of Arctic soils (Ellis 1960). These soils exhibit little development in the sense that horizons are practically absent except for the accumulation of a thin organic horizon. Where the plant cover is well developed, vegetation serves to insulate the soil, and permafrost occurs near the soil surface. Sometimes in soils developed in un-consolidated fine-textured parent materials a thin zone of gray color occurs over the permafrost, the color reflecting the occurrence of reducing or an-aerobic conditions.

Feather Mineral Patterns. The largest colony of nesting Ross' geese is found at Karrak Lake (lat. 67° 15', long. 100° 15', Ryder 1969) (Figure *137*). This lake, called Kangawan Lake by Hanson et al. (1956), drains into a tributary of the Simpson River (Figure *137*). We were able to establish the feather mineral pattern of the geese in this colony (Figure *142*) particularly well as a result of a large series of wings kindly collected for us by John P. Ryder, then of the Canadian Wildlife Service.

With respect to their nutritional experience the previous summer while growing flight feathers, the sample of Ross' geese from Karrak Lake can be regarded as unusually homogeneous. This assessment is attributed to the fact that geese tend to return with exceptional fidelity to their birthplaces or breeding areas, the females sometimes using their nest scrapes of the previous year. After the eggs hatch, the adults and broods scatter out over the surrounding tundra although their movements may tend to be mainly downstream toward the coast.

Patterns B, C, D, E, and F (Figures *143–47*) resemble in varying degrees pattern A (Figure *142*) and doubtless represent geese raised in similar geological areas. In fact, pattern F (Figure *147*) represents a Ross' goose banded along the Simpson River, and hence, this goose in all likelihood nested in the Karrak Lake colony.

Patterns G, H, I, and J (Figures *148–51*) are believed to represent geese from the Lake Arlone colony (Figure *141*) or from colonies nesting and feeding in country of somewhat similar geology. This judgment is based on the fact that a banded bird from the Lake Arlone colony is included in each of the two-bird averages represented by patterns G (Figure *148*) and J (Figure *151*). Conversely, a bird banded near Lake Arlone is included in pattern B (Figure *143*) and a goose banded along the Simpson River is included in pattern H (Figure *149*). It is possible that these latter two individuals represent "crossovers" from their earlier colonies. These two major pattern groups differ primarily in respect to sodium (averages, 577 versus 254 ppm) and by their respective zinc and copper values. The significance of the relatively high copper values and the sex differential in the values found for the Karrak Lake geese are treated in detail in Chapter 7. The difference in sodium values between the two series may represent the relative extent to which these groups reach coastal areas in their postnesting feeding movements on foot, but much more likely, it can be at-

tributed to the probable presence of clay-rich sediments in the vicinity of the Karrak Lake area that are high in sodium content, comparable to the soil sample from the bank of Laine Creek (Table 12).

The high levels of sodium and magnesium in feathers from the Karrak Lake population are especially significant in the context of data (Table 12) for water extracts of soils of the region. The soils of clay loam texture clearly have retained salinity associated with former marine subemergence and magnesium and sodium contents are notably high. In the absence of analyses of plants from this area we can infer that plant tissue levels of these elements will be enriched correspondingly. This locality, situated at considerable distance from con-

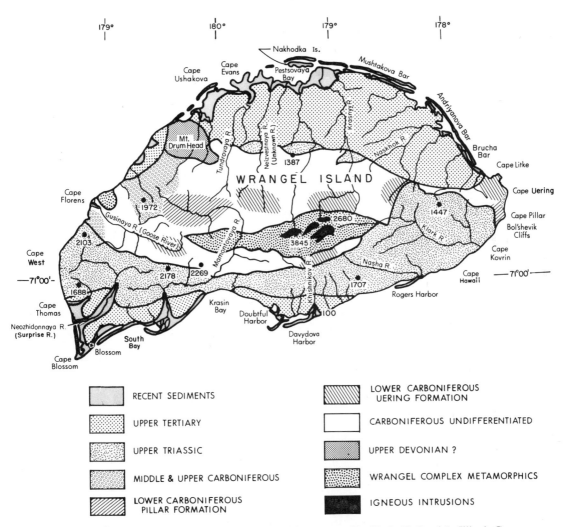

110. Geologic map of Wrangel Island, U.S.S.R. (Map prepared by Neely H. Bostick, Illinois Geological Survey; after Ivanov 1973, Gnibidenko 1968, and Tilman et al. 1964)

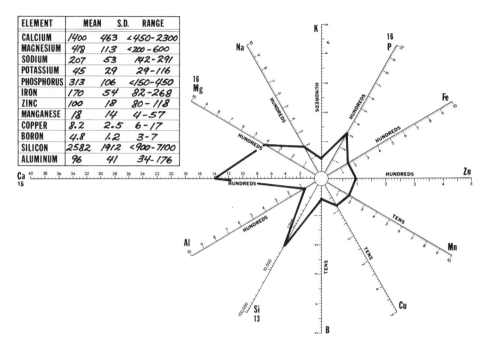

ELEMENT	MEAN	S.D.	RANGE
CALCIUM	1400	463	<450-2300
MAGNESIUM	418	113	<200-600
SODIUM	207	53	142-291
POTASSIUM	45	29	29-116
PHOSPHORUS	313	106	<150-450
IRON	170	54	82-268
ZINC	100	18	80-118
MANGANESE	18	14	4-57
COPPER	8.2	2.5	6-17
BORON	4.8	1.2	3-7
SILICON	2582	1912	<900-7100
ALUMINUM	96	41	34-176

111. Feather mineral pattern of seventeen lesser snow geese banded on Wrangel Island, U.S.S.R., and shot in the United States.

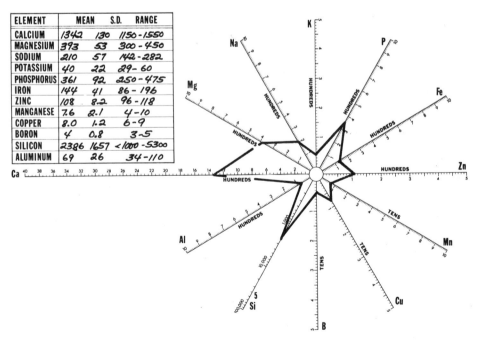

ELEMENT	MEAN	S.D.	RANGE
CALCIUM	1342	130	1150-1550
MAGNESIUM	393	53	300-450
SODIUM	210	57	142-282
POTASSIUM	40	22	29-60
PHOSPHORUS	361	92	250-475
IRON	144	41	86-196
ZINC	108	8.2	96-118
MANGANESE	7.6	2.1	4-10
COPPER	8.0	1.2	6-9
BORON	4	0.8	3-5
SILICON	2386	1657	<1000-5300
ALUMINUM	69	26	34-110

112. Feather mineral pattern A of seven (41 percent) of seventeen lesser snow geese banded on Wrangel Island, U.S.S.R., and shot in the United States.

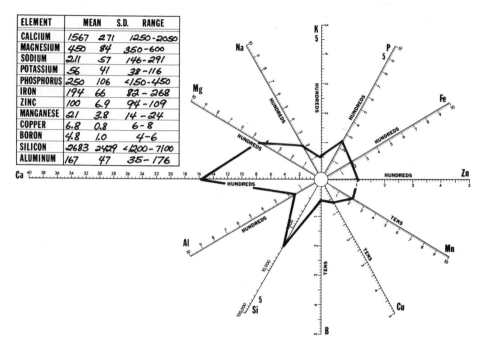

ELEMENT	MEAN	S.D.	RANGE
CALCIUM	1567	271	1250-2050
MAGNESIUM	450	84	350-600
SODIUM	211	57	146-291
POTASSIUM	56	41	38-116
PHOSPHORUS	250	106	<150-450
IRON	194	66	82-268
ZINC	100	6.9	94-109
MANGANESE	21	3.8	14-24
COPPER	6.8	0.8	6-8
BORON	4.8	1.0	4-6
SILICON	2683	2429	<1200-7100
ALUMINUM	167	47	35-176

113. Feather mineral pattern B of six (35 percent) of seventeen lesser snow geese banded on Wrangel Island, U.S.S.R., and shot in the United States.

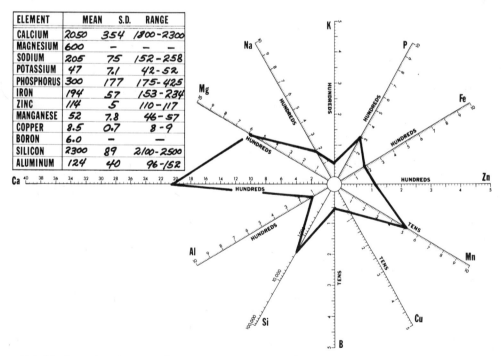

ELEMENT	MEAN	S.D.	RANGE
CALCIUM	2050	354	1800-2300
MAGNESIUM	600	—	—
SODIUM	205	75	152-258
POTASSIUM	47	7.1	42-52
PHOSPHORUS	300	177	175-425
IRON	194	57	153-234
ZINC	114	5	110-117
MANGANESE	52	7.8	46-57
COPPER	8.5	0.7	8-9
BORON	6.0	—	—
SILICON	2300	89	2100-2500
ALUMINUM	124	40	96-152

114. Feather mineral pattern C of two (12 percent) of seventeen lesser snow geese banded on Wrangel Island, U.S.S.R., and shot in the United States.

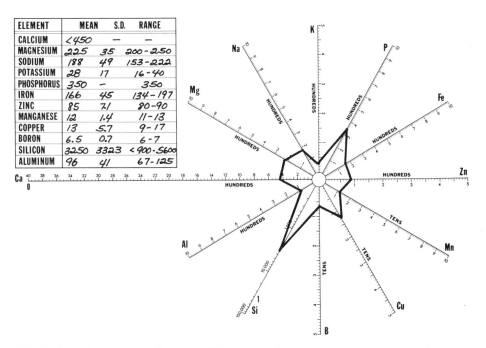

ELEMENT	MEAN	S.D.	RANGE
CALCIUM	<450	—	—
MAGNESIUM	225	35	200-250
SODIUM	188	49	153-222
POTASSIUM	28	17	16-40
PHOSPHORUS	350	—	350
IRON	166	45	134-197
ZINC	85	7.1	80-90
MANGANESE	12	1.4	11-13
COPPER	13	5.7	9-17
BORON	6.5	0.7	6-7
SILICON	3250	3323	<900-5600
ALUMINUM	96	41	67-125

115. Feather mineral pattern D of two (12 percent) of seventeen lesser snow geese banded on Wrangel Island, U.S.S.R., and shot in the United States. This resembles Anderson River feather mineral pattern except for low Mg value.

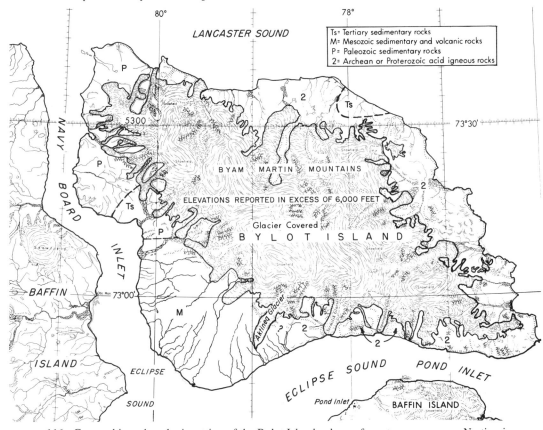

116. Geographic and geologic setting of the Bylot Island colony of greater snow geese. Nesting is restricted to the southwest triangle of the island.

90

117. Glacier and adjacent uplands, Bylot Island, N.W.T.

118. Upland slopes, the nesting habitat of the Bylot Island colony. View looking southward across Navy Board Inlet. A portion of Borden Peninsula is seen on the horizon.

119. A marshy area of coastal plain of Bylot Island. Ground patterns and polygonal ponds are formed by freeze and thaw cycles.

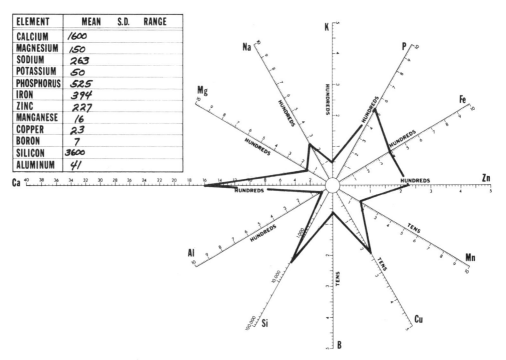

ELEMENT	MEAN	S.D.	RANGE
CALCIUM	1600		
MAGNESIUM	150		
SODIUM	263		
POTASSIUM	50		
PHOSPHORUS	525		
IRON	394		
ZINC	227		
MANGANESE	16		
COPPER	23		
BORON	7		
SILICON	3600		
ALUMINUM	41		

120. Feather mineral pattern A representing a greater snow goose collected at Pond Inlet, Baffin Island.

92

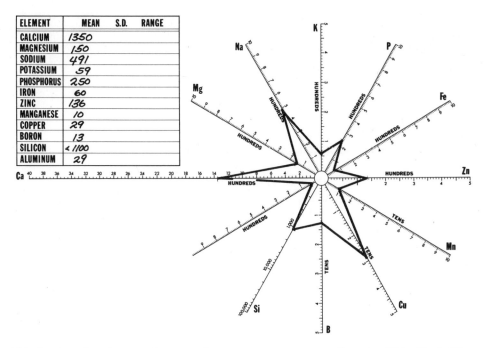

ELEMENT	MEAN	S.D.	RANGE
CALCIUM	1350		
MAGNESIUM	150		
SODIUM	491		
POTASSIUM	59		
PHOSPHORUS	250		
IRON	60		
ZINC	136		
MANGANESE	10		
COPPER	29		
BORON	13		
SILICON	<1100		
ALUMINUM	29		

121. Feather mineral pattern B representing a greater snow goose collected at Slidre Fiord, Ellesmere Island, N.W.T.

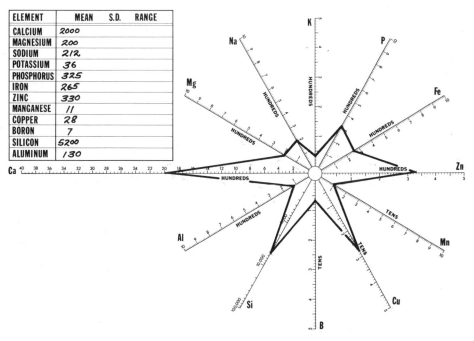

ELEMENT	MEAN	S.D.	RANGE
CALCIUM	2000		
MAGNESIUM	200		
SODIUM	212		
POTASSIUM	36		
PHOSPHORUS	325		
IRON	265		
ZINC	330		
MANGANESE	11		
COPPER	28		
BORON	7		
SILICON	5200		
ALUMINUM	130		

122. Feather mineral pattern C representing a greater snow goose collected at Croker Bay, Devon Island, N.W.T.

93

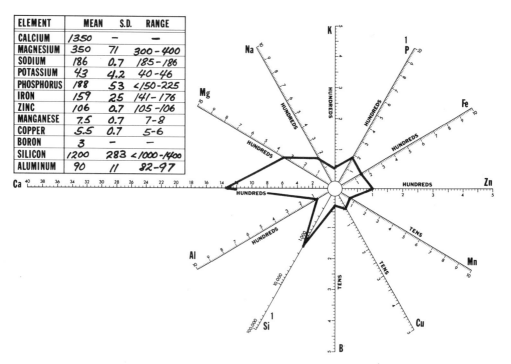

ELEMENT	MEAN	S.D.	RANGE
CALCIUM	1350	–	–
MAGNESIUM	350	71	300-400
SODIUM	186	0.7	185-186
POTASSIUM	43	4.2	40-46
PHOSPHORUS	188	53	<150-225
IRON	159	25	141-176
ZINC	106	0.7	105-106
MANGANESE	7.5	0.7	7-8
COPPER	5.5	0.7	5-6
BORON	3	–	–
SILICON	1200	283	<1000-1400
ALUMINUM	90	11	82-97

123. Feather mineral pattern D representing two greater snow geese bagged in the vicinity of Cape Tourmente, Quebec.

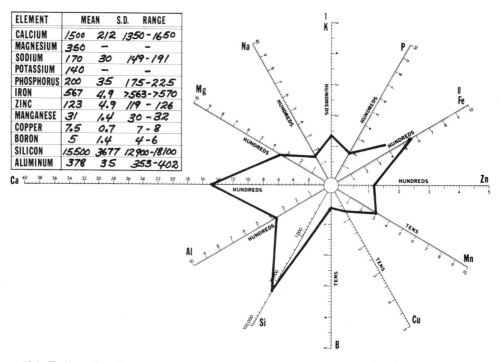

ELEMENT	MEAN	S.D.	RANGE
CALCIUM	1500	212	1350-1650
MAGNESIUM	350	–	–
SODIUM	170	30	149-191
POTASSIUM	140	–	–
PHOSPHORUS	200	35	175-225
IRON	567	4.9	>563->570
ZINC	123	4.9	119-126
MANGANESE	31	1.4	30-32
COPPER	7.5	0.7	7-8
BORON	5	1.4	4-6
SILICON	15500	3677	12900-18100
ALUMINUM	378	35	353-402

124. Feather mineral pattern E representing two greater snow geese bagged in the vicinity of Cape Tourmente, Quebec.

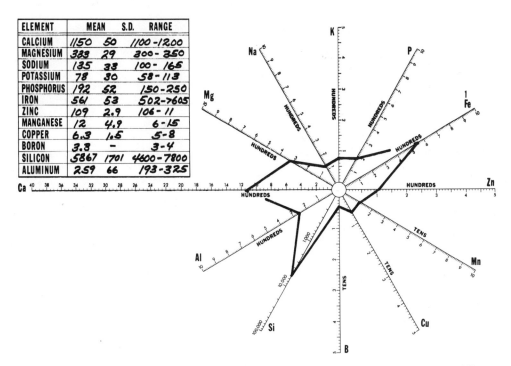

ELEMENT	MEAN	S.D.	RANGE
CALCIUM	1150	50	1100-1200
MAGNESIUM	323	29	300-350
SODIUM	135	33	100-165
POTASSIUM	78	30	58-113
PHOSPHORUS	192	52	150-250
IRON	561	53	502-7605
ZINC	109	2.9	106-11
MANGANESE	12	4.9	6-15
COPPER	6.3	1.5	5-8
BORON	3.3	–	3-4
SILICON	5867	1701	4600-7800
ALUMINUM	259	66	193-325

125. Feather mineral pattern F representing three greater snow geese bagged in the vicinity of Cape Tourmente, Quebec.

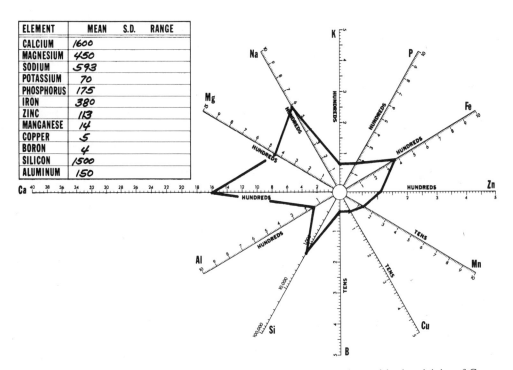

ELEMENT	MEAN	S.D.	RANGE
CALCIUM	1600		
MAGNESIUM	450		
SODIUM	593		
POTASSIUM	70		
PHOSPHORUS	175		
IRON	380		
ZINC	113		
MANGANESE	14		
COPPER	5		
BORON	4		
SILICON	1500		
ALUMINUM	150		

126. Feather mineral pattern G representing a greater snow goose bagged in the vicinity of Cape Tourmente, Quebec.

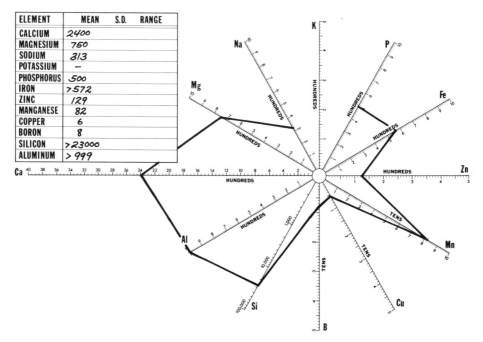

ELEMENT	MEAN	S.D.	RANGE
CALCIUM	2400		
MAGNESIUM	750		
SODIUM	313		
POTASSIUM	—		
PHOSPHORUS	500		
IRON	>572		
ZINC	129		
MANGANESE	82		
COPPER	6		
BORON	8		
SILICON	>23000		
ALUMINUM	>999		

127. Feather mineral pattern H representing a greater snow goose bagged in the vicinity of Cape Tourmente, Quebec.

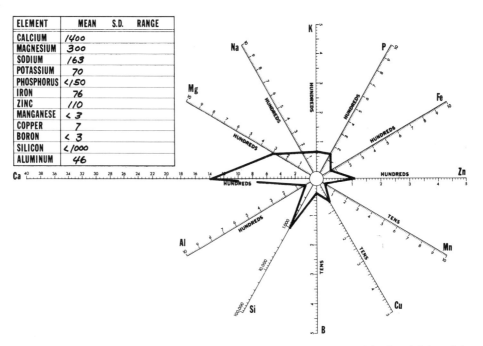

ELEMENT	MEAN	S.D.	RANGE
CALCIUM	1400		
MAGNESIUM	300		
SODIUM	163		
POTASSIUM	70		
PHOSPHORUS	<150		
IRON	76		
ZINC	110		
MANGANESE	<3		
COPPER	7		
BORON	<3		
SILICON	<1000		
ALUMINUM	46		

128. Feather mineral pattern I representing a greater snow goose bagged in the vicinity of the Squaw Creek N.W.R., Missouri.

96

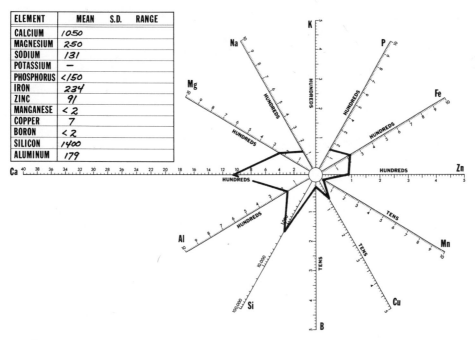

ELEMENT	MEAN	S.D.	RANGE
CALCIUM	1050		
MAGNESIUM	250		
SODIUM	131		
POTASSIUM	—		
PHOSPHORUS	<150		
IRON	234		
ZINC	91		
MANGANESE	<2		
COPPER	7		
BORON	<2		
SILICON	1400		
ALUMINUM	179		

129. Feather mineral pattern J representing a greater snow goose bagged in the vicinity of the Squaw Creek N.W.R., Missouri.

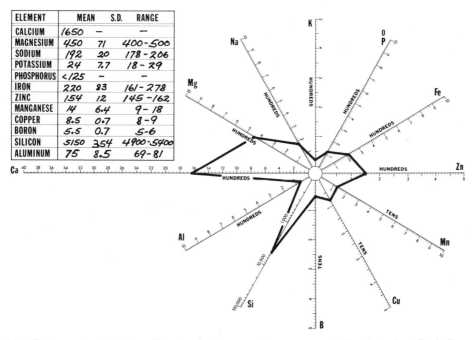

ELEMENT	MEAN	S.D.	RANGE
CALCIUM	1650	—	—
MAGNESIUM	450	71	400-500
SODIUM	192	20	178-206
POTASSIUM	24	7.7	18-29
PHOSPHORUS	<125	—	—
IRON	220	93	161-278
ZINC	154	12	145-162
MANGANESE	14	6.4	9-18
COPPER	8.5	0.7	8-9
BORON	5.5	0.7	5-6
SILICON	5150	354	4900-5400
ALUMINUM	75	8.5	69-81

130. Feather mineral pattern K representing a greater snow goose found dead at Back Bay, Virginia.

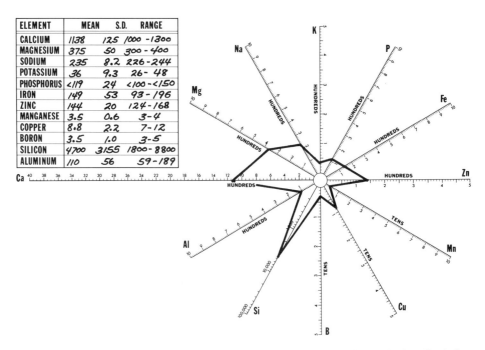

ELEMENT	MEAN	S.D.	RANGE
CALCIUM	1138	125	1000 -1300
MAGNESIUM	375	50	300 - 400
SODIUM	235	8.2	226 -244
POTASSIUM	36	9.3	26- 48
PHOSPHORUS	<119	24	<100-<150
IRON	149	53	93 -196
ZINC	144	20	124-168
MANGANESE	3.5	0.6	3-4
COPPER	8.8	2.2	7-12
BORON	3.5	1.0	3-5
SILICON	4700	3155	1800- 8800
ALUMINUM	110	56	59 -189

131. Feather mineral pattern L representing four greater snow geese found dead at Back Bay, Virginia.

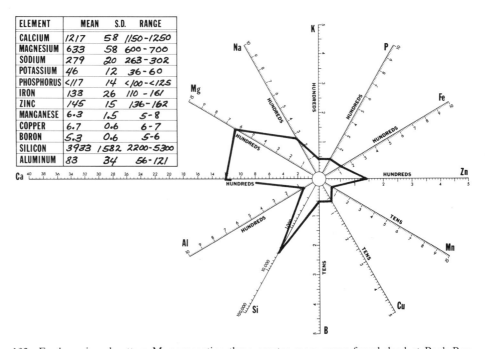

ELEMENT	MEAN	S.D.	RANGE
CALCIUM	1217	58	1150-1250
MAGNESIUM	633	58	600 -700
SODIUM	279	20	263 -302
POTASSIUM	46	12	36-60
PHOSPHORUS	<117	14	<100-<125
IRON	133	26	110 -161
ZINC	145	15	136-162
MANGANESE	6.3	1.5	5-8
COPPER	6.7	0.6	6-7
BORON	5.3	0.6	5-6
SILICON	3933	1582	2200-5300
ALUMINUM	83	34	56-121

132. Feather mineral pattern M representing three greater snow geese found dead at Back Bay, Virginia.

98

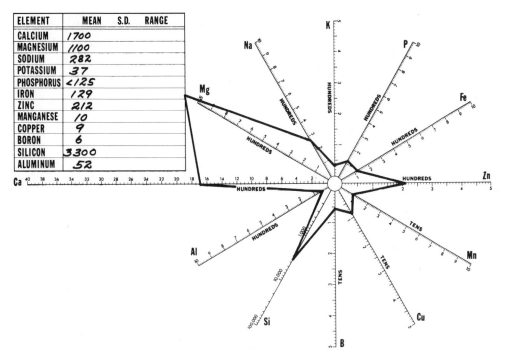

ELEMENT	MEAN	S.D.	RANGE
CALCIUM	1700		
MAGNESIUM	1100		
SODIUM	282		
POTASSIUM	37		
PHOSPHORUS	<125		
IRON	129		
ZINC	212		
MANGANESE	10		
COPPER	9		
BORON	6		
SILICON	3300		
ALUMINUM	52		

133. Feather mineral pattern N representing a greater snow goose found dead at Back Bay, Virginia.

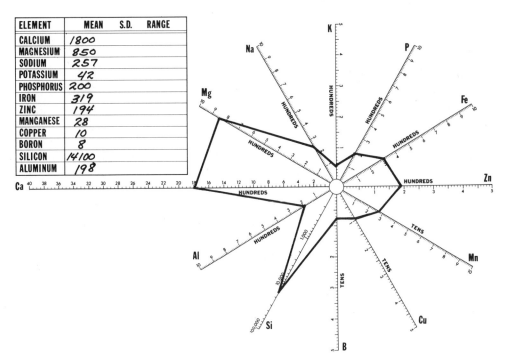

ELEMENT	MEAN	S.D.	RANGE
CALCIUM	1800		
MAGNESIUM	850		
SODIUM	257		
POTASSIUM	42		
PHOSPHORUS	200		
IRON	319		
ZINC	194		
MANGANESE	28		
COPPER	10		
BORON	8		
SILICON	14100		
ALUMINUM	198		

134. Feather mineral pattern O representing a greater snow goose found dead at Back Bay, Virginia.

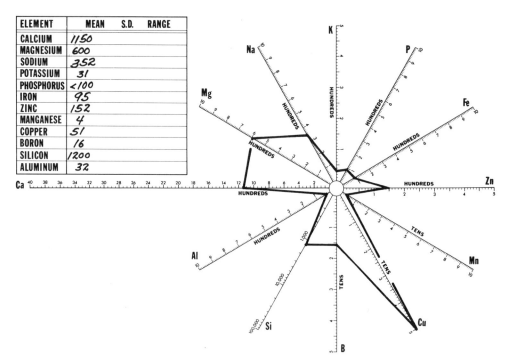

ELEMENT	MEAN	S.D.	RANGE
CALCIUM	1150		
MAGNESIUM	600		
SODIUM	352		
POTASSIUM	31		
PHOSPHORUS	<100		
IRON	95		
ZINC	152		
MANGANESE	4		
COPPER	51		
BORON	16		
SILICON	1200		
ALUMINUM	32		

135. Feather mineral pattern P representing a greater snow goose found dead at Back Bay, Virginia.

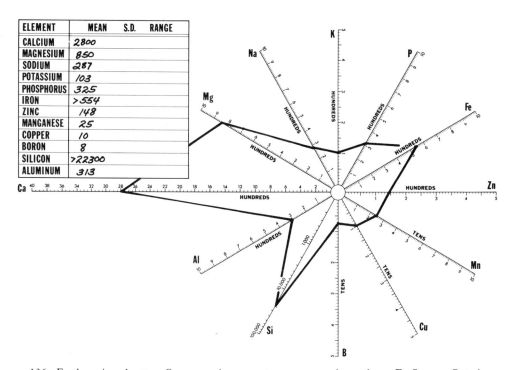

ELEMENT	MEAN	S.D.	RANGE
CALCIUM	2800		
MAGNESIUM	850		
SODIUM	287		
POTASSIUM	103		
PHOSPHORUS	325		
IRON	>554		
ZINC	148		
MANGANESE	25		
COPPER	10		
BORON	8		
SILICON	>22300		
ALUMINUM	313		

136. Feather mineral pattern Q representing a greater snow goose bagged near Ft. Severn, Ontario.

temporary marine influence, represents an interesting case of circulation of elements related to geologic events rather than to units in time. The soluble ion chemistry of the sandy-textured soils represents what may be characteristic of coarse-textured and organic matter-rich soils of the lowland. In these latter soils sodium and magnesium have leached relative to calcium and the soils are decidely acid.

Pattern K (Figure *152*), representing six geese, is markedly different from the other two pattern groups, totaling ten and twelve birds, respectively. The origin of these Ross' geese is unknown, but

we assume it lies in the Queen Maud Gulf lowlands.

Patterns L and M (Figures *153–54*) are of Ross' geese shot near the Bosque del Apache N.W.R. in the Rio Grande Valley. They are different from the other three pattern groups but resemble each other sufficiently to suggest that all four geese were from the same colony and feeding area. These distinctive patterns and the isolated nature of the wintering grounds of these geese suggest that the geese are from a breeding area of equally distinctive geology, well isolated from most or all of the other colonies in the Queen Maud Gulf lowlands.

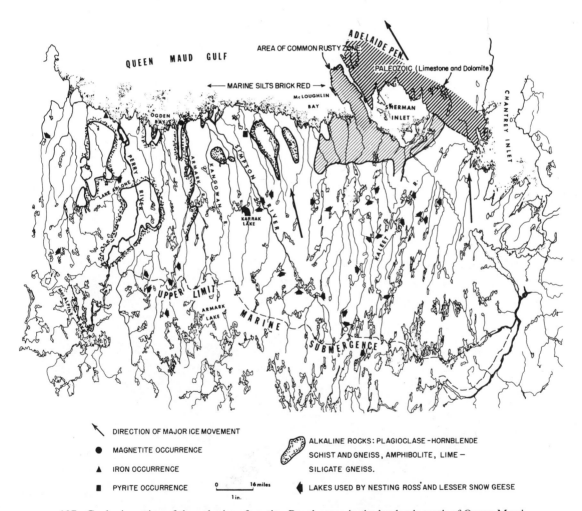

137. Geologic setting of the colonies of nesting Ross' geese in the lowlands south of Queen Maud Gulf (colony sites after Ryder 1969). Note locations of Lake Arlone and Karrak Lake. Acid rocks, mostly granites, occupy the area not indicated by the stipple convention for alkaline rocks (i.e., not shown completely between contacts). Geology was drawn from Craig (1961), Heywood (1961), and Fraser (1964).

12. Ionic contents of saturation paste extracts and potassium exchangeable by ammonium acetate of soils in the vicinity of nest colonies of Ross' geese south of Queen Maud Gulf

Locality and Site of Samples	Soil Texture	No.	pH	Ionic Content of Soil Water—Means and Ranges (ppm)						Cl me/.
				Ca	Mg	Na	K	K [a]	P	
Bank of Laine Creek [b]	Clay loam	3	5.7 5.5–6.0	9.0 7.0–10.0	14.3 14.0–15.0	79.6 75.0–87.0	13.7 12.5–15.0	171.7 165–180	14.8 13.0–16.0	1.9: 1.76–2
Nesting island in Karrak Lake	Sandy loam	6	4.7 4.3–5.1	8.4 7.0–13.5	6.3 4.5–7.0	11.6 7.5–18.0	10.6 7.0–19.0	28.3 15–45	7.9 4.0–19.0	0.3(0.08–C
Moraine bordering Karrak Lake	Sandy loam	4	4.4 4.3–4.5	8.1 7.0–10.0	7.4 6.0–8.0	9.0 7.5–11.0	17.3 13.0–20.0	82.5 55–110	8.9 6.5–10.5	0.5: 0.39–C
Mainland near Karrak Lake	Mostly fine sandy loam	9	4.8 4.3–5.2	5.6 3.5–8.0	3.3 2.0–5.0	13.2 10.0–15.5	4.1 2.0–13.5	30 [c] 5–75	2.9 1.5–8.5	0.3(0.24–C

Reference: Data courtesy of John P. Ryder and the Saskatchewan Soil Testing Laboratory.
[a] Exchangeable, dry soil basis.   [b] Tributary of Perry River. Flows near and drains Lake Arlone.
[c] Mean of three samples; other samples below detectable limit.

138. The lower Perry River area before breakup as seen from "Radio Hill" (67°32′N, 102°03′W), a ridge of Precambrian rock fifteen miles south of the seacoast.

139. View of the lower Perry River at breakup, as seen from "Radio Hill." View is eastward. Extensive flooding of the kind observable in this photograph characterizes many coastal plains areas in the mid-Arctic regions.

140. Characteristic landforms in the lower Perry River coastal plain. These "crag and tail" ridges are believed to be a series of dipping norite or diabase sills intruded into paragneiss country rock. (Reprinted from Hanson et al. 1956)

141. Lake Arlone (Lat. 67°22′, Long. 102°10′). The islands in this lake (A, B1, B2, C, D, and E) support one of the largest colonies of Ross' geese in the area of former marine submergence south of Queen Maud Gulf. In 1949, 260 pairs nested on these islands; in 1963, 769 pairs and in 1964, 906 pairs nested on these islands (Ryder 1967).

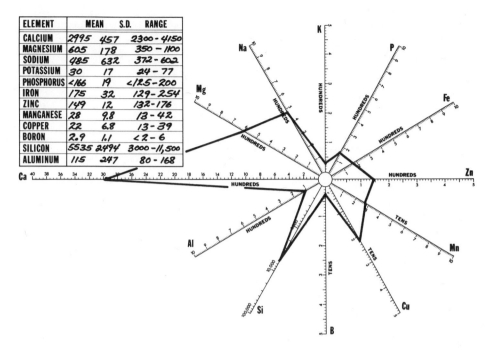

ELEMENT	MEAN	S.D.	RANGE
CALCIUM	2995	457	2300-4150
MAGNESIUM	605	178	350-1100
SODIUM	485	632	372-602
POTASSIUM	30	17	24-77
PHOSPHORUS	<166	19	<125-200
IRON	175	32	129-254
ZINC	149	12	132-176
MANGANESE	28	9.8	13-42
COPPER	22	6.8	13-39
BORON	2.9	1.1	<2-6
SILICON	5535	2494	3000-11,500
ALUMINUM	115	247	80-168

142. Feather mineral pattern A representing twenty adult Ross' geese collected in the vicinity of Karrak Lake, N.W.T.

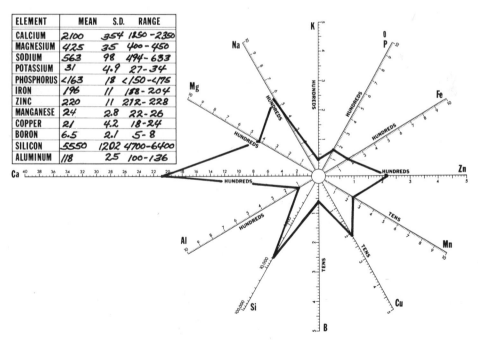

ELEMENT	MEAN	S.D.	RANGE
CALCIUM	2100	354	1250-2350
MAGNESIUM	425	35	400-450
SODIUM	563	98	494-633
POTASSIUM	31	4.9	27-34
PHOSPHORUS	<163	18	<150-<175
IRON	196	11	188-204
ZINC	220	11	212-228
MANGANESE	24	2.8	22-26
COPPER	21	4.2	18-24
BORON	6.5	2.1	5-8
SILICON	5550	1202	4700-6400
ALUMINUM	118	25	100-136

143. Feather mineral pattern B of two Ross' geese bagged in the vicinity of the Sacramento N.W.R., California. One had been banded near Lake Arlone, N.W.T.

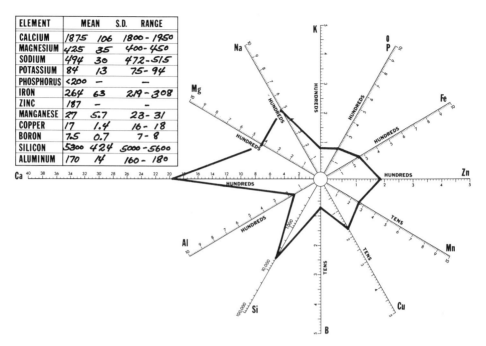

ELEMENT	MEAN	S.D.	RANGE
CALCIUM	1875	106	1800-1950
MAGNESIUM	425	35	400-450
SODIUM	494	30	472-515
POTASSIUM	84	13	75-94
PHOSPHORUS	<200	—	—
IRON	264	63	219-308
ZINC	187	—	—
MANGANESE	27	5.7	23-31
COPPER	17	1.4	16-18
BORON	7.5	0.7	7-8
SILICON	5300	424	5000-5600
ALUMINUM	170	14	160-180

144. Feather mineral pattern C of two Ross' geese bagged in the vicinity of the Sacramento N.W.R., California. One had been banded near Kindersley, Saskatchewan.

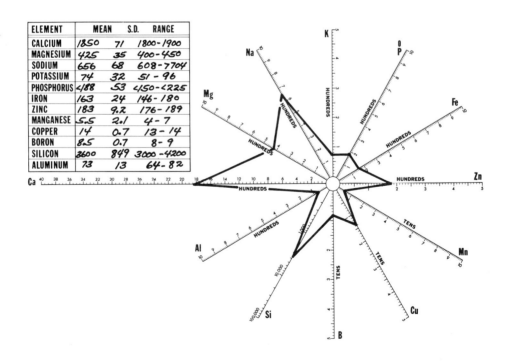

ELEMENT	MEAN	S.D.	RANGE
CALCIUM	1850	71	1800-1900
MAGNESIUM	425	35	400-450
SODIUM	656	68	608-7704
POTASSIUM	74	32	51-96
PHOSPHORUS	<188	53	<150-<225
IRON	163	24	146-180
ZINC	183	9.2	176-189
MANGANESE	5.5	2.1	4-7
COPPER	14	0.7	13-14
BORON	8.5	0.7	8-9
SILICON	3600	849	3000-4200
ALUMINUM	73	13	64-82

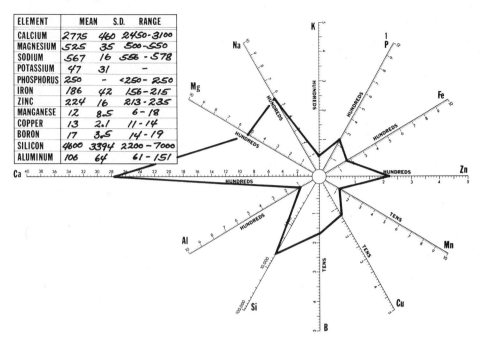

ELEMENT	MEAN	S.D.	RANGE
CALCIUM	2775	460	2450-3100
MAGNESIUM	525	35	500-550
SODIUM	567	16	556-578
POTASSIUM	47	31	-
PHOSPHORUS	250	-	<250-250
IRON	186	42	156-215
ZINC	224	16	213-235
MANGANESE	12	8.5	6-18
COPPER	13	2.1	11-14
BORON	17	3.5	14-19
SILICON	4600	3394	2200-7000
ALUMINUM	106	64	61-151

146. Feather mineral pattern E of two Ross' geese; one shot near Kindersley, Saskatchewan, and the other in Utah.

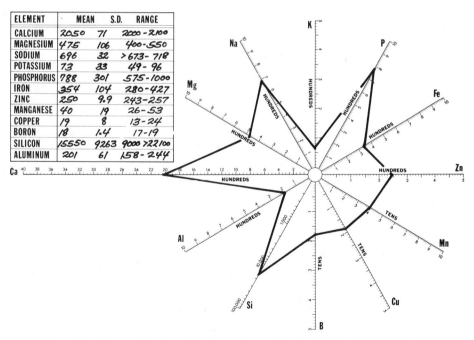

ELEMENT	MEAN	S.D.	RANGE
CALCIUM	2050	71	2000-2100
MAGNESIUM	475	106	400-550
SODIUM	696	32	>673-718
POTASSIUM	73	33	49-96
PHOSPHORUS	788	301	575-1000
IRON	354	104	280-427
ZINC	250	9.9	243-257
MANGANESE	40	19	26-53
COPPER	19	8	13-24
BORON	18	1.4	17-19
SILICON	15550	9263	9000->22100
ALUMINUM	201	61	158-244

147. Feather mineral pattern F of two Ross' geese bagged in the vicinity of Kindersley, Saskatchewan. One of these had been banded along the Simpson River, N.W.T.

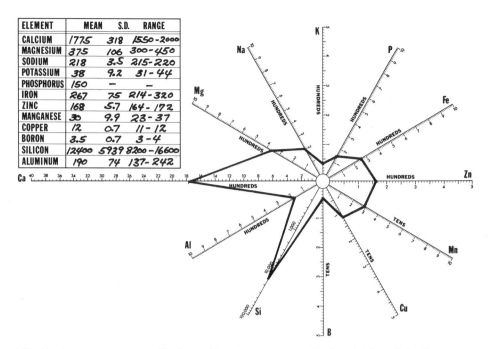

ELEMENT	MEAN	S.D.	RANGE
CALCIUM	1775	318	1550-2000
MAGNESIUM	375	106	300-450
SODIUM	218	3.5	215-220
POTASSIUM	38	9.2	31-44
PHOSPHORUS	150	—	—
IRON	267	75	214-320
ZINC	168	5.7	164-172
MANGANESE	30	9.9	23-37
COPPER	12	0.7	11-12
BORON	3.5	0.7	3-4
SILICON	12400	5939	8200-16600
ALUMINUM	190	74	137-242

148. Feather mineral pattern G of two Ross' geese bagged in the vicinity of the Sacramento N.W.R., California. One had been banded in the vicinity of Lake Arlone, N.W.T.

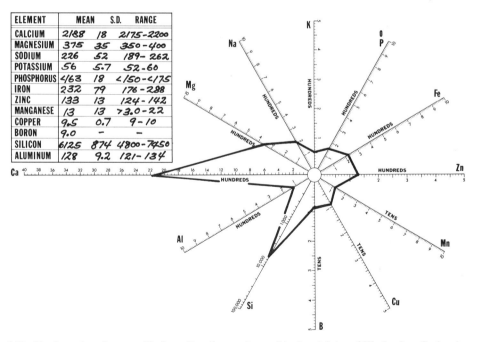

ELEMENT	MEAN	S.D.	RANGE
CALCIUM	2188	18	2175-2200
MAGNESIUM	375	35	350-400
SODIUM	226	52	189-262
POTASSIUM	56	5.7	52-60
PHOSPHORUS	<163	18	<150-<175
IRON	232	79	176-288
ZINC	133	13	124-142
MANGANESE	13	13	>3.0-22
COPPER	9.5	0.7	9-10
BORON	9.0	—	—
SILICON	6125	874	4800-7450
ALUMINUM	128	9.2	121-134

149. Feather mineral pattern H of two Ross' geese bagged in the vicinity of Kindersley, Saskatchewan. One had been banded along the Simpson River, N.W.T.

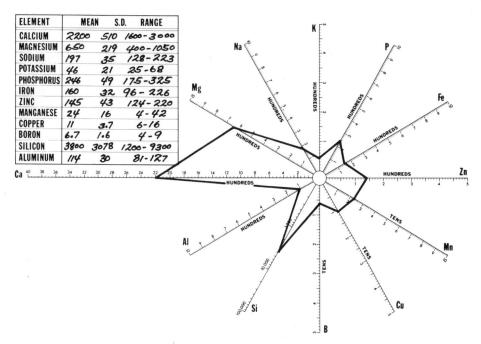

ELEMENT	MEAN	S.D.	RANGE
CALCIUM	2200	510	1600-3000
MAGNESIUM	650	219	400-1050
SODIUM	197	35	128-223
POTASSIUM	46	21	25-68
PHOSPHORUS	246	49	175-325
IRON	160	32	96-226
ZINC	145	43	124-220
MANGANESE	24	16	4-42
COPPER	11	3.7	6-16
BORON	6.7	1.6	4-9
SILICON	3800	3078	1200-9300
ALUMINUM	114	30	81-127

150. Feather mineral pattern I of six Ross' geese, four of which were shot near Kindersley, Saskatchewan, and one each near the Tule Lake and Sacramento national wildlife refuges in California. The Ross' goose shot at Tule Lake had been banded in the vicinity of Lake Arlone, N.W.T.

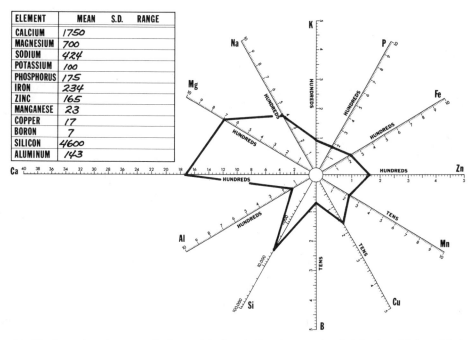

ELEMENT	MEAN	S.D.	RANGE
CALCIUM	1750		
MAGNESIUM	700		
SODIUM	424		
POTASSIUM	100		
PHOSPHORUS	175		
IRON	234		
ZINC	165		
MANGANESE	23		
COPPER	17		
BORON	7		
SILICON	4600		
ALUMINUM	143		

151. Feather mineral pattern J of an adult female Ross' goose banded in the vicinity of Lake Arlone, N.W.T., and shot near the Sacramento N.W.R., California.

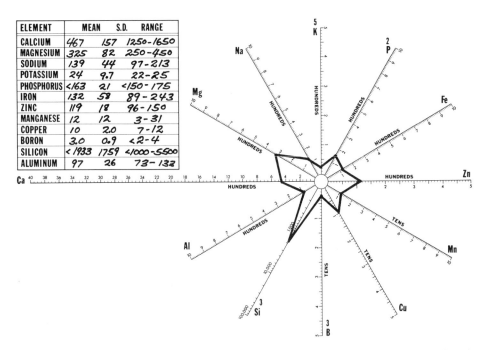

ELEMENT	MEAN	S.D.	RANGE
CALCIUM	467	157	1250-1650
MAGNESIUM	325	82	250-450
SODIUM	139	44	97-213
POTASSIUM	24	9.7	22-25
PHOSPHORUS	<163	21	<150-175
IRON	132	58	89-243
ZINC	119	18	96-150
MANGANESE	12	12	3-31
COPPER	10	20	7-12
BORON	3.0	0.9	<2-4
SILICON	<1933	1759	<1000-5500
ALUMINUM	97	26	73-132

152. Feather mineral pattern K of six Ross' geese, one of which was bagged at Ft. Severn, Ontario, four at Tule Lake N.W.R., California, and one at Sacramento N.W.R., California.

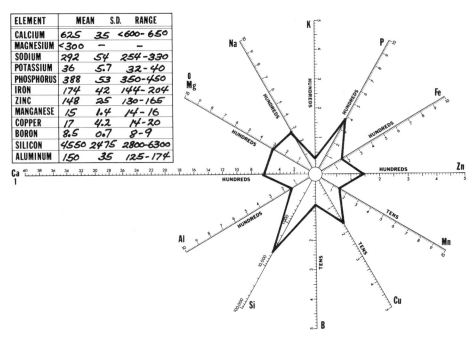

ELEMENT	MEAN	S.D.	RANGE
CALCIUM	625	35	<600-650
MAGNESIUM	<300	-	-
SODIUM	292	54	254-330
POTASSIUM	36	5.7	32-40
PHOSPHORUS	388	53	350-450
IRON	174	42	144-204
ZINC	148	25	130-165
MANGANESE	15	1.4	14-16
COPPER	17	4.2	14-20
BORON	8.5	0.7	8-9
SILICON	4550	2475	2800-6300
ALUMINUM	150	35	125-174

153. Feather mineral pattern L of two Ross' geese shot in the vicinity of the Bosque del Apache N.W.R., New Mexico.

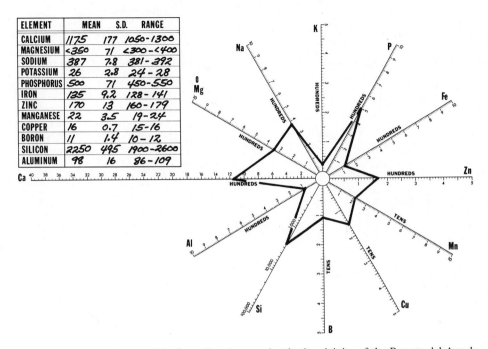

ELEMENT	MEAN	S.D.	RANGE
CALCIUM	1175	177	1050-1300
MAGNESIUM	<350	71	<300-<400
SODIUM	387	7.8	381-392
POTASSIUM	26	2.8	24-28
PHOSPHORUS	500	71	450-550
IRON	135	9.2	128-141
ZINC	170	13	160-179
MANGANESE	22	3.5	19-24
COPPER	16	0.7	15-16
BORON	11	1.4	10-12
SILICON	2250	495	1900-2600
ALUMINUM	98	16	86-109

154. Feather mineral pattern M of two Ross' geese shot in the vicinity of the Bosque del Apache N.W.R., New Mexico.

111

CHAPTER FIVE

Differences in Feather Mineral Patterns and Origins of Migrating and Wintering Geese

DIFFERENCES AMONG COLONIES

Eastern Arctic Colonies

Despite the number and diversity of intracolony feather mineral patterns, these differences were not sufficient to blur basic elemental differences between colonies that result from differential inputs of minerals into the ecosystems of the breeding and feeding areas. From tables of the distribution of values of t tests it is discernible that the four major colony groupings in the Hudson Bay area can be distinguished from one another on the basis of quantitative differences in one or more elements (Table 13).

The Southampton Island and McConnell River colonies were each distinguishable from the other four colonies on the basis of four or five elements, Baffin Island by two to five minerals, and Cape Churchill by one to four elements. The Cape Churchill and Cape Henrietta Maria colonies are most alike, only zinc being significantly different.

It is of interest to note which elements most frequently differed significantly between colonies. Of the ten minerals involved, calcium differed significantly in seven comparisons; zinc in five; iron in four; potassium, phosphorus, manganese, aluminum, boron, and silicon in three instances; and copper in two. Magnesium and sodium values did not differ significantly between any of the five colonies (Table 14).

Western Arctic Colonies

In accordance with the greater differences and complexities among the geologies of the colony areas and the mineralogies of the drainage basins in which they are situated, the total of significant differences in feather mineral patterns among western Arctic colonies was much greater than among eastern Arctic colonies. Significant colony differences were found for all twelve elements in the western Arctic as compared with ten in the eastern Arctic (Tables 13–15). Four to ten elements differed significantly between the four major western Arctic populations. Four of the five colonies in the Hudson Bay area are located on limestones. The feather mineral characteristics of the western Arctic colonies, on the other hand, are related to 1) deltaic islands that received their sediments primarily from areas underlain by Mesozoic rocks (Kendall Island colony) and from Precambrian and Cretaceous strata (Anderson River colony), or 2) are located on major islands with distinctive geologies of their own—the Banks Island colony being located on Cenozoic deposits and the Wrangel Island colony being influenced by an array of rock types ranging in age from late Paleozoic to early Mesozoic.

ORIGINS OF MIGRANT AND WINTERING POPULATIONS

At the outset of this study, we decided that if our investigations were limited only to describing the mineral profiles of each colony and determining the significant elemental differences between them, we would have achieved limited goals. If our research were to be useful to and to be employed readily by conservation agencies, we should have to confront, evaluate and solve problems of identifying the origins of unbanded geese from as many sectors of the continent as possible.

Determinations of origins must be made ultimately from comparisons of overall mineral patterns and significant elemental differences between colonies, but to make such comparisons without a computer would often be laborious, especially when populations of birds from similar geological backgrounds were involved. However, in the cases of Canada geese from Akimiski Island, James Bay, the Belcher Islands, the southern Yukon, and southern coastal Alaska-British Columbia, values for one or two elements would serve to establish their origins. For the determination of origins of geese less readily distinguished, the use of the dis-

112

13. Elements which differed significantly ($P < .05$) between colonies of blue and lesser snow geese in the eastern Arctic

	Baffin Island	Southampton Island	Cape Henrietta Maria	McConnell River	Cape Churchill
Baffin Island (n = 49)		Fe, Zn, Si, Al	Ca, P, Zn	Ca, P, Mn, Cu, B	Ca, K
Southampton Island (n = 23)	Fe, Zn, Si, Al		Ca, Fe, Zn, Si, Al	Fe, Mn, Cu, B, Al	Ca, K, Zn, Si
Cape Henrietta Maria (n = 24)	Ca, P, Zn	Ca, Fe, Zn, Si, Al		Ca, P, Zn, Mn	Zn
McConnell River (n = 38)	Ca, P, Mn Cu, B	Fe, Mn, Cu, B, Al	Ca, P, Zn, Mn		Ca, K, Fe, B
Cape Churchill (n = 4)	Ca, K	Ca, K, Zn, Si	Zn	Ca, K, Fe, B	
Total frequency of elemental difference	14	18	13	18	11

criminant function proved to be the ideal solution, but its use is restricted to samples that at least equal in number the number of basic parameters measured. Thus, because of limited sample sizes, we were able to use this test only for geese from the Hudson Bay colonies and eastern North America and not for populations from the western sectors of the continent.

An evaluation of the adequacy of the technique would be to treat statistically the individuals making up colony samples as geese of unknown origins. Results of this procedure are given in Table 16. About 85 percent of the individuals making up the Baffin Island and Cape Henrietta Maria sam-

14. Relative importance of elements exhibiting significant quantitative differences between colonies of blue and lesser snow geese

	Frequency			
	Eastern Arctic		Western Arctic	
Element	n	%	n	%
Ca	7	19.5	8	12.7
Mg			8	12.7
Na			4	6.3
K	3	8.3	8	12.7
P	3	8.3	3	4.8
Fe	4	11.1	5	8.0
Zn	5	14.0	6	9.5
Mn	3	8.3	2	3.2
Cu	2	5.6	6	9.5
B	3	8.3	4	6.3
Si	3	8.3	6	9.5
Al	3	8.3	3	4.8
Total	36	100.0	63	100.0

ples were correctly identified as to origin, and 70 percent of the geese constituting the Southampton Island and McConnell River series were correctly identified. Southampton Island geese were most often "confused" as Baffin Island geese by the computer; McConnell River geese, with highly varied patterns, were most often incorrectly identified as Baffin Island geese. These findings suggest two major conclusions: 1) a population containing some individuals that have "nondescript" feather mineral patterns that exhibit some basic similarities to the pattern of another colony inhabiting an area similar in bedrock geology and soils would be difficult to distinguish with a high rate of success, and 2) conversely, a colony or population that produces many widely varying mineral profiles because of the diverse mineral regimes the individuals in it encounter, would have a mean pattern that would be so "elastic," because of high variance for most elemental values, that many of its component members could not be distinguished readily from members of another colony, and many members of the latter would fit comfortably within the statistical confines of the more variable colony.

The problem of identifying individuals from two or more populations is greatly simplified if the populations concerned breed on diverse geologies strongly skewed for a few elements. To test this thesis, as well as to secure confidence that the technique was sound and could be used fruitfully, we tested the ability of the discriminant function to classify component members of the reference samples of seventeen races (many undescribed) of Canada geese. Of the 388 geese treated as un-

15. Elements that differed significantly $(P < .05)$ between colonies of blue and lesser snow geese in the western Arctic

	Banks Island	Anderson River	Kendall Island	Wrangel Island	Mackenzie River delta*
Banks Island (n = 3)		Ca, Mg, K, Zn, Cu	Mg, Na, K, Cu, B, Si	Ca, Mg, Fe, Zn, Cu	Ca, Mg, K, Fe, Cu, Si
Anderson River (n = 3)	Ca, Mg, K, Zn, Cu		Ca, Mg, Na, K, P, Zn, B	Ca, Mg, Na, K	Ca, Mg, K, P, Fe, Zn, Si, Al
Kendall Island (n = 7)	Mg, Na, K, Cu, B, Si	Ca, Mg, Na, K, P, Zn, B		Ca, Mg, Na, K, P, Zn, Mn, Cu, B, Si	Ca, Mg, Na, K, Fe, Mn, Cu, Si, Al
Wrangel Island (n = 17)	Ca, Mg, Fe, Zn, Cu	Ca, Mg, Na, K	Ca, Mg, Na, K, P, Zn, Mn, Cu, B, Si		K, Fe, Zn, B, Si, Al
Mackenzie River delta* (n = 11)	Ca, Mg, K, Fe, Cu, Si	Ca, Mg, K, P, Fe, Zn, Si, Al	Ca, Mg, Na, K, Fe, Mn, Cu, Si, Al	K, Fe, Zn, B, Si, Al	
Total frequency of elemental difference	22	24	32	25	29

* West shore, first landfall for Banks Island geese in fall migration.

knowns, 92 percent were correctly identified as to origins. This high score of correct identification relates, of course, as indicated above, to the unique nature of the ecosystem from which each of these populations originates and the diversity of the mineral input into the nutrient chain of these geese. Eight of these populations were identified at the 100 percent level and three at about the 95 percent level. Few taxonomists working on races of most species of birds can equal this level of differentiation, and none could remotely approach it with Canada geese in view of the erroneous state of published taxonomic reports (Hanson, unpublished).

It seemed apparent from studies of the population characteristics of the mainland colonies of

blue and lesser snow geese in the Hudson Bay area that immigration from other colonies explained in part the rapid growth of small colonies (e.g., the Cape Churchill colony). It also appeared that immigration was possibly a contributing factor, although it is now believed to be a slight one, to an apparently continuing shift toward the blue phase in the geese of the Cape Henrietta Maria colony (Hanson et al. 1972:10–15). The color-phase ratios in this colony indicated that it originates from Baffin Island stocks, and the color composition of the Cape Churchill colony suggested that it was founded chiefly from the McConnell River stocks.

Sporadic nestings by small numbers of blue and lesser snow geese along the south coast of Hudson Bay, the northwest coast of James Bay, and on

16. Origins of banded blue and lesser snow geese from four colonies in Hudson Bay as indicated by the discriminant function test from mineral analyses of the primary feathers

Colony of Origin	Total Number of Recoveries	Computer-Designated Colony							
		Baffin Island		Southampton Island		Cape Henrietta Maria		McConnell River	
		n	%	n	%	n	%	n	%
Baffin Island	49	41	83.6	2	4.1	2	4.1	4	8.2
Southampton Island	23	5	21.7	16	69.7	1	4.3	1	4.3
Cape Henrietta Maria	24	2	8.3	1	4.2	21	87.5	0	0.0
McConnell River	38	7	18.4	2	5.3	2	5.3	27	71.0

Akimiski Island are also indicative of the pioneering traits of these geese and provide evidence that some immigration to and emigration from major colonies, although limited in extent, is constantly taking place. This exchange of geese between colonies may explain in part the failure of the discriminant function to classify correctly a higher percentage of the geese banded in various colonies, as the composite data of these same banded geese were used for the colony models (Table 16). Put another way, the intervention of one or more years between the year when a snow goose was banded and the fall in which it was shot and subsequently its primaries were used as a reference sample diminishes in some measure its reliability as a representative of the colony in which it was banded. Thus, a goose that was banded on Baffin Island as an immature and spent its second summer on Cape Henrietta Maria is representative, in respect to its feather minerals, of the latter colony rather than the former. The data in Table 17 provide a basis for at least some subjective appreciation of the relative reliability of our basic reference samples. Conversely, the adherence of Canada goose populations to established breeding areas accounts in large measure for the evolution of the thirty or more races of this species (Hanson, unpublished).

Designations of origins of migrant and wintering geese derived both from feather mineral analyses of geese in hunters' bags and from band recovery analyses should be related to total kills and total populations, area by area, if a true perspective of the relative use of migration routes by geese of various colonies is to be gained. This has been done in part, as will be discussed, for available data from the Ontario coasts of Hudson and James bays (see Tables 22–25). It is not the purpose of this report to delineate migration routes, nor are the data at hand adequate for this purpose. Rather, if waterfowl management is to progress on a sound basis, our ultimate goal should be to describe the dispersal of goose colony populations, using the approaches suggested in Tables 22–25. Because of the inadequacies of our data from the standpoint of sample sizes and geographic representations, our following discussions of these aspects of the data are necessarily brief. But it is important to point out, especially to those who contemplate using feather mineral data and the discriminant function test to determine the birth or summering place of geese of unknown origin, that a relatively high percentage (weighted mean of samples—53 percent) of the "unknowns" we submitted to chemical and mathematical analyses were designated at P levels of 95 percent or better and that the great majority of each sample (weighted mean of samples—77 percent) were determined at levels of P that exceeded 70 percent.

Mississippi and Central Flyway Populations

Feather mineral patterns of blue and lesser snow geese shot on the coastal marshes of Hudson and James bays, near the mouths of the Severn and Kapiskau rivers, respectively, are shown in Figures *155–64*. Patterns for migrant populations at two points on the Great Plains—the Sand Lake N.W.R., South Dakota and the Squaw Creek N.W.R., extreme northwest Missouri, are presented in Figures *165–73*. Finally, results of analyses of feathers from geese at four localities on the Gulf Coast of Louisiana—Pass A Loutre, Grand Chenier, Johnson's Bayou, and the Sabine

17. Intervals between years blue and lesser snow geese were banded on their breeding grounds and years they were shot and reference primary feathers collected

Years Between Banding and Collection of Primary Feathers	Colony									
	Baffin Island		Cape Henrietta Maria		Southampton Island		McConnell River		Cape Churchill	
	n	%	n	%	n	%	n	%	n	%
0	32	54.2	24	100.0			21	55.2	4	100.0
1	18	30.5					2	5.3		
2					2	8.7	6	15.8		
3					2	8.7	5	13.2		
4–5					3	13.1	1	2.6		
6–7	6	10.2			9	39.1	2	5.3		
8–9	3	5.1			6	26.1	1	2.6		
10–11					1	4.3				
Total or percent	59	100.0	24	100.0	23	100.0	38	100.0	4	100.0

N.W.R.—(Figure *174*) are shown in Figures *175–202*. The computer designated origins of geese whose feather mineral patterns were analyzed are indicated in each legend. In some instances geese from more than one colony may have been averaged into the pattern groupings, all of which were subjectively determined. Each pattern has been matched to the colony of origin (as designated by the discriminant function) of the majority of the geese whose feather minerals made up the pattern.

The designations of the origins of migrant and wintering blue and lesser snow geese, based on the discriminant function, cannot be evaluated adequately unless the color phase composition of the bagged geese from which the wing feather samples were obtained is taken into consideration. Only in the case of the Baffin Island geese are color phases represented in the samples of wing feathers of geese banded on their breeding grounds and shot by hunters in the fall similar to ratios of color phases in the colonies from which the geese originated (Table 18). Also, except for the sample from the Sabine N.W.R., the samples listed in Table 18, although random, are of inadequate size to provide a satisfactory portrayal of the winter distribution of the geese from these colonies. It is also unfortunate that we did not have a wing feather collection from Texas. Nevertheless, some analyses of the data in the light of previous findings are justified.

Lemieux and Heyland (1967:686–87) made a study of the recoveries of 5,535 blue geese and 5,328 lesser snow geese banded along the Koukdjuak River, Baffin Island (Figure 27). They found that the southward migration routes followed by the lesser snow geese to their wintering grounds were to the west of those taken by the blue geese banded in the same area. Thus, 35 percent of direct recoveries of banded lesser snow geese were from the Mississippi Flyway states as compared with 47 percent of the blue goose recoveries. In contrast, 52 percent of the lesser snow goose recoveries were from the Great Plains states (Central Flyway) as compared with 18 percent of the blue geese. Few banded lesser snow geese were shot at the south end of James Bay in comparison with the percentage of banded blue geese taken there; instead, the banded lesser snow geese tended to drift westward along the south coast of Hudson Bay or southwestward from the upper portions of the west coast of James Bay after crossing the latter from the east coast. Although all studies have clearly shown the preponderance of blue geese taken at the south end of James Bay, band recovery rates for

the two color phases do not accurately represent the ratios of color phases in the nesting flocks, as the Indians who hunt this sector of the bay shoot blue geese in preference to lesser snow geese (Hanson et al. 1972:32).

The majority of the geese from the Cape Henrietta Maria colony also drift westward along the south coast of Hudson Bay as they feed in the fall before heading south (Table 19 and Harry G. Lumsden, personal communication, 1971). The geese from the Southampton Island colonies migrate southward along the west coast of Hudson Bay and feed along the south coast, where they are shot from York Factory to Cape Henrietta Maria (Cooch 1961:87). Also intermingled with the geese that feed on the south coast of Hudson Bay are birds from the McConnell River colony, many of which follow the coastline to the south end of James Bay (Tables 19, 23). In light of this partial intermingling of colony populations in migration and on the wintering grounds and the marked tendency of white-phase birds to drift to the west of blue-phase birds, it is of interest to note that 26 percent of the geese shot in the vicinity of the mouth of the Severn River were blue geese as compared with 100 percent found in the small random sample from the marshes at the mouth of the Kapiskau River (Table 20). Thirty percent of the wing feather sample taken at Ft. Severn and derived from Baffin Island was from blue geese as compared with the all-blue-phase sample primarily derived from Baffin Island and collected at the Kapiskau marsh (Table 20).

Our limited data from the Great Plains and the Gulf Coast of Louisiana essentially confirm a westward drift of white-phase geese. At the Sand Lake N.W.R., South Dakota, geese from the Southampton Island and Cape Henrietta Maria colonies appear to make up the majority of the fall concentrations, but the representation of blue-phase birds is low in comparison with their relative numbers on their breeding grounds (Tables 18–20).

At the Squaw Creek N.W.R., geese from Baffin Island were most importantly represented in the small sample that we collected (but not necessarily in the migrant population) (Figure *241*). The major migration of part of this colony, at least in the past, has been down the eastern sector of the Mississippi Flyway; it is not surprising, therefore, in view of the westward drift of lesser snow geese, that blue geese comprised only a small portion of the Baffin Island contingent at the Squaw Creek N.W.R. (Tables 19–20).

The color differential in the shot sample of Cape

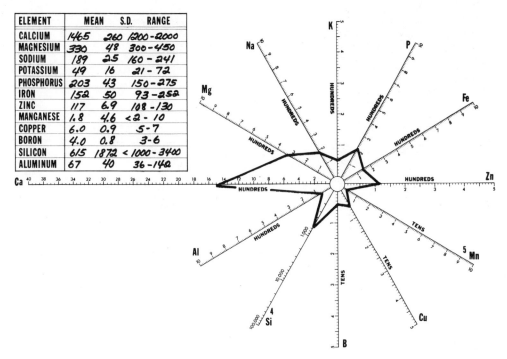

ELEMENT	MEAN	S.D.	RANGE
CALCIUM	1465	260	1200-2000
MAGNESIUM	330	48	300-450
SODIUM	189	25	160-241
POTASSIUM	49	16	21-72
PHOSPHORUS	203	43	150-275
IRON	152	50	93-252
ZINC	117	6.9	108-130
MANGANESE	1.8	4.6	<2-10
COPPER	6.0	0.9	5-7
BORON	4.0	0.8	3-6
SILICON	615	1872	<1000-3400
ALUMINUM	67	40	36-142

155. Feather mineral pattern A representing ten (35 percent) of twenty-nine blue and lesser snow geese bagged near the mouth of the Severn River, Ontario.

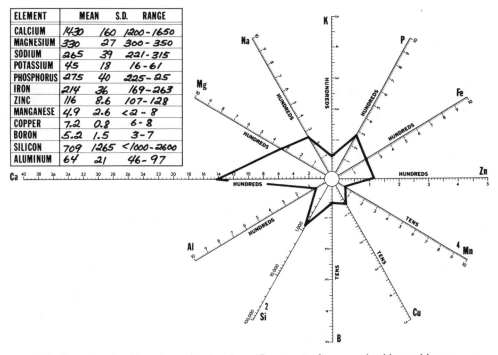

ELEMENT	MEAN	S.D.	RANGE
CALCIUM	1430	160	1200-1650
MAGNESIUM	330	27	300-350
SODIUM	265	39	221-315
POTASSIUM	45	18	16-61
PHOSPHORUS	275	40	225-25
IRON	214	36	169-263
ZINC	116	8.6	107-128
MANGANESE	4.9	2.6	<2-8
COPPER	7.2	0.8	6-8
BORON	5.2	1.5	3-7
SILICON	709	1265	<1000-2600
ALUMINUM	64	21	46-97

156. Feather mineral pattern B representing five (17 percent) of twenty-nine blue and lesser snow geese bagged near the mouth of the Severn River, Ontario.

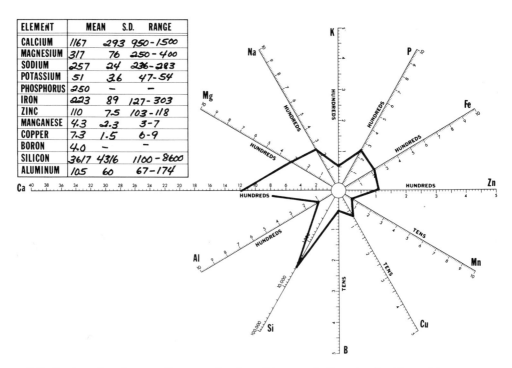

ELEMENT	MEAN	S.D.	RANGE
CALCIUM	1167	293	950-1500
MAGNESIUM	317	76	250-400
SODIUM	257	24	236-283
POTASSIUM	51	3.6	47-54
PHOSPHORUS	250	–	–
IRON	223	89	127-303
ZINC	110	7.5	103-118
MANGANESE	4.3	2.3	3-7
COPPER	7.3	1.5	6-9
BORON	4.0	–	–
SILICON	3617	4316	1100-8600
ALUMINUM	105	60	67-174

157. Feather mineral pattern C representing three (11 percent) of twenty-nine blue and lesser snow geese bagged near the mouth of the Severn River, Ontario.

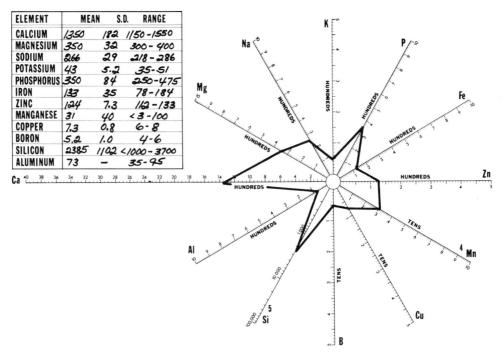

ELEMENT	MEAN	S.D.	RANGE
CALCIUM	1350	182	1150-1550
MAGNESIUM	350	32	300-400
SODIUM	266	29	218-286
POTASSIUM	43	5.2	35-51
PHOSPHORUS	350	84	250-475
IRON	133	35	78-184
ZINC	124	7.3	112-133
MANGANESE	31	40	<3-100
COPPER	7.3	0.8	6-8
BORON	5.2	1.0	4-6
SILICON	2385	1102	<1000-3700
ALUMINUM	73	–	35-95

158. Feather mineral pattern D representing six (21 percent) of twenty-nine blue and lesser snow geese bagged near the mouth of the Severn River, Ontario.

118

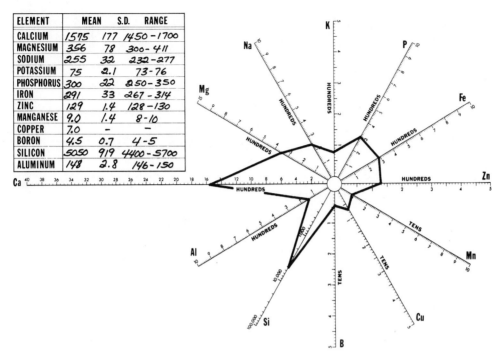

ELEMENT	MEAN	S.D.	RANGE
CALCIUM	1575	177	1450-1700
MAGNESIUM	356	78	300-411
SODIUM	255	32	232-277
POTASSIUM	75	2.1	73-76
PHOSPHORUS	300	22	250-350
IRON	291	33	267-314
ZINC	129	1.4	128-130
MANGANESE	9.0	1.4	8-10
COPPER	7.0	–	–
BORON	4.5	0.7	4-5
SILICON	5050	919	4400-5700
ALUMINUM	148	2.8	146-150

159. Feather mineral pattern E representing two (7 percent) of twenty-nine blue and lesser snow geese bagged near the mouth of the Severn River, Ontario.

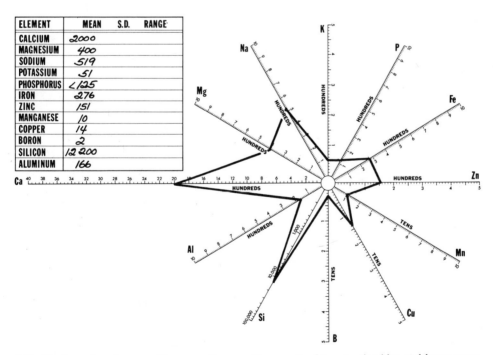

ELEMENT	MEAN	S.D.	RANGE
CALCIUM	2000		
MAGNESIUM	400		
SODIUM	519		
POTASSIUM	51		
PHOSPHORUS	<125		
IRON	276		
ZINC	151		
MANGANESE	10		
COPPER	14		
BORON	2		
SILICON	12200		
ALUMINUM	166		

160. Feather mineral pattern F representing one (3 percent) of twenty-nine blue and lesser snow geese bagged near the mouth of the Severn River, Ontario.

119

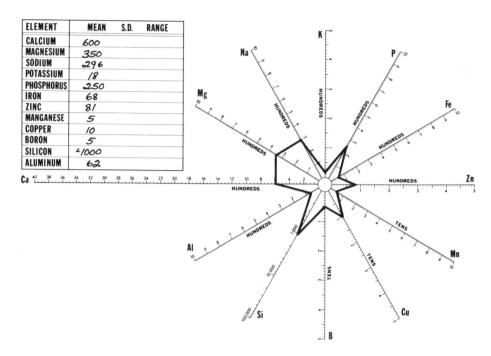

ELEMENT	MEAN	S.D.	RANGE
CALCIUM	600		
MAGNESIUM	350		
SODIUM	296		
POTASSIUM	18		
PHOSPHORUS	250		
IRON	68		
ZINC	81		
MANGANESE	5		
COPPER	10		
BORON	5		
SILICON	<1000		
ALUMINUM	62		

161. Feather mineral pattern G representing one (3 percent) of twenty-nine blue and lesser snow geese bagged near the mouth of the Severn River, Ontario.

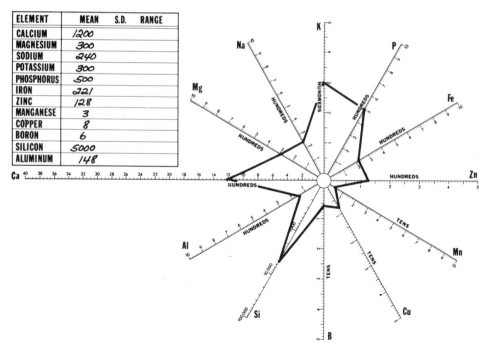

ELEMENT	MEAN	S.D.	RANGE
CALCIUM	1200		
MAGNESIUM	300		
SODIUM	240		
POTASSIUM	300		
PHOSPHORUS	500		
IRON	221		
ZINC	128		
MANGANESE	3		
COPPER	8		
BORON	6		
SILICON	5000		
ALUMINUM	148		

162. Feather mineral pattern H representing one (3 percent) of twenty-nine blue and lesser snow geese bagged near the mouth of the Severn River, Ontario.

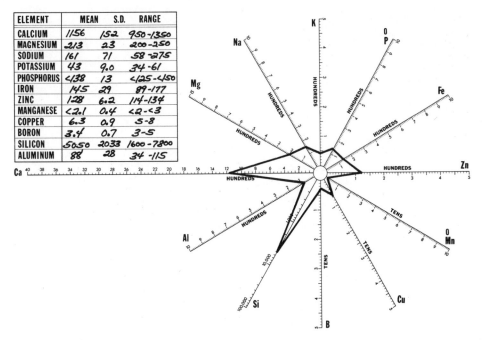

ELEMENT	MEAN	S.D.	RANGE
CALCIUM	1156	152	950-1350
MAGNESIUM	213	23	200-250
SODIUM	161	71	58-275
POTASSIUM	43	9.0	34-61
PHOSPHORUS	<138	13	<125-<150
IRON	145	29	89-177
ZINC	128	6.2	114-134
MANGANESE	<2.1	0.4	<2-<3
COPPER	6.3	0.9	5-8
BORON	3.4	0.7	3-5
SILICON	5050	2033	1600-7800
ALUMINUM	88	28	34-115

163. Feather mineral pattern A representing eight (53 percent) of fifteen blue geese bagged near the mouth of the Kapiskau River, Ontario.

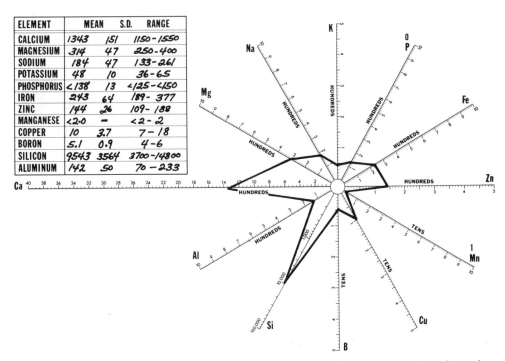

ELEMENT	MEAN	S.D.	RANGE
CALCIUM	1343	151	1150-1550
MAGNESIUM	314	47	250-400
SODIUM	184	47	133-261
POTASSIUM	48	10	36-65
PHOSPHORUS	<138	13	<125-<150
IRON	243	64	189-377
ZINC	144	26	109-188
MANGANESE	<2.0	—	<2-2
COPPER	10	3.7	7-18
BORON	5.1	0.9	4-6
SILICON	9543	3564	3700-14800
ALUMINUM	142	50	70-233

164. Feather mineral pattern B representing seven (47 percent) of fifteen blue geese bagged near the mouth of the Kapiskau River, Ontario.

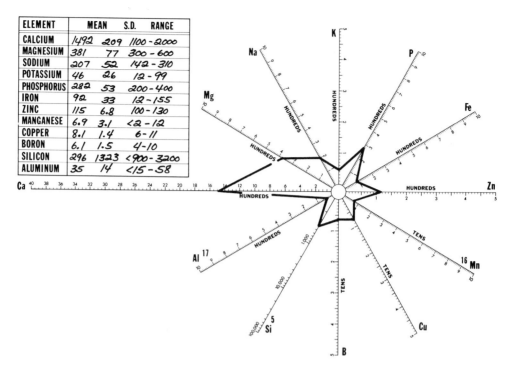

ELEMENT	MEAN	S.D.	RANGE
CALCIUM	1492	209	1100-2000
MAGNESIUM	381	77	300-600
SODIUM	207	52	142-310
POTASSIUM	46	26	12-99
PHOSPHORUS	282	53	200-400
IRON	92	33	12-155
ZINC	115	6.8	100-130
MANGANESE	6.9	3.1	<2-12
COPPER	8.1	1.4	6-11
BORON	6.1	1.5	4-10
SILICON	296	1323	<900-3200
ALUMINUM	35	14	<15-58

165. Feather mineral pattern A representing eighteen (78 percent) of twenty-three blue and lesser snow geese bagged in the vicinity of Sand Lake N.W.R., South Dakota.

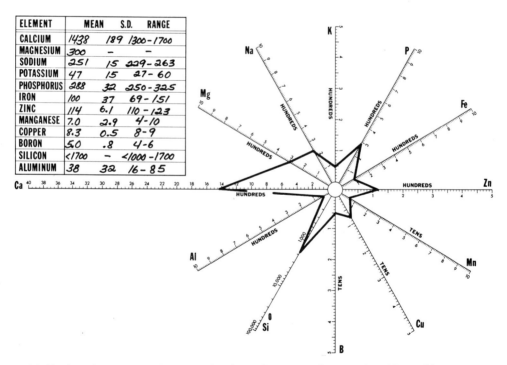

ELEMENT	MEAN	S.D.	RANGE
CALCIUM	1438	189	1300-1700
MAGNESIUM	300	-	-
SODIUM	251	15	229-263
POTASSIUM	47	15	27-60
PHOSPHORUS	288	32	250-325
IRON	100	37	69-151
ZINC	114	6.1	110-123
MANGANESE	7.0	2.9	4-10
COPPER	8.3	0.5	8-9
BORON	5.0	.8	4-6
SILICON	<1700	-	<1000-1700
ALUMINUM	38	32	16-85

166. Feather mineral pattern B representing four (18 percent) of twenty-three blue and lesser snow geese bagged in the vicinity of Sand Lake N.W.R., South Dakota.

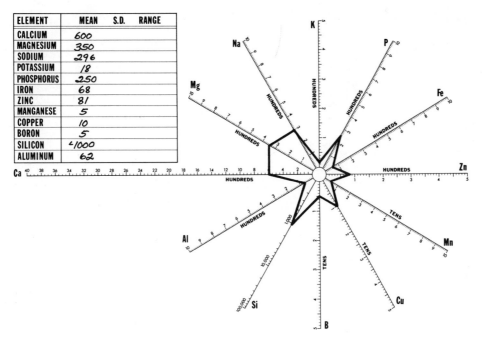

ELEMENT	MEAN	S.D.	RANGE
CALCIUM	600		
MAGNESIUM	350		
SODIUM	296		
POTASSIUM	18		
PHOSPHORUS	250		
IRON	68		
ZINC	81		
MANGANESE	5		
COPPER	10		
BORON	5		
SILICON	<1000		
ALUMINUM	62		

167. Feather mineral pattern C representing one (4 percent) of twenty-three blue and lesser snow geese bagged in the vicinity of Sand Lake N.W.R., South Dakota.

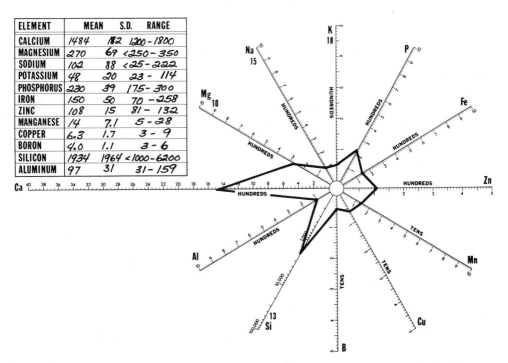

ELEMENT	MEAN	S.D.	RANGE
CALCIUM	1484	182	1200-1800
MAGNESIUM	270	69	<250-350
SODIUM	102	88	<25-222
POTASSIUM	48	20	23-114
PHOSPHORUS	230	39	175-300
IRON	150	50	70-258
ZINC	108	15	81-132
MANGANESE	14	7.1	5-28
COPPER	6.3	1.7	3-9
BORON	4.0	1.1	3-6
SILICON	1934	1964	<1000-6200
ALUMINUM	97	31	31-159

168. Feather mineral pattern A representing nineteen (62 percent) of thirty-one blue and lesser snow geese bagged in the vicinity of Squaw Creek N.W.R., Missouri.

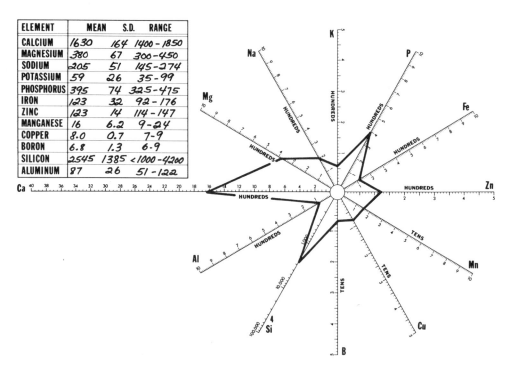

ELEMENT	MEAN	S.D.	RANGE
CALCIUM	1630	164	1400 - 1850
MAGNESIUM	380	67	300 - 450
SODIUM	205	51	145 - 274
POTASSIUM	59	26	35 - 99
PHOSPHORUS	395	74	325 - 475
IRON	123	32	92 - 176
ZINC	123	14	114 - 147
MANGANESE	16	6.2	9 - 24
COPPER	8.0	0.7	7 - 9
BORON	6.8	1.3	6 - 9
SILICON	2545	1385	<1000 - 4200
ALUMINUM	87	26	51 - 122

169. Feather mineral pattern B representing five (16 percent) of thirty-one blue and lesser snow geese bagged in the vicinity of Squaw Creek N.W.R., Missouri.

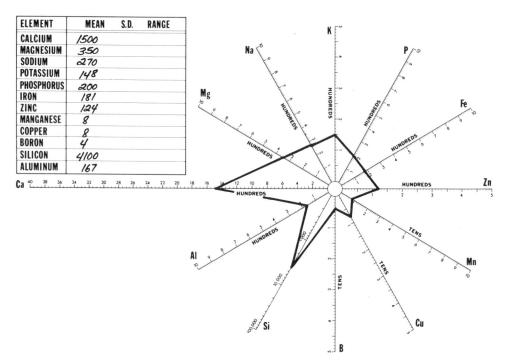

ELEMENT	MEAN	S.D.	RANGE
CALCIUM	1500		
MAGNESIUM	350		
SODIUM	270		
POTASSIUM	148		
PHOSPHORUS	200		
IRON	181		
ZINC	124		
MANGANESE	8		
COPPER	8		
BORON	4		
SILICON	4100		
ALUMINUM	167		

170. Feather mineral pattern C representing one (3 percent) of thirty-one blue and lesser snow geese bagged in the vicinity of Squaw Creek N.W.R., Missouri.

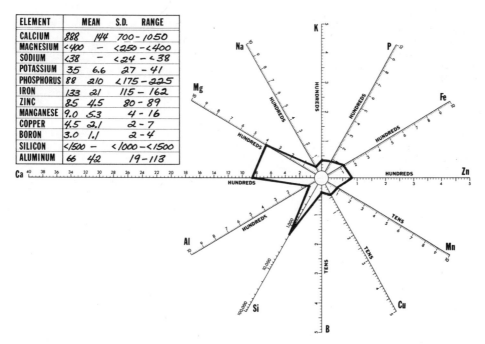

ELEMENT	MEAN	S.D.	RANGE
CALCIUM	888	144	700 – 1050
MAGNESIUM	<400	–	<250 – <400
SODIUM	<38	–	<24 – <38
POTASSIUM	35	6.6	27 – 41
PHOSPHORUS	88	210	<175 – 225
IRON	133	21	115 – 162
ZINC	85	4.5	80 – 89
MANGANESE	9.0	5.3	4 – 16
COPPER	4.5	2.1	2 – 7
BORON	3.0	1.1	2 – 4
SILICON	<1500	–	<1000 – <1500
ALUMINUM	66	42	19 – 113

171. Feather mineral pattern D representing four (13 percent) of thirty-one blue and lesser snow geese bagged in the vicinity of Squaw Creek N.W.R., Missouri.

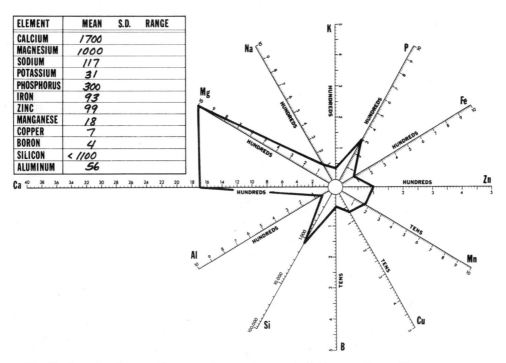

ELEMENT	MEAN	S.D.	RANGE
CALCIUM	1700		
MAGNESIUM	1000		
SODIUM	117		
POTASSIUM	31		
PHOSPHORUS	300		
IRON	93		
ZINC	99		
MANGANESE	18		
COPPER	7		
BORON	4		
SILICON	<1100		
ALUMINUM	56		

172. Feather mineral pattern E representing one (3 percent) of thirty-one blue and lesser snow geese bagged in the vicinity of Squaw Creek N.W.R., Missouri.

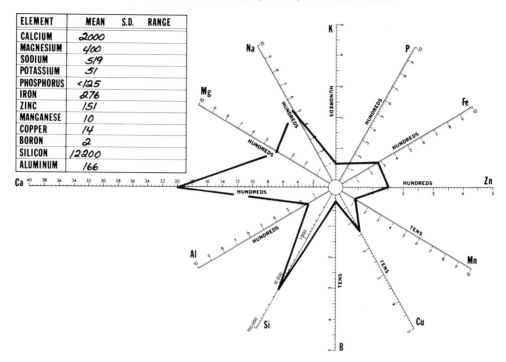

ELEMENT	MEAN	S.D.	RANGE
CALCIUM	2000		
MAGNESIUM	400		
SODIUM	519		
POTASSIUM	51		
PHOSPHORUS	<125		
IRON	276		
ZINC	151		
MANGANESE	10		
COPPER	14		
BORON	2		
SILICON	12200		
ALUMINUM	166		

173. Feather mineral pattern F representing one (3 percent) of thirty-one blue and lesser snow geese bagged in the vicinity of Squaw Creek N.W.R., Missouri.

Henrietta Maria geese largely paralleled that of Baffin Island geese although a greater proportion of the Cape Henrietta Maria birds tended to follow migration routes to the west of those used by Baffin Island stocks. At Pass A Loutre, at the eastern end of the wintering range on the Gulf Coast (Figure *174*), 52 percent of the shot sample of geese presumably originated from Cape Henrietta Maria, and 91 percent of these were blue-phase geese (Tables 19–20).

18. Percent of blue geese by colony in samples of geese analyzed for feather minerals

Colony	Source of Sample	Number in Sample	Blue Geese n	%
Baffin Island, N.W.T.	Wing sample[a]	59	49	83
	Banded Sample[b]	2,418	2,128	88[b]
		7,361	5,585	76[d]
Cape Henrietta Maria, Ont.	Wing sample[a]	24	8	33
	Aerial photographs[c]	3,734	2,768	74
Southampton Island, N.W.T.	Wing sample[a]	23	13	57
	Ground counts	?	?	35[e]
		?	?	30[f]
McConnell River N.W.T.	Wing sample[a]	38	16	42
	Aerial photographs[c]	12,383	3,098	25

[a] Of geese banded on their breeding grounds.
[b] Banded at Bluegoose Prairie, 1967 (from Kerbes 1969).
[c] From aerial photography for years of wing collection (data from Hanson et al. 1972).
[d] Banded at Koukdjuak Plain, 1968 (from Kerbes 1969).
[e] For East Bay (from Cooch 1961:75).
[f] For Boas River (from Cooch 1961:75).

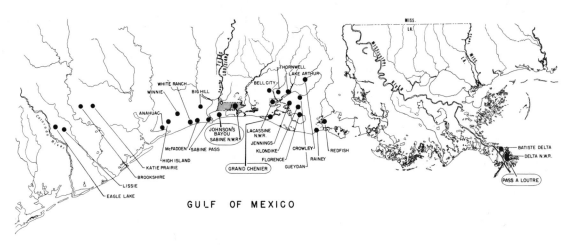

174. Present-day wintering range of blue and lesser snow geese along the Gulf coast of southern Louisiana and Texas. Wing feathers were analyzed from geese collected at encircled localities. Also shown are principal points at which winter inventory observations are made by state and federal personnel.

In the vicinity of Sabine N.W.R., western Louisiana (Figure *174*), Cape Henrietta Maria geese comprised only 16 percent of a fairly adequate sample, and of these, 69 percent were of the blue phase (Tables 19–20). Because of the immense numbers of geese nesting in the Baffin Island colony, it is not unreasonable that geese of this colony would account for a larger portion of the wintering population at the Sabine N.W.R. than would the Cape Henrietta Maria flock, which is possibly less than one-tenth the size of the former. Only about 60 percent of the geese presumably originating from Baffin Island and wintering at the Sabine N.W.R. were blue geese—a percentage well below the prevalence of this color phase on Baffin Island.

Similarly, at the Sabine N.W.R. the represen-

tation of blue geese in the contingents of blues and snows from the Southampton Island colonies and the McConnell River colonies was well above the percentage of these geese on their respective breeding grounds (Tables 16, 18, 20). However, the color-phase composition of the flocks in the western localities of the wintering range along the coastal region of east Texas—Katie Prairie, Lissie, and Eagle Lake (Figure *174*)—is 75 percent lesser snow geese (Lynch 1973:36). The lesser snow goose population that nests in the central Canadian Arctic could not account for the numbers of blue and lesser snow geese seen at these Texas localities. Cooch (1961:87) has pointed out that the greatest number of band recoveries of Southampton Island geese on the Gulf Coast are made in east Texas. The McConnell River geese essentially du-

19. Indicated colony sources of unbanded blue and lesser snow geese shot in the Mississippi and Central flyways, 1966–67, based on minerals in the primary feathers

Locality of Sample of Unknown Origin	Number in Sample	Indicated Colony of Origin									
		Baffin Island		Southampton Island		Cape Henrietta Maria		McConnell River		Unknown	
		n	%	n	%	n	%	n	%	n	%
Severn River, Ont.	27	10	37.6	9	33.3	4	14.8	4	14.8	0	0.0
Kapiskau River, Ont.	15	13	86.6	1	6.7	0	0	1	6.7	0	0.0
Sand Lake, N.W.R., S. Dak.	23	3	13.0	5	21.7	13	56.6	2	8.7	0	0.0
Squaw Creek N.W.R., Mo.	31	17	54.8	6	19.4	5	16.1	1	3.2	2	6.5
Pass A Loutre, La.	21	1	4.8	5	23.8	11	52.4	2	9.5	2	9.5
Johnson's Bayou, La.	3	0	—	0	—	0	—	3	100.0	0	0.0
Grand Chenier, La.	17	1	5.9	6	35.3	3	17.6	7	4.1	0	0.0
Sabine N.W.R., La.	82	42	51.2	8	9.8	13	15.9	19	23.1	0	0.0

20. Color phase composition of samples of migrating and wintering populations of blue and lesser snow geese

					Indicated Origin							
					Baffin Island				Cape Henrietta Maria			
			Total		Total		Blue		Total		Blue	
Locality (in the vicinity of)	Year	Number in Sample	Blue Geese n	%	n	%	n	%	n	%	n	%
Ft. Severn, Ont.	1966	27	7	25.9	10	37.1	nd[a]		4	14.8	nd	
Kapiskau River, Ont.	1967	16	16	100.0	13	81.2	13	100.0	0	0.0	0	0.0
Sand Lake N.W.R., S.Dak.	1966	23	2	8.7	3	13.1	1	33.3	13	56.5	1	7.7
Squaw Creek N.W.R., Mo.	1966	31	9	29.0	17	54.8	2	11.8	5	16.1	3	60.0
Pass A Loutre, La.	1967	21[b]	19	95.2	1	4.8	1	100.0	11	52.4	10	90.9
Grand Chenier, La.	1967	17	16	94.1	1	5.9	1	100.0	3	17.6	3	100.0
Sabine N.W.R., La.	1967	82	56	68.3	42	51.2	25	59.5	13	15.8	9	69.2
Johnson's Bayou, La.	1967	3	1	33.3	0	0.0	0	0.0	0	0.0	0	0.0
Total or percent		220[c]	126	57.7	87	39.9			49	22.5		

Note: Origins were determined by use of the discriminant function test.   [b] Origin of two birds undetermined.
[a] Unknown.   [c] Base line percentages computed from 218.

plicate these movements although their migration routes may be somewhat to the west of those of the Southampton Island birds. A more pronounced westward movement of the white-phase birds of these two colonies readily explains color ratios observed in east Texas.

If our determinations of the origins of the hunter-killed geese from feather analyses are seemingly not in agreement with assumed distribu-

tion patterns based on band recoveries and observed color ratios, it should be pointed out that dramatic shifts in the blue and snow goose populations wintering on the Gulf Coast have occurred in recent years. In this respect, we can refer to no better authority than John J. Lynch (personal communication, 12 January 1973):

"Incidentally these [Mississippi River] Delta Snows & Blues have been declining in numbers in

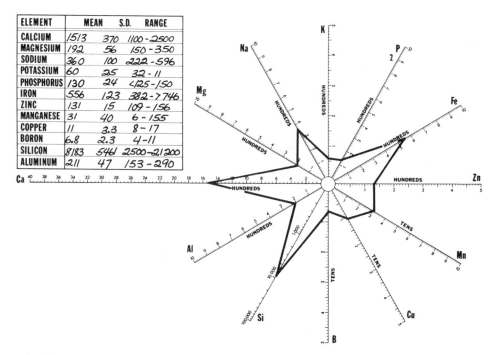

ELEMENT	MEAN	S.D.	RANGE
CALCIUM	1513	370	1100-2500
MAGNESIUM	192	56	150-350
SODIUM	360	100	222-596
POTASSIUM	60	25	32-11
PHOSPHORUS	130	24	<125-150
IRON	556	123	382->746
ZINC	131	15	109-156
MANGANESE	31	40	6-155
COPPER	11	3.3	8-17
BORON	6.8	2.3	4-11
SILICON	8183	5461	2500-21200
ALUMINUM	211	47	153-290

175. Feather mineral pattern A representing twelve (52 percent) of twenty-three blue and lesser snow geese bagged in the vicinity of Pass A Loutre, Louisiana.

Indicated Origin							
Southampton Island				McConnell River			
Total		Blue		Total		Blue	
n	%	n	%	n	%		%
9	33.3	nd		4	14.8	nd	
2	12.5	2	100.0	1	6.3	1	100.0
5	21.7	0	0.0	2	8.7	0	0.0
7	22.6	4	57.1	2	6.5	0	0.0
5	23.8	5	100.0	2	9.5	2	100.0
6	35.3	6	100.0	7	41.2	6	85.7
8	9.8	5	62.5	19	23.2	17	89.5
0	0.0	0	0.0	3	100.0	2	66.7
42	19.3			40	18.3		

recent years. A decade or more ago we could find 60,000 to 100,000+ of these geese in SE Louisiana, and at one time we thought we had 300,000 there (ground guesstimates, Lynch, O'Neil & Lay, Jour. Wildl. Mgt. 11:1, 50–76, January 1947). But now we have to work real hard to locate as many as 30,000 geese in that southeastern region. Recent changes in wintering conditions might have brought about some local shifting of birds here, but would not explain their decline in total numbers. The latter could well reflect the drop in productivity of Blues, occasioned by the recent cooling off of *nesting* grounds in the eastern Arctic.

"Those old Baffin-nesters that may have moved to the newly-developing nest colonies on the west side of Hudson Bay, and their progeny, could now be coming south via more westerly flight corridors. In so doing, they would be apt to show up in Gulf winter quarters at or west of Vermilion Bay, or might even get involved to some extent in the interrupted fall migration of "Great Plains" blue and snows (in which case, color-ratios would quickly be obscured). Our efforts to identify via color and age those blue and snow flocks that come to wintering-grounds between Lacassine NWR and High Island, Texas, have run into all kinds of problems. Added to these problems are the blues and snows that no longer come this far south."

Cooch (1961:87) has postulated for the Southampton Island stocks an eastward shift of blue geese and a westward shift of lesser snow geese along the Gulf Coast. In brief, the picture may be more simple—a westward drift of all populations in migration before and after they reach the Gulf Coast marshes, but a more pronounced drift by the white-phase birds.

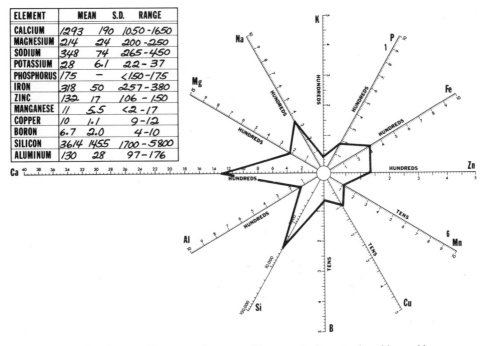

ELEMENT	MEAN	S.D.	RANGE
CALCIUM	1293	190	1050-1650
MAGNESIUM	214	24	200-250
SODIUM	348	74	265-450
POTASSIUM	28	6.1	22-37
PHOSPHORUS	175	–	<150-175
IRON	318	50	257-380
ZINC	132	17	106-150
MANGANESE	11	5.5	<2-17
COPPER	10	1.1	9-12
BORON	6.7	2.0	4-10
SILICON	3614	1455	1700-5800
ALUMINUM	130	28	97-176

176. Feather mineral pattern B representing seven (31 percent) of twenty-three blue and lesser snow geese bagged in the vicinity of Pass A Loutre, Louisiana.

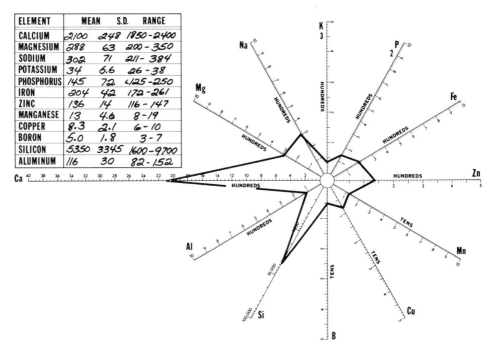

ELEMENT	MEAN	S.D.	RANGE
CALCIUM	2100	248	1850-2400
MAGNESIUM	288	63	200-350
SODIUM	302	71	211-384
POTASSIUM	34	6.6	26-38
PHOSPHORUS	145	72	<125-250
IRON	204	42	172-261
ZINC	136	14	116-147
MANGANESE	13	4.6	8-19
COPPER	8.3	2.1	6-10
BORON	5.0	1.8	3-7
SILICON	5350	3345	1600-9700
ALUMINUM	116	30	82-152

177. Feather mineral pattern C representing four (17 percent) of twenty-three blue and lesser snow geese bagged in the vicinity of Pass A Loutre, Louisiana.

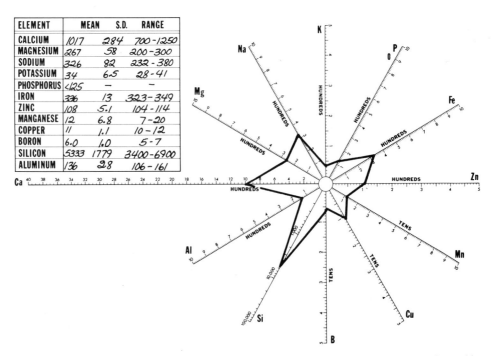

ELEMENT	MEAN	S.D.	RANGE
CALCIUM	1017	284	700-1250
MAGNESIUM	267	58	200-300
SODIUM	326	82	232-380
POTASSIUM	34	6.5	28-41
PHOSPHORUS	<125	–	–
IRON	336	13	323-349
ZINC	108	5.1	104-114
MANGANESE	12	6.8	7-20
COPPER	11	1.1	10-12
BORON	6.0	1.0	5-7
SILICON	5333	1779	3400-6900
ALUMINUM	136	28	106-161

178. Feather mineral pattern representing 100 percent of three blue and lesser snow geese bagged in the vicinity of Johnson's Bayou, Louisiana.

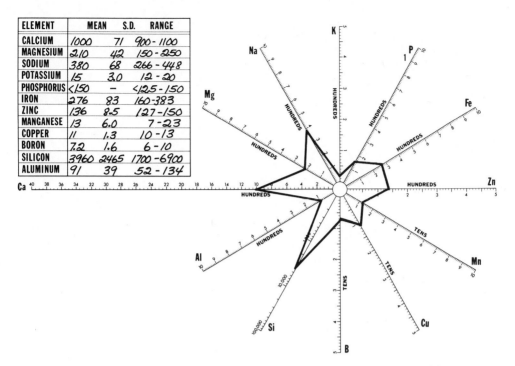

ELEMENT	MEAN	S.D.	RANGE
CALCIUM	1000	71	900-1100
MAGNESIUM	210	42	150-250
SODIUM	380	68	266-448
POTASSIUM	15	3.0	12-20
PHOSPHORUS	<150	–	<125-150
IRON	276	83	160-383
ZINC	136	8.5	127-150
MANGANESE	13	6.0	7-23
COPPER	11	1.3	10-13
BORON	7.2	1.6	6-10
SILICON	3960	2465	1700-6900
ALUMINUM	91	39	52-134

179. Feather mineral pattern A representing eleven (69 percent) of sixteen blue and lesser snow geese bagged in the vicinity of Grand Chenier, Louisiana.

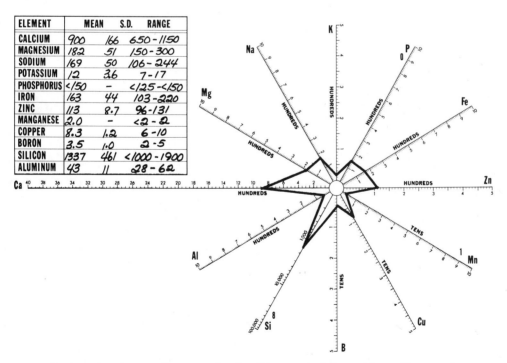

ELEMENT	MEAN	S.D.	RANGE
CALCIUM	900	166	650-1150
MAGNESIUM	182	51	150-300
SODIUM	169	50	106-244
POTASSIUM	12	3.6	7-17
PHOSPHORUS	<150	–	<125-<150
IRON	163	44	103-220
ZINC	113	8.7	96-131
MANGANESE	2.0	–	<2-2
COPPER	8.3	1.2	6-10
BORON	3.5	1.0	2-5
SILICON	1337	461	<1000-1900
ALUMINUM	43	11	28-62

180. Feather mineral pattern B representing five (31 percent) of sixteen blue and lesser snow geese bagged in the vicinity of Grand Chenier, Louisiana.

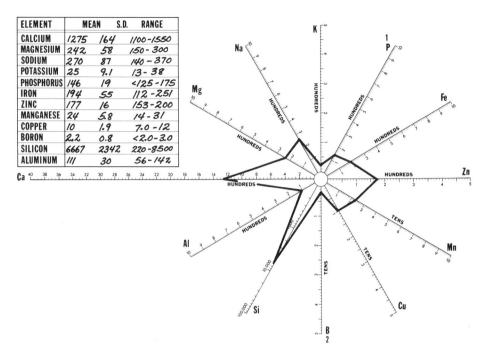

ELEMENT	MEAN	S.D.	RANGE
CALCIUM	1275	164	1100-1550
MAGNESIUM	242	58	150-300
SODIUM	270	87	140-370
POTASSIUM	25	9.1	13-38
PHOSPHORUS	146	19	<125-175
IRON	194	55	112-251
ZINC	177	16	153-200
MANGANESE	24	5.8	14-31
COPPER	10	1.9	7.0-12
BORON	2.2	0.8	<2.0-3.0
SILICON	6667	2342	220-8500
ALUMINUM	111	30	56-142

181. Feather mineral pattern A representing six (7.3 percent) of eighty-two blue and lesser snow geese bagged in the vicinity of Sabine N.W.R., Louisiana.

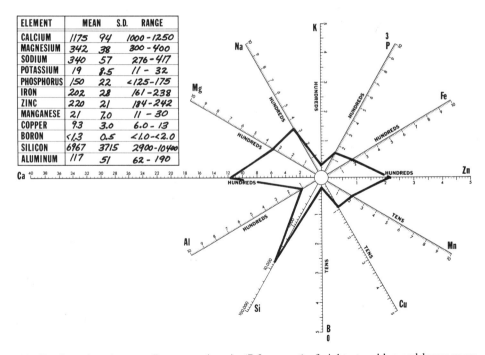

ELEMENT	MEAN	S.D.	RANGE
CALCIUM	1175	94	1000-1250
MAGNESIUM	342	38	300-400
SODIUM	340	57	276-417
POTASSIUM	19	8.5	11-32
PHOSPHORUS	150	22	<125-175
IRON	202	28	161-238
ZINC	220	21	184-242
MANGANESE	21	7.0	11-30
COPPER	9.3	3.0	6.0-13
BORON	<1.3	0.5	<1.0-<2.0
SILICON	6967	3715	2900-10400
ALUMINUM	117	51	62-190

182. Feather mineral pattern B representing six (7.3 percent) of eighty-two blue and lesser snow geese bagged in the vicinity of Sabine N.W.R., Louisiana.

132

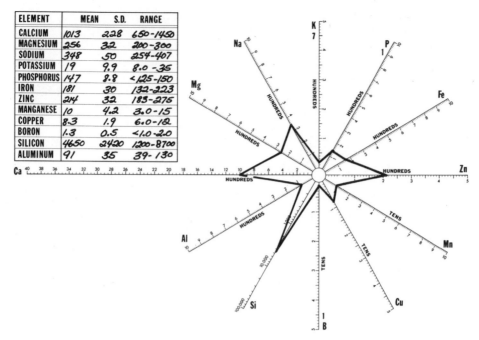

ELEMENT	MEAN	S.D.	RANGE
CALCIUM	1013	228	650-1450
MAGNESIUM	256	32	200-300
SODIUM	348	50	254-407
POTASSIUM	19	9.9	8.0-35
PHOSPHORUS	147	8.8	<125-150
IRON	181	30	132-223
ZINC	214	32	183-275
MANGANESE	10	4.2	3.0-15
COPPER	8.3	1.9	6.0-12
BORON	1.3	0.5	<1.0-2.0
SILICON	4650	2420	1200-8700
ALUMINUM	91	35	39-130

183. Feather mineral pattern C representing eight (9.8 percent) of eighty-two blue and lesser snow geese bagged in the vicinity of Sabine N.W.R., Louisiana.

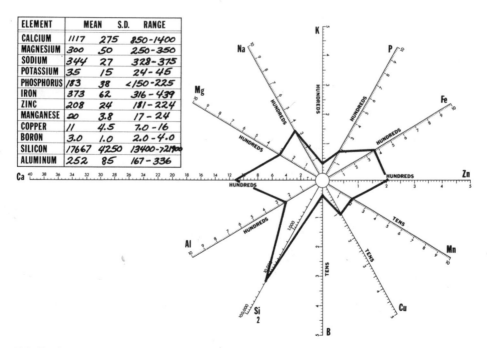

ELEMENT	MEAN	S.D.	RANGE
CALCIUM	1117	275	850-1400
MAGNESIUM	300	50	250-350
SODIUM	344	27	328-375
POTASSIUM	35	15	24-45
PHOSPHORUS	183	38	<150-225
IRON	373	62	316-439
ZINC	208	24	181-224
MANGANESE	20	3.8	17-24
COPPER	11	4.5	7.0-16
BORON	3.0	1.0	2.0-4.0
SILICON	17667	4250	13400-72190
ALUMINUM	252	85	167-336

184. Feather mineral pattern D representing three (3.7 percent) of eighty-two blue and lesser snow geese bagged in the vicinity of Sabine N.W.R., Louisiana.

133

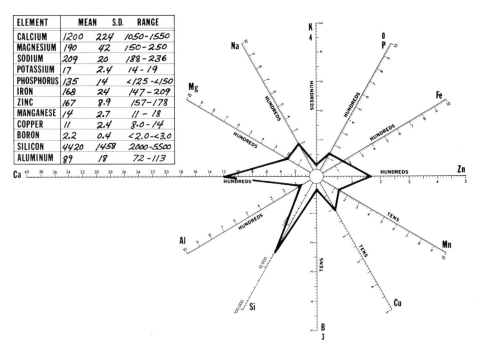

ELEMENT	MEAN	S.D.	RANGE
CALCIUM	1200	224	1050-1550
MAGNESIUM	190	42	150-250
SODIUM	209	20	188-236
POTASSIUM	17	2.4	14-19
PHOSPHORUS	135	14	<125-<150
IRON	168	24	147-209
ZINC	167	8.9	157-178
MANGANESE	14	2.7	11-18
COPPER	11	2.4	8.0-14
BORON	2.2	0.4	<2.0-<3.0
SILICON	4420	1458	2000-5500
ALUMINUM	89	18	72-113

185. Feather mineral pattern E representing five (6.1 percent) of eighty-two blue and lesser snow geese bagged in the vicinity of sabine N.W.R., Louisiana.

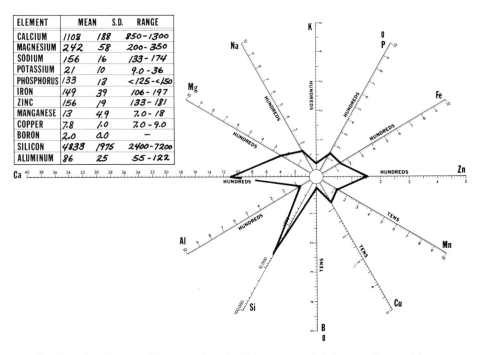

ELEMENT	MEAN	S.D.	RANGE
CALCIUM	1108	188	850-1300
MAGNESIUM	242	58	200-350
SODIUM	156	16	133-174
POTASSIUM	21	10	9.0-36
PHOSPHORUS	133	13	<125-<150
IRON	149	39	106-197
ZINC	156	19	133-181
MANGANESE	13	4.9	7.0-18
COPPER	7.8	1.0	7.0-9.0
BORON	2.0	0.0	—
SILICON	4833	1975	2400-7200
ALUMINUM	86	25	55-122

186. Feather mineral pattern F representing six (7.3 percent) of eighty-two blue and lesser snow geese bagged in the vicinity of Sabine N.W.R., Louisiana.

134

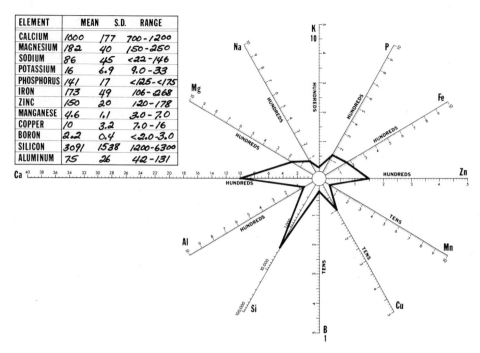

ELEMENT	MEAN	S.D.	RANGE
CALCIUM	1000	177	700-1200
MAGNESIUM	182	40	150-250
SODIUM	86	45	<22-146
POTASSIUM	16	6.9	9.0-33
PHOSPHORUS	141	17	<125-<175
IRON	173	49	106-268
ZINC	150	20	120-178
MANGANESE	4.6	1.1	3.0-7.0
COPPER	10	3.2	7.0-16
BORON	2.2	0.4	<2.0-3.0
SILICON	3091	1538	1200-6300
ALUMINUM	75	26	42-131

187. Feather mineral pattern G representing eleven (13.5 percent) of eighty-two blue and lesser snow geese bagged in the vicinity of Sabine N.W.R., Louisiana.

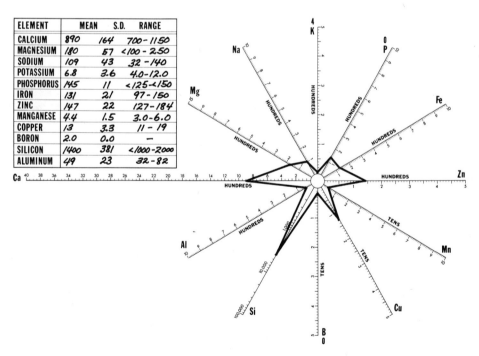

ELEMENT	MEAN	S.D.	RANGE
CALCIUM	890	164	700-1150
MAGNESIUM	180	57	<100-250
SODIUM	109	43	32-140
POTASSIUM	6.8	3.6	4.0-12.0
PHOSPHORUS	145	11	<125-<150
IRON	131	21	97-150
ZINC	147	22	127-184
MANGANESE	4.4	1.5	3.0-6.0
COPPER	13	3.3	11-19
BORON	2.0	0.0	—
SILICON	1400	381	<1000-2000
ALUMINUM	49	23	32-82

188. Feather mineral pattern H representing five (6.1 percent) of eighty-two blue and lesser snow geese bagged in the vicinity of Sabine N.W.R., Louisiana.

135

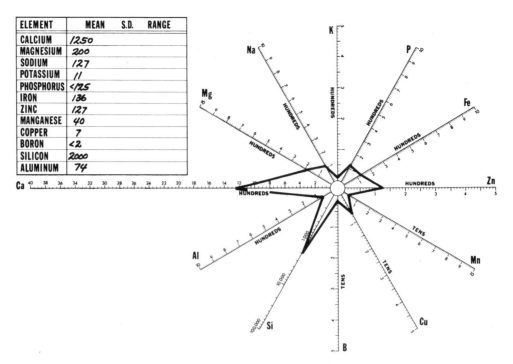

ELEMENT	MEAN	S.D.	RANGE
CALCIUM	1250		
MAGNESIUM	200		
SODIUM	127		
POTASSIUM	11		
PHOSPHORUS	<125		
IRON	136		
ZINC	127		
MANGANESE	40		
COPPER	7		
BORON	<2		
SILICON	2000		
ALUMINUM	74		

189. Feather mineral pattern I representing one (1.2 percent) of eighty-two blue and lesser snow geese bagged in the vicinity of Sabine N.W.R., Louisiana.

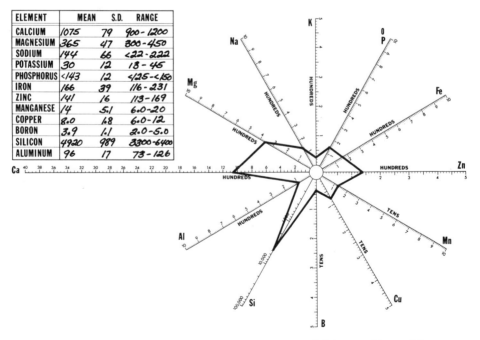

ELEMENT	MEAN	S.D.	RANGE
CALCIUM	1075	79	900-1200
MAGNESIUM	365	47	300-450
SODIUM	144	66	<22-222
POTASSIUM	30	12	13-45
PHOSPHORUS	<143	12	<125-<150
IRON	166	39	116-231
ZINC	141	16	113-169
MANGANESE	14	5.1	6.0-20
COPPER	8.0	1.8	6.0-12
BORON	3.9	1.1	2.0-5.0
SILICON	4920	989	3300-6400
ALUMINUM	96	17	73-126

190. Feather mineral pattern J representing ten (12.3 percent) of eighty-two blue and lesser snow geese bagged in the vicinity of Sabine N.W.R., Louisiana.

136

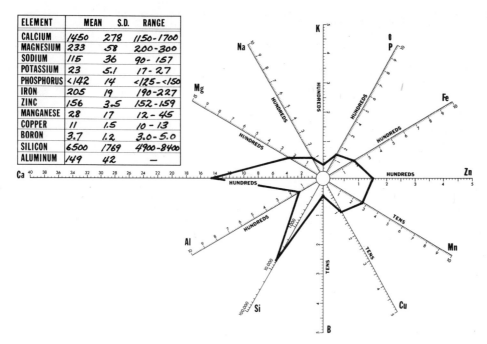

ELEMENT	MEAN	S.D.	RANGE
CALCIUM	1450	278	1150-1700
MAGNESIUM	233	58	200-300
SODIUM	115	36	90-157
POTASSIUM	23	5.1	17-27
PHOSPHORUS	<142	14	<125-<150
IRON	205	19	190-227
ZINC	156	3.5	152-159
MANGANESE	28	17	12-45
COPPER	11	1.5	10-13
BORON	3.7	1.2	3.0-5.0
SILICON	6500	1769	4900-8400
ALUMINUM	149	42	—

191. Feather mineral pattern K representing three (3.7 percent) of eighty-two blue and lesser snow geese bagged in the vicinity of the Sabine N.W.R., Louisiana.

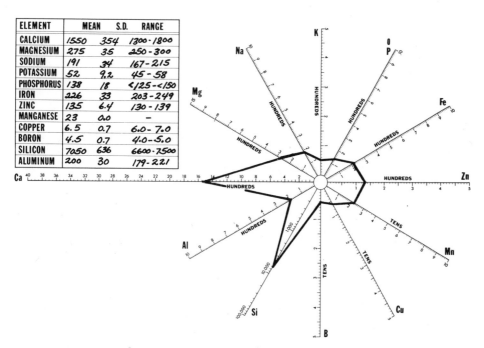

ELEMENT	MEAN	S.D.	RANGE
CALCIUM	1550	354	1300-1800
MAGNESIUM	275	35	250-300
SODIUM	191	34	167-215
POTASSIUM	52	9.2	45-58
PHOSPHORUS	138	18	<125-<150
IRON	226	33	203-249
ZINC	135	6.4	130-139
MANGANESE	23	0.0	—
COPPER	6.5	0.7	6.0-7.0
BORON	4.5	0.7	4.0-5.0
SILICON	7050	636	6600-7500
ALUMINUM	200	30	179-221

192. Feather mineral pattern L representing two (2.4 percent) of eighty-two blue and lesser snow geese bagged in the vicinity of Sabine N.W.R., Louisiana.

137

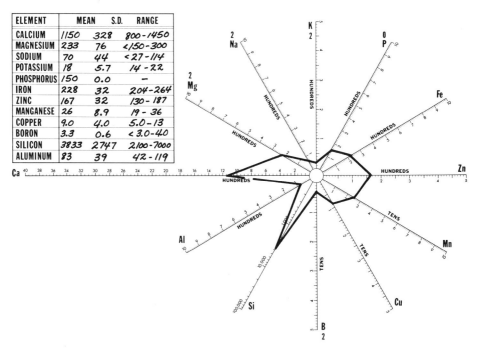

ELEMENT	MEAN	S.D.	RANGE
CALCIUM	1150	328	800-1450
MAGNESIUM	233	76	<150-300
SODIUM	70	44	<27-114
POTASSIUM	18	5.7	14-22
PHOSPHORUS	150	0.0	–
IRON	228	32	204-264
ZINC	167	32	130-187
MANGANESE	26	8.9	19-36
COPPER	9.0	4.0	5.0-13
BORON	3.3	0.6	<3.0-40
SILICON	3833	2747	2100-7000
ALUMINUM	83	39	42-119

193. Feather mineral pattern M representing three (3.7 percent) of eighty-two blue and lesser snow geese bagged in the vicinity of Sabine N.W.R., Louisiana.

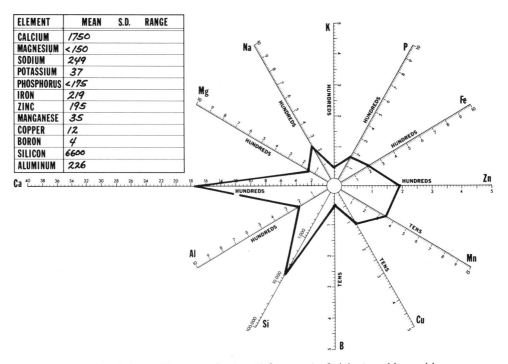

ELEMENT	MEAN	S.D.	RANGE
CALCIUM	1750		
MAGNESIUM	<150		
SODIUM	249		
POTASSIUM	37		
PHOSPHORUS	<175		
IRON	219		
ZINC	195		
MANGANESE	35		
COPPER	12		
BORON	4		
SILICON	6600		
ALUMINUM	226		

194. Feather mineral pattern N representing one (1.2 percent) of eighty-two blue and lesser snow geese bagged in the vicinity of Sabine N.W.R., Louisiana.

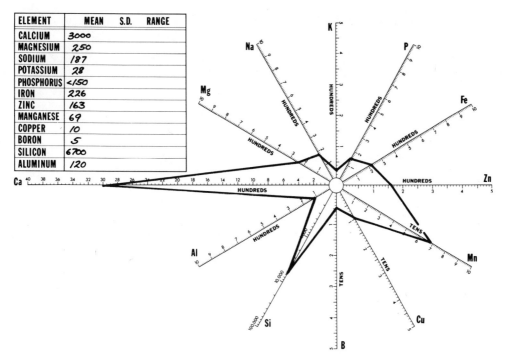

ELEMENT	MEAN	S.D.	RANGE
CALCIUM	3000		
MAGNESIUM	250		
SODIUM	187		
POTASSIUM	28		
PHOSPHORUS	<150		
IRON	226		
ZINC	163		
MANGANESE	69		
COPPER	10		
BORON	5		
SILICON	6700		
ALUMINUM	120		

195. Feather mineral pattern O representing one (1.2 percent) of eight-two blue and lesser snow geese bagged in the vicinity of Sabine N.W.R., Louisiana.

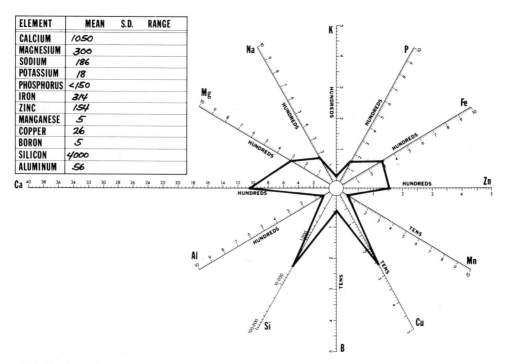

ELEMENT	MEAN	S.D.	RANGE
CALCIUM	1050		
MAGNESIUM	300		
SODIUM	186		
POTASSIUM	18		
PHOSPHORUS	<150		
IRON	314		
ZINC	154		
MANGANESE	5		
COPPER	26		
BORON	5		
SILICON	4000		
ALUMINUM	56		

196. Feather mineral pattern P representing one (1.2 percent) of eighty-two blue and lesser snow geese bagged in the vicinity of Sabine N.W.R., Louisiana.

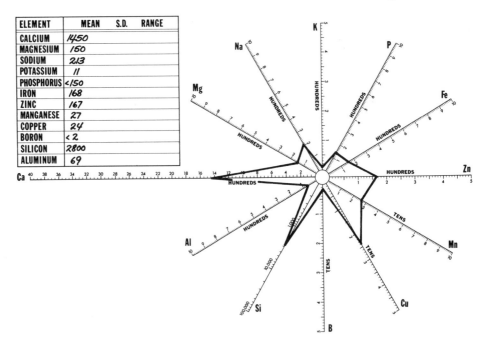

ELEMENT	MEAN	S.D.	RANGE
CALCIUM	1450		
MAGNESIUM	150		
SODIUM	213		
POTASSIUM	11		
PHOSPHORUS	<150		
IRON	168		
ZINC	167		
MANGANESE	27		
COPPER	24		
BORON	<2		
SILICON	2800		
ALUMINUM	69		

197. Feather mineral pattern Q representing one (1.2 percent) of eighty-two blue and lesser snow geese bagged in the vicinity of Sabine N.W.R., Louisiana.

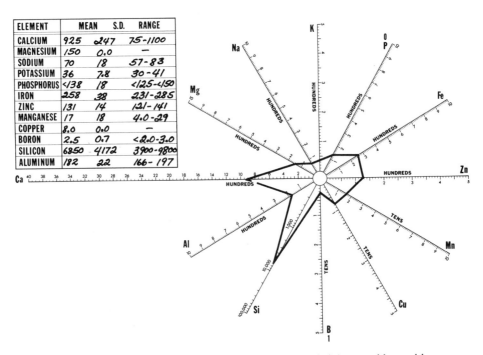

ELEMENT	MEAN	S.D.	RANGE
CALCIUM	925	247	75-1100
MAGNESIUM	150	0.0	—
SODIUM	70	18	57-83
POTASSIUM	36	7.8	30-41
PHOSPHORUS	<138	18	<125-<150
IRON	258	38	231-285
ZINC	131	14	121-141
MANGANESE	17	18	4.0-29
COPPER	8.0	0.0	—
BORON	2.5	0.7	<2.0-3.0
SILICON	6850	4172	3900-9800
ALUMINUM	182	22	166-197

198. Feather mineral pattern R representing two (2.4 percent) of eighty-two blue and lesser snow geese bagged in the vicinity of Sabine N.W.R., Louisiana.

140

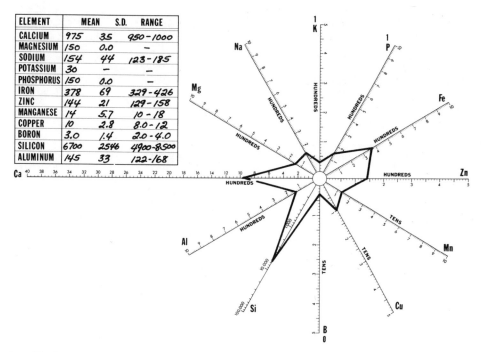

ELEMENT	MEAN	S.D.	RANGE
CALCIUM	975	35	950-1000
MAGNESIUM	150	0.0	−
SODIUM	154	44	123-185
POTASSIUM	30	−	−
PHOSPHORUS	150	0.0	−
IRON	378	69	329-426
ZINC	144	21	129-158
MANGANESE	14	5.7	10-18
COPPER	10	2.8	8.0-12
BORON	3.0	1.4	2.0-4.0
SILICON	6700	2546	4900-8500
ALUMINUM	145	33	122-168

199. Feather mineral pattern S representing two (2.4 percent) of eighty-two blue and lesser snow geese bagged in the vicinity of Sabine N.W.R., Louisiana.

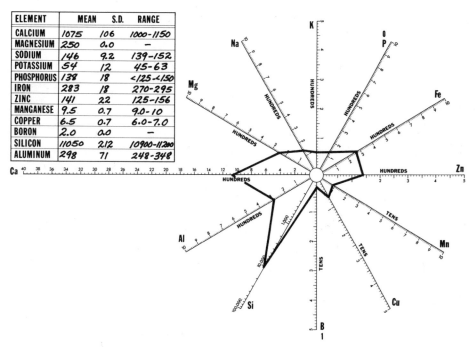

ELEMENT	MEAN	S.D.	RANGE
CALCIUM	1075	106	1000-1150
MAGNESIUM	250	0.0	−
SODIUM	146	9.2	139-152
POTASSIUM	54	12	45-63
PHOSPHORUS	138	18	<125-<150
IRON	283	18	270-295
ZINC	141	22	125-156
MANGANESE	9.5	0.7	9.0-10
COPPER	6.5	0.7	6.0-7.0
BORON	2.0	0.0	−
SILICON	11050	212	10900-11200
ALUMINUM	298	71	248-348

200. Feather mineral pattern T representing two (2.4 percent) of eighty-two blue and lesser snow geese bagged in the vicinity of Sabine N.W.R., Louisiana.

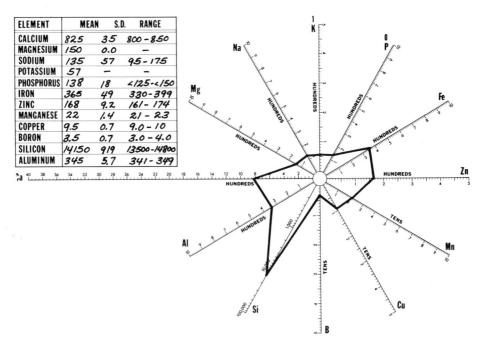

ELEMENT	MEAN	S.D.	RANGE
CALCIUM	825	35	800-850
MAGNESIUM	150	0.0	-
SODIUM	135	57	95-175
POTASSIUM	57	-	-
PHOSPHORUS	138	18	<125-<150
IRON	365	49	330-399
ZINC	168	9.2	161-174
MANGANESE	22	1.4	21-23
COPPER	9.5	0.7	9.0-10
BORON	3.5	0.7	3.0-4.0
SILICON	14150	919	13500-14800
ALUMINUM	345	5.7	341-349

201. Feather mineral pattern U representing two (2.4 percent) of eighty-two blue and lesser snow geese bagged in the vicinity of Sabine N.W.R., Louisiana.

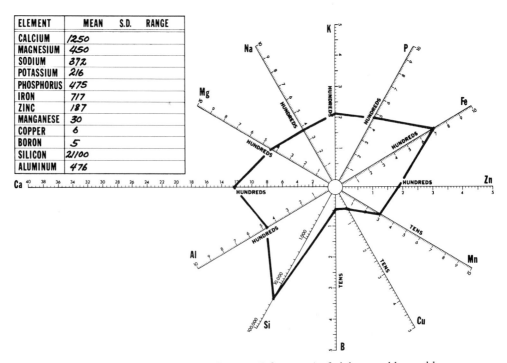

ELEMENT	MEAN	S.D.	RANGE
CALCIUM	1250		
MAGNESIUM	450		
SODIUM	372		
POTASSIUM	216		
PHOSPHORUS	475		
IRON	717		
ZINC	187		
MANGANESE	30		
COPPER	6		
BORON	5		
SILICON	21100		
ALUMINUM	476		

202. Feather mineral pattern V representing one (1.2 percent) of eighty-two blue and lesser snow geese bagged in the vicinity of Sabine N.W.R., Louisiana.

Pacific Flyway Populations

Primary feathers of migrant and wintering lesser snow geese from eleven localities in western North America were analyzed. Feather mineral patterns summarizing the results are shown in these figures: for Kindersley, Saskatchewan, Figures *203–5;* for the Bear River marshes, Utah, Figures *206–8;* for Fallon-Carson Sink area, Nevada, Figures *209–14;* for Bosque del Apache N.W.R., New Mexico, Figure *215;* for Westham Island, Frazier River delta, British Columbia, Figures *216–17;* for the Skagit River delta, Washington, Figures *218–19;* for Summer Lake, Oregon, Figures *220–28;* and for Tule Lake N.W.R., Figures *229–30;* Sacramento N.W.R., Figures *231–33;* Gray Lodge state refuge, Figures *234–38;* and Imperial state refuge, Figures *239–43.*

The origins of migrant and wintering lesser snow goose populations in western North America, based on comparison of group patterns of feather minerals with patterns of geese collected on or near their breeding grounds, are summarized in Table 21. In general, these subjective judgments of origin based on pattern similarities are in agreement with the dispersal of the various colonies as known from band recoveries.

As will be emphasized below, feather mineral data should, when possible, be evaluated in the light of dispersal patterns inferred from band recoveries. For example, it is reasonable to assume

that geese from Banks Island may constitute a major portion of the flocks passing through the Kindersley area of southwestern Saskatchewan. It is less to be expected that Wrangel Island geese contribute to the kill there (Table 21), but recoveries of lesser snow geese banded on Wrangel Island have shown that some components of this population do migrate south through western sectors of the Great Plains.

Recoveries of lesser snow geese banded at the delta of the Anderson River have revealed that the geese of this colony migrate southward through Utah and winter in the Salton Sea area of California (Nagel 1969:44). However, the Anderson River colony consists of only about four to five thousand birds, a number that cannot account for all the lesser snow geese that use the Bear River marshes in autumn. Only 1,365 lesser snow geese in the Banks Island colony have been banded—300 in 1955 and 1,065 in 1961; therefore band recoveries could not draw attention to the presence of geese from this colony in Utah in more recent years. Moreover, a segment of the Banks Island population does migrate through Utah (Nagel 1969:40), and feather mineral patterns of a limited sample of geese indicate that Banks Island geese make up a high proportion of the flights stopping in Utah en route to the Salton Sea area (Table 21 and Figures *206–8*).

Feather mineral patterns of lesser snow geese bagged in the Fallon-Carson Sink area of Nevada

21. Origin of samples of migrant and wintering populations of lesser snow geese in western North America

Locality	Physiographic Region or Area	Number in Sample	Wrangel Island		Banks Island		Kendall Island		Queen Maud Gulf Lowlands	
			n	%	n	%	n	%	n	%
Kindersley, Sask.	Western Great Plains	15	2	13	13	87				
Bear River marshes, Utah	Great Basin	15			15	100				
Fallon-Carson Sink area, Nev.	Great Basin	21			21	100				
Bosque del Apache N.W.R., N.Mex.	Rio Grande Valley	4							4	100
Westham Island, Frazier River delta, B.C.	Pacific Coast	6	6	100						
Skagit River delta, Wash.	Pacific Coast	16	16	100						
Summer Lake Refuge, Ore.	Great Basin	29	25	86	4	14				
Tule Lake N.W.R., Calif.	Great Basin	21	21	100						
Sacramento N.W.R., Calif.	Sacramento Valley	7	5	71			2	29		
Gray Lodge Refuge, Calif.	Sacramento Valley	27			27	100				
Imperial Refuge, Calif.	Imperial Valley	27			27	100				

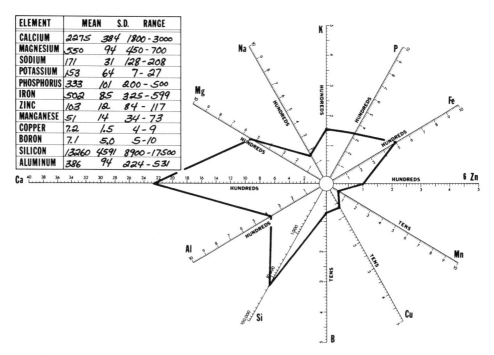

ELEMENT	MEAN	S.D.	RANGE
CALCIUM	2275	384	1800-3000
MAGNESIUM	550	94	450-700
SODIUM	171	31	128-208
POTASSIUM	153	64	7-27
PHOSPHORUS	333	101	200-500
IRON	502	85	325-599
ZINC	103	12	84-117
MANGANESE	51	14	34-73
COPPER	7.2	1.5	4-9
BORON	7.1	5.0	5-10
SILICON	13260	4591	8900-17500
ALUMINUM	386	94	224-531

203. Feather mineral pattern A representing three (20 percent) of fifteen lesser snow geese bagged in the vicinity of Kindersley, Saskatchewan. Assumed origin: Banks Island.

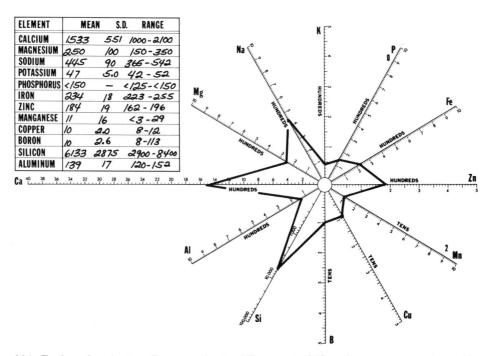

ELEMENT	MEAN	S.D.	RANGE
CALCIUM	1533	551	1000-2100
MAGNESIUM	250	100	150-350
SODIUM	445	90	366-542
POTASSIUM	47	5.0	42-52
PHOSPHORUS	<150	—	<125-<150
IRON	234	18	223-255
ZINC	184	19	162-196
MANGANESE	11	16	<3-29
COPPER	10	2.0	8-12
BORON	10	2.6	8-113
SILICON	6133	2875	2900-8400
ALUMINUM	139	17	120-152

204. Feather mineral pattern B representing ten (67 percent) of fifteen lesser snow geese bagged in the vicinity of Kindersley, Saskatchewan. Assumed origin: Banks Island.

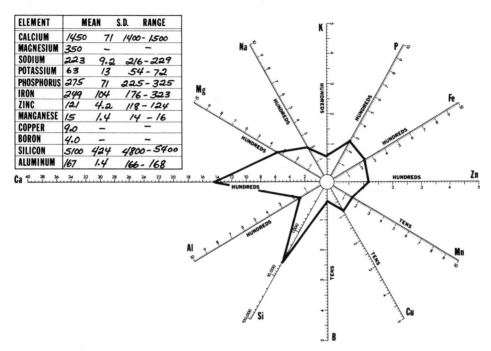

ELEMENT	MEAN	S.D.	RANGE
CALCIUM	1450	71	1400-1500
MAGNESIUM	350	—	—
SODIUM	223	9.2	216-229
POTASSIUM	63	13	54-72
PHOSPHORUS	275	71	225-325
IRON	249	104	176-323
ZINC	121	4.2	118-124
MANGANESE	15	1.4	14-16
COPPER	9.0	—	—
BORON	4.0	—	—
SILICON	5100	424	4800-5400
ALUMINUM	167	1.4	166-168

205. Feather mineral pattern C representing two (13 percent) of fifteen lesser snow geese bagged in the vicinity of Kindersley, Saskatchewan. Assumed origin: Wrangel Island.

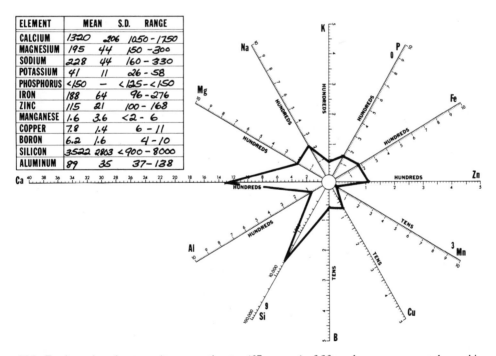

ELEMENT	MEAN	S.D.	RANGE
CALCIUM	1320	206	1050-1750
MAGNESIUM	195	44	150-300
SODIUM	228	44	160-330
POTASSIUM	41	11	26-58
PHOSPHORUS	<150	—	<125-<150
IRON	188	64	96-276
ZINC	115	21	100-168
MANGANESE	1.6	3.6	<2-6
COPPER	7.8	1.4	6-11
BORON	6.2	1.6	4-10
SILICON	3522	2803	<900-8000
ALUMINUM	89	35	37-138

206. Feather mineral pattern A representing ten (67 percent) of fifteen lesser snow geese bagged in the Bear River marshes, Utah. Assumed origin: Banks Island.

145

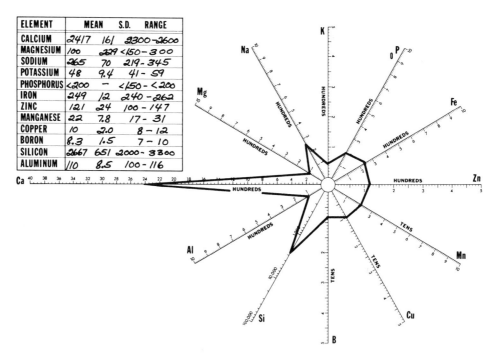

ELEMENT	MEAN	S.D.	RANGE
CALCIUM	2417	161	2300-2600
MAGNESIUM	100	229	<150-300
SODIUM	265	70	219-345
POTASSIUM	48	9.4	41-59
PHOSPHORUS	<200	-	<150-<200
IRON	249	12	240-262
ZINC	121	24	100-147
MANGANESE	22	7.8	17-31
COPPER	10	2.0	8-12
BORON	8.3	1.5	7-10
SILICON	2667	651	2000-3300
ALUMINUM	110	8.5	100-116

207. Feather mineral pattern B representing three (20 percent) of fifteen lesser snow geese bagged in the Bear River marshes, Utah. Assumed origin: Banks Island.

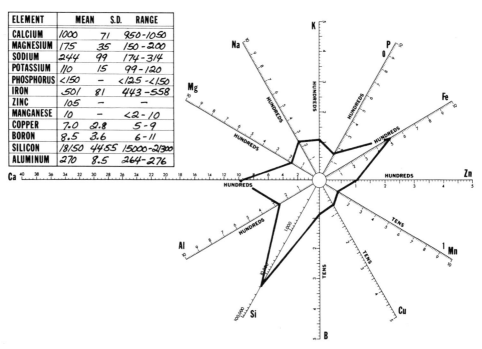

ELEMENT	MEAN	S.D.	RANGE
CALCIUM	1000	71	950-1050
MAGNESIUM	175	35	150-200
SODIUM	244	99	174-314
POTASSIUM	110	15	99-120
PHOSPHORUS	<150	-	<125-<150
IRON	501	81	443-558
ZINC	105	-	-
MANGANESE	10	-	<2-10
COPPER	7.0	2.8	5-9
BORON	8.5	3.6	6-11
SILICON	18150	4455	15000-21300
ALUMINUM	270	8.5	264-276

208. Feather mineral pattern C representing two (13 percent) of fifteen lesser snow geese bagged in the Bear River marshes, Utah. Assumed origin: Banks Island.

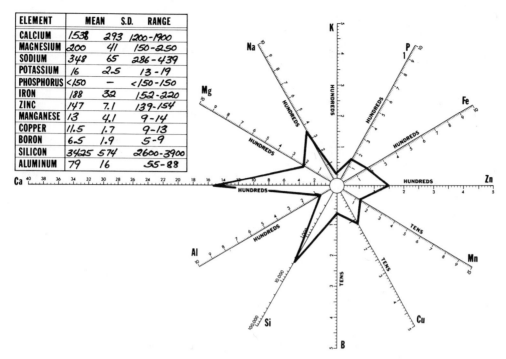

ELEMENT	MEAN	S.D.	RANGE
CALCIUM	1538	293	1200-1900
MAGNESIUM	200	41	150-250
SODIUM	348	65	286-439
POTASSIUM	16	2.5	13 - 19
PHOSPHORUS	<150	—	<150-150
IRON	188	32	152-220
ZINC	147	7.1	139-154
MANGANESE	13	4.1	9-14
COPPER	11.5	1.7	9-13
BORON	6.5	1.9	5-9
SILICON	3425	574	2600-3900
ALUMINUM	79	16	55-88

209. Feather mineral pattern A representing four (19 percent) of twenty-one lesser snow geese bagged in the vicinity of Fallon and Carson Sink, Nevada. Assumed origin: Banks Island.

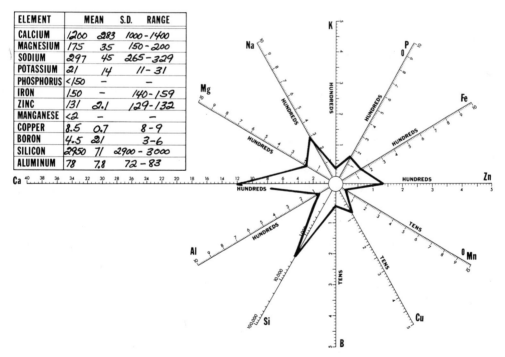

ELEMENT	MEAN	S.D.	RANGE
CALCIUM	1200	283	1000-1400
MAGNESIUM	175	35	150-200
SODIUM	297	45	265-329
POTASSIUM	21	14	11-31
PHOSPHORUS	<150	—	—
IRON	150	—	140-159
ZINC	131	2.1	129-132
MANGANESE	<2	—	
COPPER	8.5	0.7	8-9
BORON	4.5	21	3-6
SILICON	2950	71	2900-3000
ALUMINUM	78	7.8	72-83

210. Feather mineral pattern B representing two (10 percent) of twenty-one lesser snow geese bagged in the vicinity of Fallon and Carson Sink, Nevada. Assumed origin: Banks Island.

147

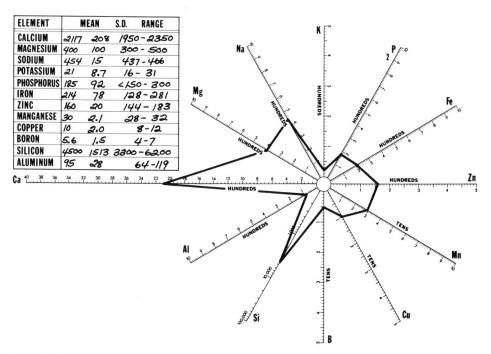

211. Feather mineral pattern C representing three (14 percent) of twenty-one lesser snow geese bagged in the vicinity of Fallon and Carson Sink, Nevada. Assumed origin: Banks Island.

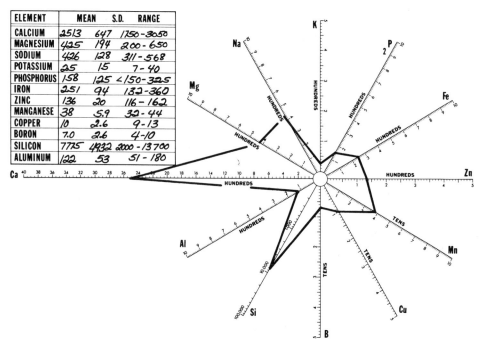

212. Feather mineral pattern D representing four (19 percent) of twenty-one lesser snow geese bagged in the vicinity of Fallon and Carson Sink, Nevada. Assumed origin: Banks Island.

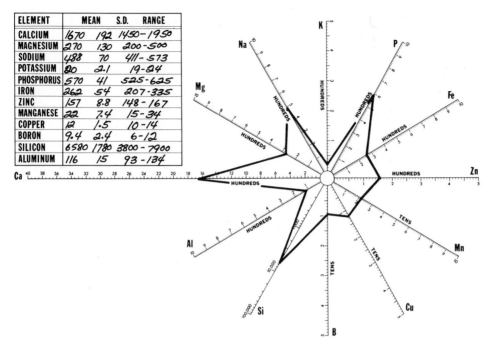

ELEMENT	MEAN	S.D.	RANGE
CALCIUM	1670	192	1450-1950
MAGNESIUM	270	130	200-500
SODIUM	488	70	411-573
POTASSIUM	20	2.1	19-24
PHOSPHORUS	570	41	525-625
IRON	262	54	207-335
ZINC	157	8.8	148-167
MANGANESE	22	7.4	15-34
COPPER	12	1.5	10-14
BORON	9.4	2.4	6-12
SILICON	6580	1780	3800-7900
ALUMINUM	116	15	93-134

213. Feather mineral pattern E representing five (24 percent) of twenty-one lesser snow geese bagged in the vicinity of Fallon and Carson Sink, Nevada. Assumed origin: Banks Island.

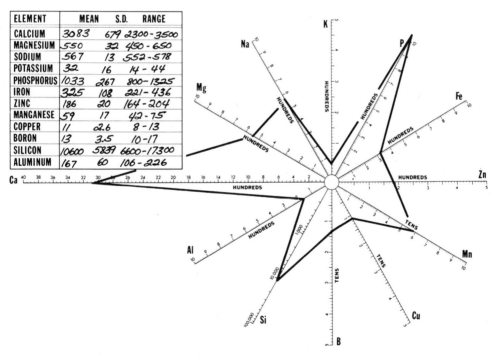

ELEMENT	MEAN	S.D.	RANGE
CALCIUM	3083	679	2300-3500
MAGNESIUM	550	32	450-650
SODIUM	567	13	552-578
POTASSIUM	32	16	14-44
PHOSPHORUS	1033	267	800-1325
IRON	325	108	221-436
ZINC	186	20	164-204
MANGANESE	59	17	42-75
COPPER	11	2.6	8-13
BORON	13	3.5	10-17
SILICON	10600	5839	6600-17300
ALUMINUM	167	60	106-226

214. Feather mineral pattern F representing three (14 percent) of twenty-one lesser snow geese bagged in the vicinity of Fallon and Carson Sink, Nevada. Assumed origin: Banks Island.

149

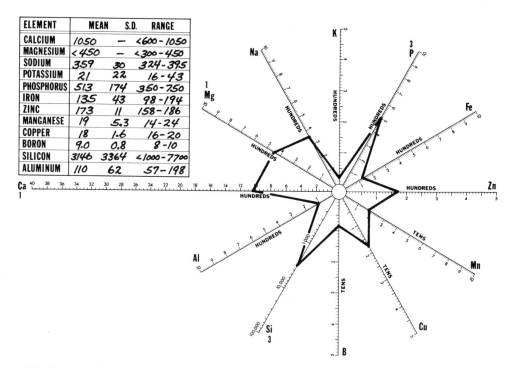

ELEMENT	MEAN	S.D.	RANGE
CALCIUM	1050	—	<600-1050
MAGNESIUM	<450	—	<300-450
SODIUM	359	30	324-395
POTASSIUM	21	22	16-43
PHOSPHORUS	513	174	350-750
IRON	135	43	98-194
ZINC	173	11	158-186
MANGANESE	19	5.3	14-24
COPPER	18	1.6	16-20
BORON	9.0	0.8	8-10
SILICON	3146	3364	<1000-7700
ALUMINUM	110	62	57-198

215. Feather mineral pattern of four lesser snow geese bagged in the vicinity of the Bosque del Apache N.W.R., New Mexico.

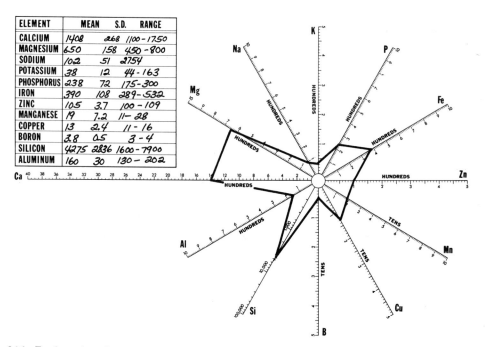

ELEMENT	MEAN	S.D.	RANGE
CALCIUM	1408	268	1100-1750
MAGNESIUM	650	158	450-800
SODIUM	102	51	2754
POTASSIUM	38	12	44-163
PHOSPHORUS	238	72	175-300
IRON	390	108	289-532
ZINC	105	3.7	100-109
MANGANESE	19	7.2	11-28
COPPER	13	2.4	11-16
BORON	3.8	0.5	3-4
SILICON	4275	2836	1600-7900
ALUMINUM	160	30	130-202

216. Feather mineral pattern A representing four (67 percent) of six lesser snow geese bagged on Westham Island, B.C. Assumed origin: Wrangel Island.

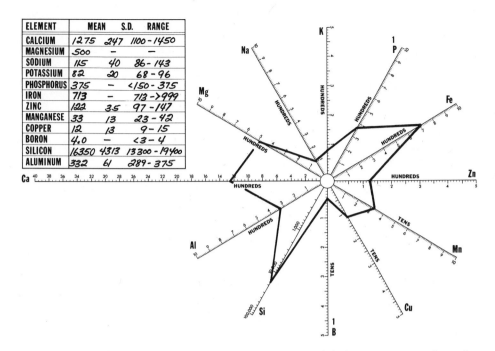

ELEMENT	MEAN	S.D.	RANGE
CALCIUM	1275	247	1100-1450
MAGNESIUM	500	-	-
SODIUM	115	40	86-143
POTASSIUM	82	20	68-96
PHOSPHORUS	375	-	<150-375
IRON	713	-	713->999
ZINC	122	35	97-147
MANGANESE	33	13	23-42
COPPER	12	13	9-15
BORON	4.0	-	<3-4
SILICON	16350	4313	13300-19400
ALUMINUM	332	61	289-375

217. Feather mineral pattern B representing two (33 percent) of six lesser snow geese bagged on Westham Island, B.C. Assumed origin: Wrangel Island.

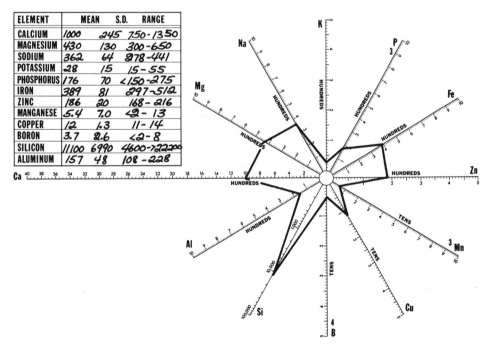

ELEMENT	MEAN	S.D.	RANGE
CALCIUM	1000	245	750-1350
MAGNESIUM	430	130	300-650
SODIUM	362	64	278-441
POTASSIUM	28	15	15-55
PHOSPHORUS	176	70	<150-275
IRON	389	81	297-512
ZINC	186	20	168-216
MANGANESE	5.4	7.0	<2-13
COPPER	12	1.3	11-14
BORON	3.7	2.6	<2-8
SILICON	11100	6990	4600-22200
ALUMINUM	157	48	108-228

218. Feather mineral pattern A representing eleven (69 percent) of sixteen lesser snow geese bagged in the Skagit River delta, Washington. Assumed origin: Wrangel Island.

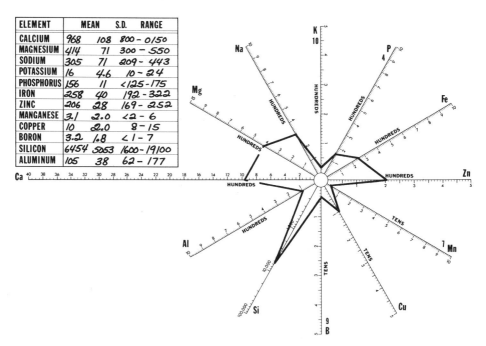

ELEMENT	MEAN	S.D.	RANGE
CALCIUM	968	108	800-0150
MAGNESIUM	414	71	300-550
SODIUM	305	71	209-443
POTASSIUM	16	4.6	10-24
PHOSPHORUS	156	11	<125-175
IRON	258	40	192-322
ZINC	206	28	169-252
MANGANESE	3.1	2.0	<2-6
COPPER	10	2.0	8-15
BORON	3.2	1.8	<1-7
SILICON	6454	5053	1600-19100
ALUMINUM	105	38	62-177

219. Feather mineral pattern B representing five (31 percent) of sixteen lesser snow geese bagged in the Skagit River delta, Washington. Assumed origin: Wrangel Island.

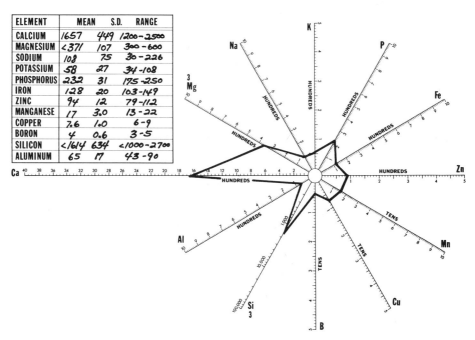

ELEMENT	MEAN	S.D.	RANGE
CALCIUM	1657	449	1200-2500
MAGNESIUM	<371	107	300-600
SODIUM	108	75	30-226
POTASSIUM	58	27	34-108
PHOSPHORUS	232	31	175-250
IRON	128	20	103-149
ZINC	94	12	79-112
MANGANESE	17	3.0	13-22
COPPER	7.6	1.0	6-9
BORON	4	0.6	3-5
SILICON	<1614	634	<1000-2700
ALUMINUM	65	17	43-90

220. Feather mineral pattern A representing seven (24 percent) of twenty-nine lesser snow geese bagged in the vicinity of Summer Lake, Oregon. Assumed origin: Wrangel Island.

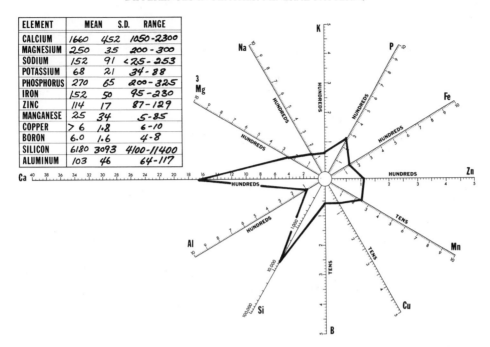

ELEMENT	MEAN	S.D.	RANGE
CALCIUM	1660	452	1050-2300
MAGNESIUM	250	35	200-300
SODIUM	152	91	<25-253
POTASSIUM	68	21	34-88
PHOSPHORUS	270	65	200-325
IRON	152	50	95-230
ZINC	114	17	87-129
MANGANESE	25	34	5-85
COPPER	>6	1.8	6-10
BORON	6.0	1.6	4-8
SILICON	6180	3093	4100-11400
ALUMINUM	103	46	64-117

221. Feather mineral pattern B representing four (14 percent) of twenty-nine lesser snow geese bagged in the vicinity of Summer Lake, Oregon. Assumed origin: Banks Island.

were assigned to the Banks Island colony, largely on the basis of their relatively extreme calcium-magnesium ratios, which were closest to those of the Banks Island geese (Table 21 and Figures *209–14*). A more adequate series than we had of wings from the Banks Island colony, judging on the basis of the kinds of variation found in other colonies, would probably encompass the patterns found in the Nevada geese. Nevertheless, it must be admitted that these patterns, especially E and F (Figures *213–14*), are unique and may represent discrete segments of the Banks Island population that follow an equally distinctive migration route.

The feather mineral patterns for four lesser snow geese bagged near the Bosque del Apache N.W.R. in the central Rio Grande valley of New Mexico are unique for western North America. It is highly significant, however, that they closely resemble patterns for four Ross' geese also taken near this refuge (Figure *215*). As the probability is great that the Ross' geese originated in the Queen Maud Gulf lowlands, it is equally probable that the lesser snow geese at the Bosque del Apache N.W.R. were from the same breeding area.

The lesser snow geese that stop at Westham Island in the Frazier River delta, British Columbia, and at the Skagit River delta, Washington, are wholly from the Wrangel Island colony (Table 21

and Figures *218–19*). According to the late Hugh Monahan (personal communication, 1966), the only bands that have been recovered at Westham Island from lesser snow geese were from geese banded on Wrangel Island.

The great majority of migrant lesser snow geese using the Summer Lake refuge in Oregon are also from the Wrangel Island colony (Table 21 and Figures *220–28*). Most of the wings of banded geese used for the model of feather minerals for this colony were collected at Summer Lake.

The migrant flocks using the Tule Lake N.W.R., California (Figures *229–30*), are apparently also mainly from the Wrangel Island colonies (Table 21 and Figures *111–15*), but assumed origins based on feather mineral patterns (Table 21 and Figures *231–33*) suggest that the main terminus of migration for this colony is the Sacramento N.W.R., California. The odds that geese from the small Kendall Island colony would be included in a small sample of geese from this refuge are minimal, but the feather mineral pattern of two geese from this refuge (Figure *233*) resembles sufficiently the model for Kendall Island to suggest that the Sacramento N.W.R. is also the winter terminus for this small colony.

Feather mineral patterns for geese bagged near the Gray Lodge and Imperial state refuges in California indicate that populations wintering on these refuges are apparently almost wholly from Banks Island (Table 21 and Figures *234–43*). If further research supports this judgment, kill quotas at these refuges—particularly at the Imperial refuge—should be set in relation to population estimates for the Banks Island colony and aerial photography appraisals of production (Hanson et al. 1972).

ORIGINS VERSUS DISPERSAL

We believe that we have established that feather mineral data can be used effectively to determine the origins of wild geese (Figure *244*). We also believe that ultimately this technique will, in part, supplant the need for large-scale banding operations—indeed, will supply more useful and more accurate information in most management situations than is obtainable from band recoveries. Moreover, determinations of origin from feather mineral data expressed as percentages can be used to calculate the numerical contributions made by various breeding populations to a local kill. As will be shown, it is impossible to do this from band recovery data alone. Nevertheless, perhaps both approaches should be used simultaneously in coordinated programs to study the dispersal of breeding populations and origins of geese killed by hunters. The primary point of this discussion is that data derived from feather analyses and band recoveries are basically different and not directly compatible; the objective here is to show how these data can be reconciled.

Recoveries of geese after they have been banded on the breeding grounds portray *dispersal,* but the recoveries made from place to place are not comparable unless corrections are made for relative hunting pressures—recoveries weighted in relation to the sizes of local kills, assuming that band reporting rates are similar. The feather mineral technique, on the other hand, is based on random samples of bagged geese—and necessarily assumes that the bagged geese are representative of the various components of the population passing through or wintering in an area. When the *origins* of the hunter-killed sample, largely unbanded, have been determined and converted to percentages, the proportion contributed by each breeding population to a local kill is known. More importantly, in regions where the total kill is known (e.g., the Ontario coasts of Hudson and James

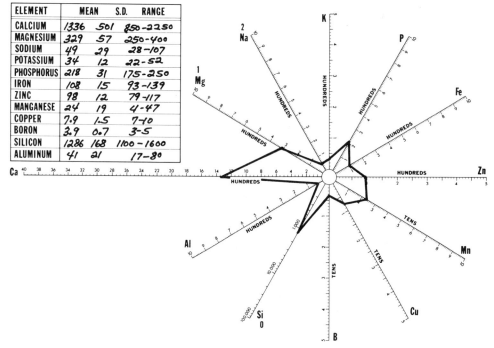

ELEMENT	MEAN	S.D.	RANGE
CALCIUM	1336	501	850-2250
MAGNESIUM	329	57	250-400
SODIUM	49	29	28-107
POTASSIUM	34	12	22-52
PHOSPHORUS	218	31	175-250
IRON	108	15	93-139
ZINC	98	12	79-117
MANGANESE	24	19	4-47
COPPER	7.9	1.5	7-10
BORON	3.9	0.7	3-5
SILICON	1286	168	1100-1600
ALUMINUM	41	21	17-80

222. Feather mineral pattern C representing seven (24 percent) of twenty-nine lesser snow geese bagged in the vicinity of Summer Lake, Oregon. Assumed origin: Wrangel Island.

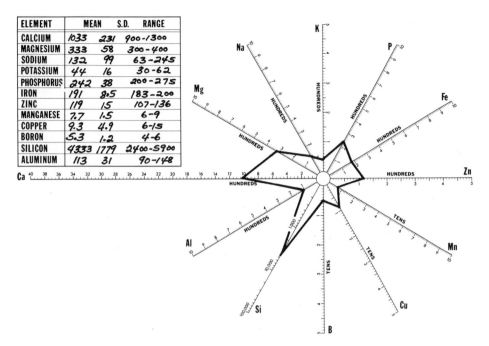

ELEMENT	MEAN	S.D.	RANGE
CALCIUM	1033	231	900-1300
MAGNESIUM	333	58	300-400
SODIUM	132	99	63-245
POTASSIUM	44	16	30-62
PHOSPHORUS	242	38	200-275
IRON	191	8.5	183-200
ZINC	119	15	107-136
MANGANESE	7.7	1.5	6-9
COPPER	9.3	4.9	6-15
BORON	5.3	1.2	4-6
SILICON	4333	1779	2400-5900
ALUMINUM	113	31	90-148

223. Feather mineral pattern D representing three (10 percent) of twenty-nine lesser snow geese bagged in the vicinity of Summer Lake, Oregon. Assumed origin: Wrangel Island.

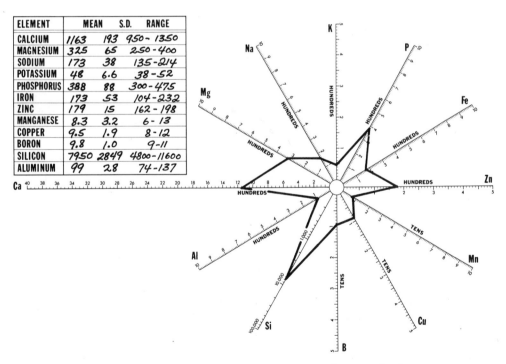

ELEMENT	MEAN	S.D.	RANGE
CALCIUM	1163	193	950-1350
MAGNESIUM	325	65	250-400
SODIUM	173	38	135-214
POTASSIUM	48	6.6	38-52
PHOSPHORUS	388	88	300-475
IRON	173	53	104-232
ZINC	179	15	162-198
MANGANESE	8.3	3.2	6-13
COPPER	9.5	1.9	8-12
BORON	9.8	1.0	9-11
SILICON	7950	2849	4800-11600
ALUMINUM	99	28	74-137

224. Feather mineral pattern E representing four (14 percent) of twenty-nine lesser snow geese bagged in the vicinity of Summer Lake, Oregon. Assumed origin: Wrangel Island.

155

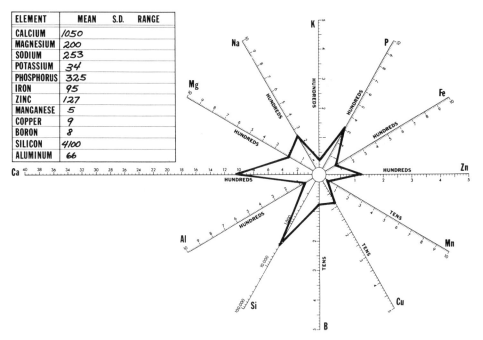

ELEMENT	MEAN	S.D.	RANGE
CALCIUM	1050		
MAGNESIUM	200		
SODIUM	253		
POTASSIUM	34		
PHOSPHORUS	325		
IRON	95		
ZINC	127		
MANGANESE	5		
COPPER	9		
BORON	8		
SILICON	4100		
ALUMINUM	66		

225. Feather mineral pattern F representing one (3.5 percent) of twenty-nine lesser snow geese bagged in the vicinity of Summer Lake, Oregon. Assumed origin: Wrangel Island.

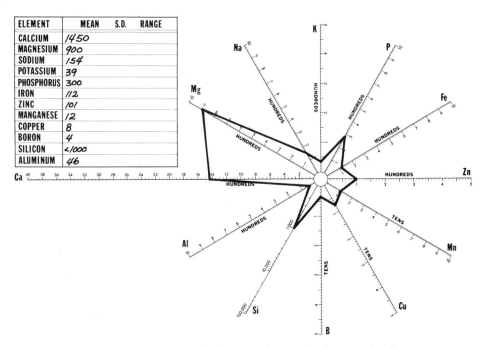

ELEMENT	MEAN	S.D.	RANGE
CALCIUM	1450		
MAGNESIUM	900		
SODIUM	154		
POTASSIUM	39		
PHOSPHORUS	300		
IRON	112		
ZINC	101		
MANGANESE	12		
COPPER	8		
BORON	4		
SILICON	<1000		
ALUMINUM	46		

226. Feather mineral pattern G representing one (3.5 percent) of twenty-nine lesser snow geese bagged in the vicinity of Summer Lake, Oregon. Assumed origin: Wrangel Island.

156

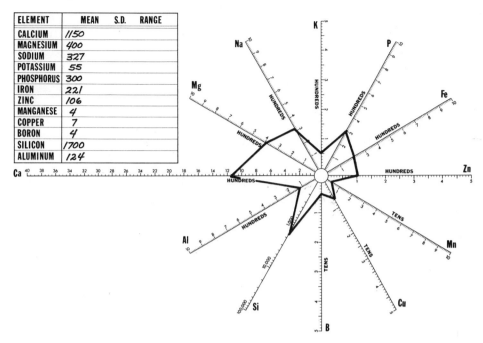

ELEMENT	MEAN	S.D.	RANGE
CALCIUM	1150		
MAGNESIUM	400		
SODIUM	327		
POTASSIUM	55		
PHOSPHORUS	300		
IRON	221		
ZINC	106		
MANGANESE	4		
COPPER	7		
BORON	4		
SILICON	1700		
ALUMINUM	124		

227. Feather mineral pattern H representing one (3.5 percent) of twenty-nine lesser snow geese bagged in the vicinity of Summer Lake, Oregon. Assumed origin: Wrangel Island.

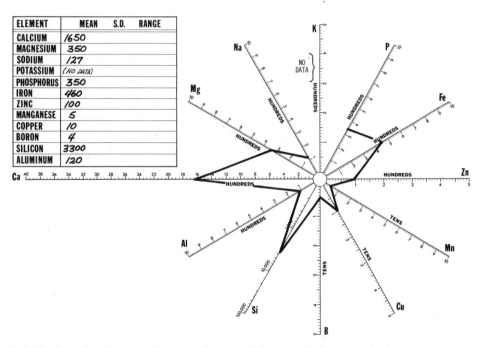

ELEMENT	MEAN	S.D.	RANGE
CALCIUM	1650		
MAGNESIUM	350		
SODIUM	127		
POTASSIUM	(NO DATA)		
PHOSPHORUS	350		
IRON	460		
ZINC	100		
MANGANESE	5		
COPPER	10		
BORON	4		
SILICON	3300		
ALUMINUM	120		

228. Feather mineral pattern I representing one (3.5 percent) of twenty-nine lesser snow geese bagged in the vicinity of Summer Lake, Oregon. Assumed origin: Wrangel Island.

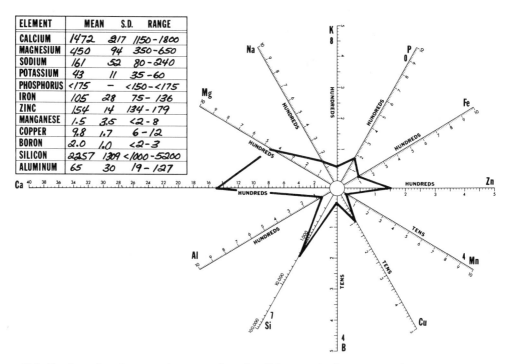

ELEMENT	MEAN	S.D.	RANGE
CALCIUM	1472	217	1150-1800
MAGNESIUM	450	94	350-650
SODIUM	161	52	80-240
POTASSIUM	43	11	35-60
PHOSPHORUS	<175	—	<150-<175
IRON	105	28	75-136
ZINC	154	14	134-179
MANGANESE	1.5	3.5	<2-8
COPPER	9.8	1.7	6-12
BORON	2.0	1.0	<2-3
SILICON	2257	1309	<1000-5200
ALUMINUM	65	30	19-127

229. Feather mineral pattern A representing nine (43 percent) of twenty-one lesser snow geese bagged in the vicinity of Tule Lake N.W.R., California. Assumed origin: Wrangel Island.

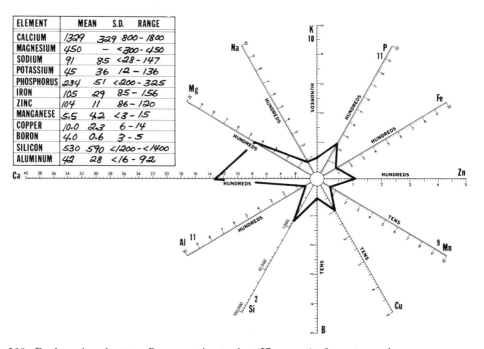

ELEMENT	MEAN	S.D.	RANGE
CALCIUM	1329	329	800-1800
MAGNESIUM	450	—	<300-450
SODIUM	91	85	<28-147
POTASSIUM	45	36	12-136
PHOSPHORUS	234	51	<200-325
IRON	105	29	85-156
ZINC	104	11	86-120
MANGANESE	5.5	4.2	<3-15
COPPER	10.0	2.3	6-14
BORON	4.0	0.6	3-5
SILICON	530	590	<1200-<1400
ALUMINUM	42	28	<16-92

230. Feather mineral pattern B representing twelve (57 percent) of twenty-one lesser snow geese bagged in the vicinity of Tule Lake N.W.R., California. Assumed origin: Wrangel Island.

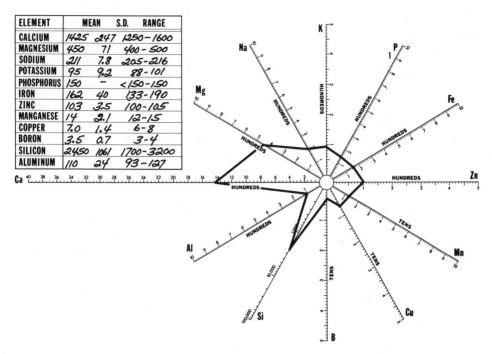

ELEMENT	MEAN	S.D.	RANGE
CALCIUM	1425	247	1250-1600
MAGNESIUM	450	71	400-500
SODIUM	211	7.8	205-216
POTASSIUM	95	9.2	88-101
PHOSPHORUS	150	–	<150-150
IRON	162	40	133-190
ZINC	103	3.5	100-105
MANGANESE	14	2.1	12-15
COPPER	7.0	1.4	6-8
BORON	3.5	0.7	3-4
SILICON	2450	1061	1700-3200
ALUMINUM	110	24	93-127

231. Feather mineral pattern A representing two (29 percent) of seven lesser snow geese bagged in the vicinity of the Sacramento N.W.R., California. Assumed origin: Wrangel Island.

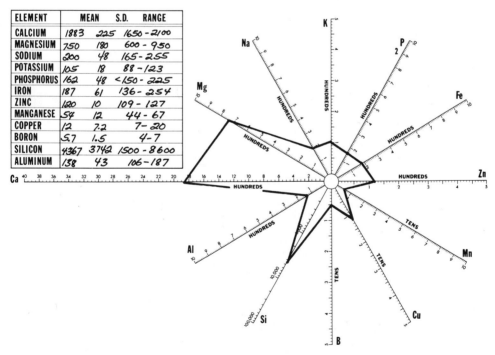

ELEMENT	MEAN	S.D.	RANGE
CALCIUM	1883	225	1650-2100
MAGNESIUM	750	180	600-950
SODIUM	200	48	165-255
POTASSIUM	105	18	88-123
PHOSPHORUS	162	48	<150-225
IRON	187	61	136-254
ZINC	120	10	109-127
MANGANESE	54	12	44-67
COPPER	12	7.2	7-20
BORON	5.7	1.5	4-7
SILICON	4367	3742	1500-8600
ALUMINUM	158	43	106-187

232. Feather mineral pattern B representing three (42 percent) of seven lesser snow geese bagged in the vicinity of the Sacramento N.W.R., California. Assumed origin: Wrangel Island.

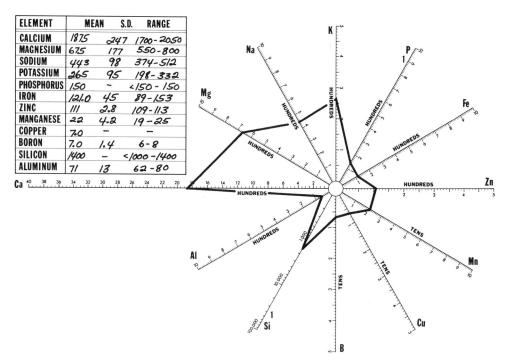

ELEMENT	MEAN	S.D.	RANGE
CALCIUM	1875	247	1700-2050
MAGNESIUM	675	177	550-800
SODIUM	443	98	374-512
POTASSIUM	265	95	198-332
PHOSPHORUS	150	–	<150-150
IRON	121.0	45	89-153
ZINC	111	2.8	109-113
MANGANESE	22	4.2	19-25
COPPER	7.0	–	–
BORON	7.0	1.4	6-8
SILICON	1400	–	<1000-1400
ALUMINUM	71	13	62-80

233. Feather mineral pattern C representing two (29 percent) of seven lesser snow geese bagged in the vicinity of the Sacramento N.W.R., California. Assumed origin: Kendall Island.

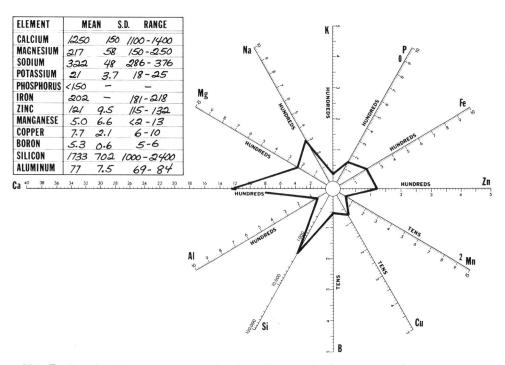

ELEMENT	MEAN	S.D.	RANGE
CALCIUM	1250	150	1100-1400
MAGNESIUM	217	58	150-250
SODIUM	322	48	286-376
POTASSIUM	21	3.7	18-25
PHOSPHORUS	<150	–	–
IRON	202	–	181-218
ZINC	121	9.5	115-132
MANGANESE	5.0	6.6	<2-13
COPPER	7.7	2.1	6-10
BORON	5.3	0.6	5-6
SILICON	1733	702	1000-2400
ALUMINUM	77	7.5	69-84

234. Feather mineral pattern A representing three (11 percent) of twenty-seven lesser snow geese bagged in the vicinity of Gray Lodge Refuge, California. Assumed origin: Banks Island.

160

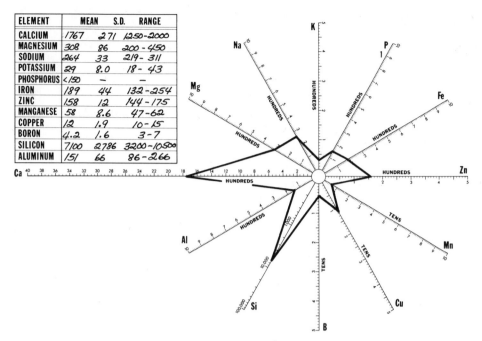

ELEMENT	MEAN	S.D.	RANGE
CALCIUM	1767	271	1250-2000
MAGNESIUM	308	86	200-450
SODIUM	264	33	219-311
POTASSIUM	29	8.0	18-43
PHOSPHORUS	<150	—	—
IRON	189	44	132-254
ZINC	158	12	144-175
MANGANESE	58	8.6	47-62
COPPER	12	1.9	10-15
BORON	4.2	1.6	3-7
SILICON	7100	2786	3200-10500
ALUMINUM	151	66	86-266

235. Feather mineral pattern B representing six (22 percent) of twenty-seven lesser snow geese bagged in the vicinity of Gray Lodge Refuge, California. Assumed origin: Banks Island.

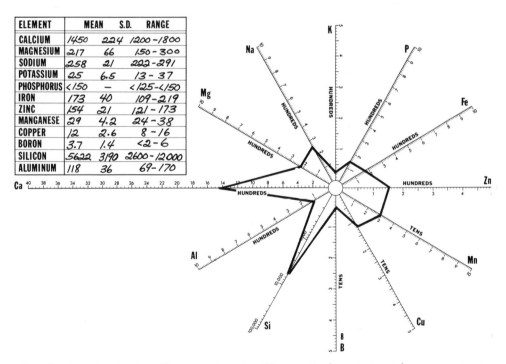

ELEMENT	MEAN	S.D.	RANGE
CALCIUM	1450	224	1200-1800
MAGNESIUM	217	66	150-300
SODIUM	258	21	222-291
POTASSIUM	25	6.5	13-37
PHOSPHORUS	<150	—	<125-<150
IRON	173	40	109-219
ZINC	154	21	121-173
MANGANESE	29	4.2	24-38
COPPER	12	2.6	8-16
BORON	3.7	1.4	<2-6
SILICON	5622	3190	2600-12000
ALUMINUM	118	36	69-170

236. Feather mineral pattern C representing nine (33 percent) of twenty-seven lesser snow geese bagged in the vicinity of Gray Lodge Refuge, California. Assumed origin: Banks Island.

161

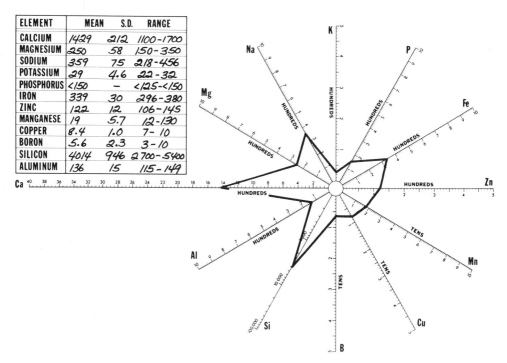

ELEMENT	MEAN	S.D.	RANGE
CALCIUM	1429	212	1100-1700
MAGNESIUM	250	58	150-350
SODIUM	359	75	218-456
POTASSIUM	29	4.6	22-32
PHOSPHORUS	<150	—	<125-<150
IRON	339	30	296-380
ZINC	122	12	106-145
MANGANESE	19	5.7	12-130
COPPER	8.4	1.0	7-10
BORON	5.6	2.3	3-10
SILICON	4014	946	2700-5400
ALUMINUM	136	15	115-149

237. Feather mineral pattern D representing seven (26 percent) of twenty-seven lesser snow geese bagged in the vicinity of Gray Lodge Refuge, California. Assumed origin: Banks Island.

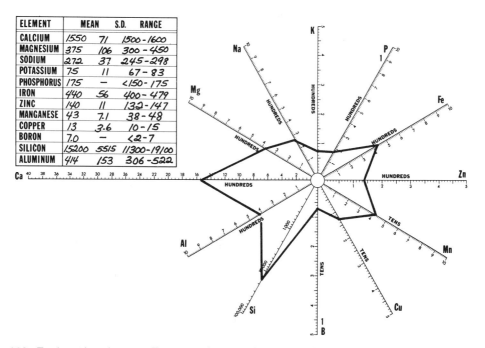

ELEMENT	MEAN	S.D.	RANGE
CALCIUM	1550	71	1500-1600
MAGNESIUM	375	106	300-450
SODIUM	272	37	245-298
POTASSIUM	75	11	67-83
PHOSPHORUS	175	—	<150-175
IRON	440	56	400-479
ZINC	140	11	132-147
MANGANESE	43	7.1	38-48
COPPER	13	3.6	10-15
BORON	7.0	—	<2-7
SILICON	15200	5515	11300-19100
ALUMINUM	414	153	306-522

238. Feather mineral pattern E representing two (8 percent) of twenty-seven lesser snow geese bagged in the vicinity of Gray Lodge Refuge, California. Assumed origin: Banks Island.

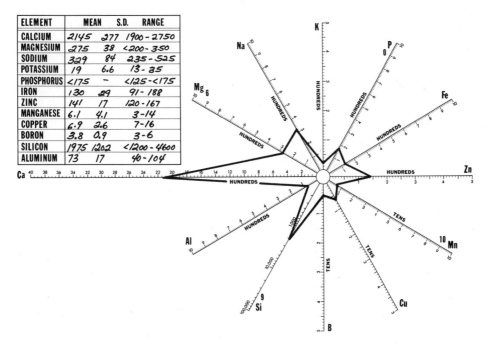

239. Feather mineral pattern A representing eleven (41 percent) of twenty-seven lesser snow geese bagged in the vicinity of the Imperial Refuge, California. Assumed origin: Banks Island.

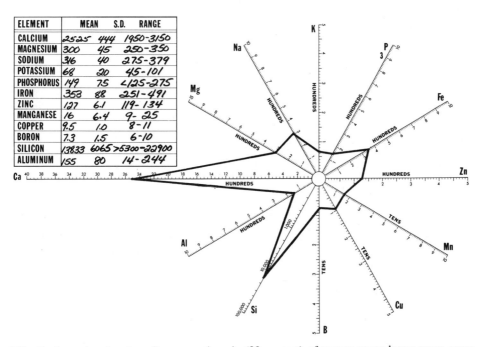

240. Feather mineral pattern B representing six (22 percent) of twenty-seven lesser snow geese bagged in the vicinity of the Imperial Refuge, California. Assumed origin: Banks Island.

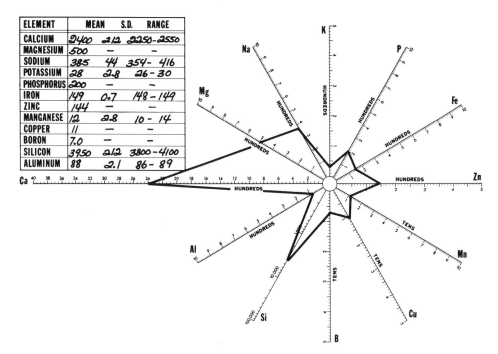

ELEMENT	MEAN	S.D.	RANGE
CALCIUM	2400	212	2250-2550
MAGNESIUM	500	—	
SODIUM	385	44	354- 416
POTASSIUM	28	2.8	26- 30
PHOSPHORUS	200	—	—
IRON	149	0.7	148- 149
ZINC	144	—	—
MANGANESE	12	2.8	10- 14
COPPER	11	—	—
BORON	7.0	—	—
SILICON	3950	212	3800-4100
ALUMINUM	88	2.1	86- 89

241. Feather mineral pattern C representing two (7 percent) of twenty-seven lesser snow geese bagged in the vicinity of the Imperial Refuge, California. Assumed origin: Banks Island.

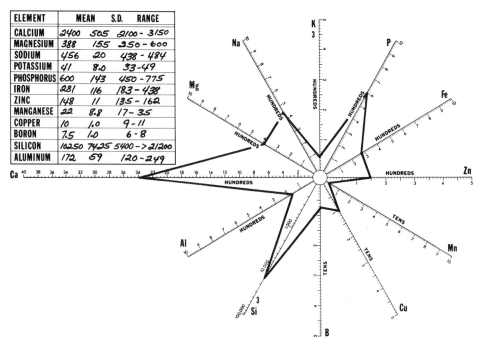

ELEMENT	MEAN	S.D.	RANGE
CALCIUM	2400	505	2100- 3150
MAGNESIUM	388	155	250- 600
SODIUM	456	20	438- 484
POTASSIUM	41	8.0	33-49
PHOSPHORUS	600	143	450 -775
IRON	281	116	183 - 438
ZINC	148	11	135 - 162
MANGANESE	22	8.8	17- 35
COPPER	10	1.0	9 - 11
BORON	7.5	1.0	6 - 8
SILICON	10250	7425	5400 - >21200
ALUMINUM	172	59	120 -249

242. Feather mineral pattern D representing four (15 percent) of twenty-seven lesser snow geese bagged in the vicinity of the Imperial Refuge, California. Assumed origin: Banks Island.

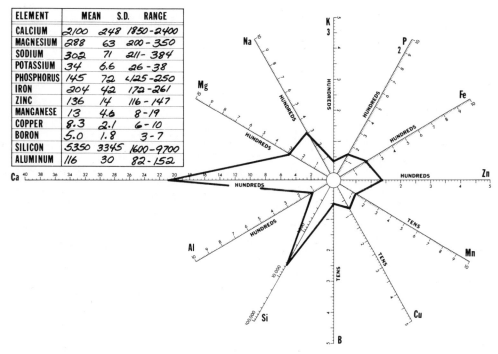

ELEMENT	MEAN	S.D.	RANGE
CALCIUM	2100	248	1850-2400
MAGNESIUM	288	63	200-350
SODIUM	302	71	211-384
POTASSIUM	34	6.6	26-38
PHOSPHORUS	145	72	<125-250
IRON	204	42	172-261
ZINC	136	14	116-147
MANGANESE	13	4.6	8-19
COPPER	8.3	2.1	6-10
BORON	5.0	1.8	3-7
SILICON	5350	3345	1600-9700
ALUMINUM	116	30	82-152

243. Feather mineral pattern E representing four (15 percent) of twenty-seven lesser snow geese bagged in the vicinity of the Imperial Refuge, California. Assumed origin: Banks Island.

bays), the true *distribution* of each breeding population in, and the numerical contribution of each population to, the overall regional kill can be determined readily.

The reconciliation of these two kinds of data is modeled from data, partly theoretical, for the coastal settlements of northern Ontario. As a result of the diligent efforts of personnel of the Ontario Ministry of Natural Resources and personnel of the Royal Canadian Mounted Police, the kill of blue and lesser snow geese is known within close limits for the settlements on Hudson and James bays, and band reporting rates for these communities are very high. The nature of these two kinds of data is represented in Figure *245.*

Results of feather mineral analyses of wing samples from Ft. Severn and the Kapiskau rivers are given in Table 22. They are regarded as inadequate in size as random samples of the flights of geese available to local hunters. For the sake of completeness, hypothetical but reasonable approximations of the relative kills of Baffin Island geese are also given for other settlements in Table 22. With the exception of the second percentage column in this model (only for the Baffin Island geese) the data in Table 22 are read and totaled

horizontally. In the first column are the percentages of geese from each of the four major breeding areas killed at each of the hunting areas. Note that these first vertically aligned percentages do not give the proportions of Baffin Island geese in kills at various settlements. However, if a random sample of wings from bagged geese were obtained and analyzed for feather minerals, the percentages representing breeding colonies applied to the total kill at each settlement can provide an estimate of the numbers of geese from each of the four colonies shot at each settlement. These estimates can, in turn, be totaled and converted to percentages (e.g., second column under Baffin Island in Table 22) which indicate the relative contribution each colony makes to the kill at each settlement. Thus, from breeding origins determined from feather analyses, both kinds of information needed for waterfowl-resource management are obtained.

Band recoveries reported from the coastal settlements of northern Ontario from 1966 to 1970 are summarized in Table 23. (As these band recoveries are a cumulative series without regard to age, sex, or exact year of recovery and, in the model presented in Table 23, are related to average annual kills, their use here in no way precludes a later

244. Blue and lesser snow geese feeding in a cornfield on the Squaw Creek N.W.R., Missouri. Color phases of major flock groupings indicate that geese from at least two colonies contribute to the fall concentration at this refuge. Feather mineral data suggest that Baffin Island geese make the major contribution to the fall concentrations at this refuge.

refined and definitive analysis of dispersal and origins.) These recoveries portray the dispersal of geese from each breeding colony, but the actual figures do not reflect the relative dispersal of the southward flights of any given colony of geese, because losses from hunting vary markedly from place to place. This problem is solved by expressing the numbers of recoveries as quotients of the kill, totaling these, and converting the quotients to percentages (e.g., columns 3 and 4, Table 23).

The relative contribution that each colony makes to the kill and to migrant flocks along each sector of the coast is then determined from *band recoveries* with reasonable accuracy. However, these estimates of the origin of the kill based on band recoveries cannot provide estimates of the converse, i.e., the percentage of geese from each of the four major nesting colonies making up the kill *at any one post*. The condensed and simplified model in Table 24 makes these relationships more apparent.

22. An incomplete, theoretical model (excepting data for Severn and Kapiskau areas) showing treatment of data indicating origins of geese based on quantitative analyses of minerals in primary feathers

Hunting Area	Average Recorded Kill 1966–70	Number in Sample	Indicated Colony of Origin													Total (%)
			Baffin Island				Cape Henrietta Maria			Southampton Island			McConnell River			
			n	%	Calculated Kill	%	n	%	Calculated Kill	n	%	Calculated Kill	n	%	Calculated Kill	
Severn River	1,922	27	10	37.1	713	3.5	4	14.8	284	9	33.3	640	4	14.8	284	100.0
Winisk and Sutton rivers	2,079			20.0	416	2.0										
Attawapiskat and Kapiskau rivers	4,926	15	13	86.6	4,266	20.7	0	0.0	0	1	6.7	330	1	6.7	330	100.0
Albany River	8,062			60.0	4,837	23.5										
J. Bluff Shoal	1,028			70.0	720	3.5										
Moose River	10,076			80.0	8,061	39.2										
Tidewater Camp and Hannah Bay	1,741			90.0	1,567	7.6										
Total	29,834				20,580											
Percent					69.0	100.0										

In short, in the way the band recovery data are presented in Table 24, they can only be read meaningfully *vertically*.

For the use which band recoveries are employed in Tables 23–25, the number of geese originally banded at each colony is irrelevant. The only important parameters are the number of band recoveries in a given period from each settlement and the origins of the banded geese. The purpose in calculating the second column of percentages (column 4 in Table 23), aside from their own intrinsic value, was to have a set of figures (column 4 in Table 25) directly comparable to the data on kills derived from primary feather analyses (column 3 in Table 25)—in short, to reconcile these two approaches to quantify the kills by colony origins. If for each settlement there is an adequate number of band recoveries of geese from all major colonies and if primary feathers from an adequate random sample of geese shot at each settlement are obtained for analyses, the two sets of percentages (columns 3 and 4 in Table 25) derived from these data should be quite similar. In practice, however, to gain more refined insights, a definitive treatment of the data would involve relating recoveries from each year to the kills of that year and to treat first-year recoveries separately from recoveries made one or more years after banding. However, to enlarge the sample, it may be desirable to combine the two band recovery series.

Ideally, to place problems of origins and dispersal into a final, definitive perspective, a coordinated program should be instituted and carried out over several years. This program would involve simultaneous, massive bandings conducted on all four major breeding colonies (actually, two on Southampton Island) combined with intensive efforts to record kills and band recoveries and to obtain a random collection of wings for mineral analyses.

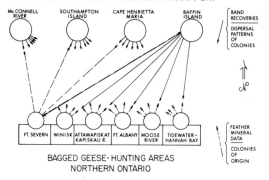

245. Diagrammatic representation of the relationship between dispersal data obtained from band recoveries and colony of origin data obtained from feather mineral analyses.

23. Band recoveries of blue and lesser snow geese banded on their breeding grounds and shot along the coasts of Hudson and James bays

Hunting Area	Average Recorded Kill 1966–70	Baffin Island			Cape Henrietta Maria		
		Number of Recoveries	Recoveries Per 1,000 Kill	Distribution of Kill by %	Number of Recoveries	Recoveries Per 1,000 Kill	Distribution of Kill by %
Severn River	1,922	10	5.20	12.43	10	5.20	16.03
Winisk and Sutton rivers	2,079	19	9.14	21.86	40	19.24	59.31
Attawapiskat and Kapiskau rivers	4,926	6	1.22	2.92	1	0.20	0.62
Albany River	8,062	27	3.35	8.01	9	1.12	3.45
N. Bluff Shoal	1,028	44	3.96	9.47	4	0.36	1.11
Moose River	10,076						
Tidewater Camp and Hannah Bay	1,741	33	18.95	45.31	11	6.32	19.48
Total or percent	29,834	139	41.82	100.00	75	32.44	100.00

24. Simplified, theoretical model of weighted band-recovery data converted to percentages to show relative contribution of each breeding colony to kills made in various hunting areas along the coasts of Hudson and James bays, Ontario

Hunting Area	Colony of Origin of Kill				
	Baffin Island (595,000)*	Cape Henrietta Maria (79,000)*	Southampton Island (208,000)*	McConnell River (520,000)*	Total (%)
Severn River	10	15	50	50	125
Winisk and Sutton rivers	20	60	10	20	110
Attawapiskat, Kapiskau, and Albany rivers	15	5	15	10	45
N. Bluff Shoal to Hannah Bay	55	20	25	20	120
Total (%)	100	100	100	100	

* Numbers of adult-plumaged birds in 1973 (from R. H. Kerbes [in press]).

Southampton Island			McConnell River		
Number of Recoveries	Recoveries Per 1,000 Kill	Distribution of Kill by %	Number of Recoveries	Recoveries Per 1,000 Kill	Distribution of Kill by %
21	10.93	53.0	20	10.41	50.03
5	2.41	11.69	6	2.89	13.89
3	0.61	2.96	4	0.81	3.89
7	0.87	4.22	8	0.99	4.76
7	0.63	3.06	6	0.54	2.59
9	5.17	25.07	9	5.17	24.84
52	20.62	100.0	53	20.81	100.0

25. Condensed versions of models of estimations of the distribution of geese from Baffin Island shot on the Ontario coasts of Hudson and James bays based on feather mineral analyses and band recoveries

Hunting Area	Combined Average Kills 1966–70 (from all colonies)	Distribution of Kill of Baffin Island Geese			
		Indicated by Analyses of Primaries		% Kill Indicated by Band Recoveries	
		Number Killed[a]	% of Average Kill[b]	Weighted (by kill)[c]	Unweighted[d]
Severn River	1,922	713	3.5	12.5	7.2
Winisk and Sutton rivers	2,079	416	2.0	22.0	13.7
Attawapiskat, Kapiskau, and Albany rivers	12,988	9,103	44.2	11.0	23.7
N. Bluff Shoal to Hannah Bay	12,845	10,348	50.3	54.5	55.4
Total or percent	29,834	20,580	100.0	100.0	100.0

[a] From column 5, Table 22.
[b] From column 6, Table 22.
[c] From column 4, Table 23.
[d] From column 2, Table 23.

CHAPTER SIX

Previous Studies of Keratin and Factors Affecting Feather Mineral Patterns

PREVIOUS STUDIES

Studies of the mineral content of feathers in relation to mineral concentrations in the ecosystem, previous to those of McCullough and Grant (1952–53) and the present authors (1968), are almost nonexistent. We are aware only of the study by Johnels and Westermark (1969), who found that the mercury content of the feathers of the goshawk (*Accipter gentilis*) was closely associated with the use of mercurial fungicides with seeds in Sweden, beginning about 1940. They concluded that the ratio of mercury in the feather to that in dry breast muscle was seven to one and that the feathers "register the level of mercury for the formative periods." Avian populations or species not associated with mercury, either because of geographic isolation or lack of food-chain involvement, did not show exceptional variation in content with time.

Kelsall (1970) reported on a preliminary study of mineral levels in white-fronted geese and three species of ducks, but samples were taken from too wide a geographical area to be meaningful. Achievement of the objectives of his study was further defeated by analyzing composite feather samples from nineteen to forty-seven individuals. Kelsall (1972) subsequently documented differences in feather mineral concentrations among different species of ducks. Aside from the factor of diet, differences among species can probably be related to species-related differences in the combinations of amino acids that comprise their feather keratin.

Dr. Harry V. Warren and his associates in the Department of Geology and Geography, University of British Columbia, have carried out many pioneering studies in biogeochemistry. Our conclusion that plant foods are the primary source of feather minerals in geese is in accordance with the comment of Warren et al. (1967:670) that "anybody who has not done biogeochemical work may be surprised at the range in the amounts of trace elements found in plants," but Warren also admonishes that "in spite of the fact that there is a tendency for the amounts of trace elements in soils to parallel those in the rocks from which they are derived, and for vegetal matter to reflect the composition of the soils on which it is growing, this parallelism must never be assumed."

Hair

Hair is homologous to feathers and, like feathers, is composed of keratin. In view of the paucity of data in the literature on the mineral content of feathers, we would be remiss if recent findings on mineral deposition in hair were not briefly summarized. Determinations of the concentration of heavy metals in hair—particularly arsenic—have long been used in forensic medicine, but only in recent years has hair been used in nutritional studies. It is evident that the mineral content of hair, as would be expected, also reflects the minerals in the ecosystem and that analyses of hair could be used to evaluate the adequacy of the mineral diet. The potential for hair-ecosystem-related studies in wildlife research management, particularly in the case of wild ruminants, would seem considerable. However, it should be noted that seasonal variation in manganese, for example, has been found in the hair of both moose and domestic cattle (Flynn and Franzmann 1974; Hartmans 1974).

Human Hair. Calcium, magnesium, copper, iron, zinc, silicon, chlorine, sulfur, and phosphorus have been found in human hair in amounts exceeding 100 ppm. Manganese, aluminum, lead, cobalt, nickel, boron, and chromium have been found in amounts varying from 1 to 100 ppm. Hair is an important excretion route for lead, selenium, and copper and a major route for the excretion of arsenic. External exposure to copper and its subsequent combination with cystine may explain high concentrations of copper in hair (Schwartz 1960:373–74).

Some studies seeking associations of hair minerals with diet or environment have arrived at negative conclusions. Thus, Bate and Dyer (1965) concluded that humans of the same age, sex, and occupation were less likely to have similar levels of hair minerals than were subjects of different ages, sexes, or occupations. After assembling data for magnesium, zinc, cadmium, copper, cobalt, nickel, chromium, and lead for a large sample of human hair representative of a wide span of age classes, Schroeder and Nason (1969) concluded that levels of minerals in human hair were not satisfactory for evaluating tissue stores. However, random samples of this nature can seldom be expected to be meaningful, particularly because of the diverse backgrounds of the individuals. Hair has been found to be a sensitive index of zinc metabolism in man (Strain and Pories 1966) and to afford a good measure of body stores of iron (Eatough et al. 1974:661).

Hair of Domestic Animals. The question of the relationship of hair color to specific patterns of mineral accumulation has been studied. Miller et al. (1965) found from studies of the black and white hair of Holstein cattle and the fawn hair of Jersey cattle that white, particularly in the tail region, and fawn hairs were the most highly mineralized. O'Mary et al. (1970) reported that the copper content of white hair from Holstein and Hereford cows fed a copper-supplemented ration was higher and less variable than was the pigmented hair. (Similar studies of human hair [Anke and Schneider 1966; Umarji and Bellare 1966] have shown that white and gray hair have higher levels of manganese than has black-pigmented hair.)

Hair of cattle, swine, and sheep has been analyzed for elements of interest in nutritional studies. However, Anke (1965) warned that bovine hair samples taken prior to, during, and after shedding were unreliable indicators of the mineral diet although Anke's and his colleague's research later had success in relating the mineral content of hair to diet. Chauvaux et al. (1965) failed to find that copper and manganese levels in hair were indicative of dietary deficiencies in cattle as indicated by analyses of these elements in liver and blood. Hartmans (1967) evaluated several washing techniques, concluding that contamination by sweat and feces largely vitiated any possibility that cattle hair might reflect mineral status.

Nevertheless, despite the problems of obtaining hair samples uncontaminated by external sources, mineral levels in the hair can be expected to reflect mineral levels in experimental diets—or minerals introduced into the body of the animal by injection. Taucins and Svilane (1965) and Eyubov (1968) found that supplemental rations of cobalt,

manganese, zinc, and copper raised the levels of these minerals in wool. Arthur (1965) concluded that the hair of guinea pigs could be used to assess molybdenum and copper interactions, which, in guinea pigs, are similar to the relationships between these elements in ruminants. Strain et al. (1971) noted that rapid uptake of ^{131}I, ^{54}Mn, ^{85}Sr, and ^{65}Zn by hair occurs in rats injected with these radioisotopes. Zinc accumulated relatively slowly, but was unique in that its radioactivity remained high eight days after injection. Reinhold et al. (1967) fed one group of rats a purified diet containing two to four ppm of zinc and a control group a diet containing thirty ppm of zinc. Both diets were low in protein content. Decreased levels of zinc in hair were noted in rats on the zinc-deficient diet within forty days of the beginning of the experiment. Rabbits injected with ^{210}Pb incorporated this isotope into their growing hair at rates of 7-fold to 137-fold greater than that found in hair in the resting phase (Jaworowski et al. 1966).

Several studies have related mineral levels in bovine hair to mineral levels in forage. Schellner (1971) found that manganese levels in the hair of cattle from eight regions in which the soils ranged from sandy to chalk could be correlated with manganese levels in the clover forages (red clover, *Trifolium pratense* and hop clover, *T. agrarium*) of these areas. No relationship between mineral levels in hair and forage could be found for zinc, a finding which, in general, is similar to our own in respect to the relative stability of zinc levels in the primary feathers of wild geese. Previously, Krolak (1968), working with bovine hair from regions in Poland either deficient or sufficient in manganese, concluded that the concentration of manganese in hair was a satisfactory indicator of nutritional levels of this element. Manganese in the hair of manganese-deficient animals ranged from 2.8 to 10.3 ppm, compared with 18.5 to 32.8 ppm in the hair of animals on forages that contained manganese at rates of 20 to 55 and 105 to 146 ppm, respectively. Werner and Anke (1960) found that levels of iron, manganese, copper, molybdenum, and cobalt in cattle hair were representative of amounts of these elements in forages from soils of diverse parent materials. Anke later (1967) reported that levels of phosphorus, manganese, and molybdenum in red hop clover grown in geochemically different areas were especially well reflected in hair concentrations. Anke established these levels as representative of dietary sufficiency: magnesium, 750 to 800 ppm; sodium, 200 ppm; zinc, 115 ppm; manganese, 4.0 ppm;

copper, 7.0 ppm; molybdenum, 0.35 ppm; and cobalt, 0.05 ppm.

GENERAL CONSIDERATIONS AND PRESENTATION OF SUMMARY DATA

The practical application of feather mineral analyses in determining origins of blue and snow geese has been presented in Chapters 3 to 5. Of primary concern here are the more basic ecosystem and metabolic relationships between these geese and the mineralogy of their breeding environments as indicated by the deposition of minerals in primary feathers. These concerns, in turn, provide a better understanding of the origins of feather minerals and the limits of their usefulness as biological tracers or avian fingerprints. If the latter, vernacular term is used, it should be remembered that feather mineral patterns are primarily a result of metabolic interactions with the environment and are not a result of genetic processes.

It will be apparent to many that in attempting to elucidate these relationships we have necessarily given superficial treatment to a host of complex subjects dealing with earth sciences and the physiology and biochemistry of plants and animals. Observations of a commonplace nature relating to ornithology may be informative to earth or plant scientists, and vice versa. The practical question is: where does one draw the line in attempting to trace the chain of events that accounts for the concentrations of elements in the ecosystem and their eventual deposition in feathers? Of necessity, because of the multidisciplinary nature of the subject and the summary nature of parts of the discussion, we have relied chiefly on major reviews rather than the journal literature, an approach not without recent precedent (Schepartz 1973:ix). There is some advantage to this restricted use of the literature, namely, the greater availability of review volumes in libraries and the benefit of the expert evaluations of the literature by the reviewers. The present report, whatever its faults, we believe, is the first of its kind. Therefore, an attempt to provide a brief sketch of the links in the nutrient chain is justified, both to give the casual reader some measure of coherent understanding of the problems as a whole, as well as to provide an introduction to the subject that will whet the appetite of those in wildlife research for more refined and penetrating studies.

An ideal organizational and sequential treatment of the various interrelationships between environmental minerals and their deposition in feathers is difficult to achieve. We decided that this goal might be best attained by introducing summary tables at the outset and then discussing in turn the origins of each of the elements in the ecosystem and some of the major factors influencing their absorption and metabolism in plants, birds, and mammals.

It is generally held that few, if any, essential minerals or trace elements are absorbed or enter into metabolic reactions in animals independently of other elements. Carried back several steps, similar relationships should hold true in the plants that enter the diets of geese, and both plants and geese should reflect, in some measure, the abundance of these elements in the water, soil, and parent rock. However, we found only limited evidence that negative interelemental influences affect the quantitative deposition of the elements studied in the primary feathers of geese. We do not, however, infer that inverse correlations between some of these elements do not exist. Examples of the complexities found in evaluating antagonistic interactions between elements have been given by Mills (1974:83):

"Changes in tissue element content or patterns of subcellular metal distribution usually provide the first evidence that such antagonistic interactions are operating. Where there are indications that an antagonist has depleted the tissue content of the target element, it must be presumed that the mode of action of the antagonist is through a restriction of absorption, an inhibition of storage within the tissue matrix, or, alternatively, an enhancement of excretion. That this is an insufficiently detailed concept to account for the mechanism of action of several practically important situations of trace-element antagonism is clearly shown by those circumstances in which the antagonist enhances tissue deposition of the target element, even though there are clearly clinical or biochemical implications that the incorporation of the target element into its functional sites is impaired. Thus, many studies of the effects of molybdenum ingestion by the rat have illustrated the development of clinical syndromes at least partially responsive to copper, even though the storage of copper in the liver is enhanced. Similarly, the development of clinical signs of zinc deficiency in copper intoxication of the pig is accompanied by an increase in liver zinc content.

"Most studies of the Cu/Zn/Cd relationship have strongly emphasized effects of these interactions upon liver composition and have largely neglected effects upon other tissues. In consequence it is often difficult to assess whether the effects of an antagonistic interaction are merely upon the distribution of elements between tissues or whether the primary effects are upon absorption or excretion.''

In the following discussions, two or more elements may sometimes be dealt with jointly, because the discussion of the relative abundance of any element in the primary feathers cannot be compartmentalized readily but must be related to the other elements that have influenced its fate en route to its terminal point in the feathers, either through enhancement of, or antagonism toward, its absorption by the plant roots or similar effects in the digestive tract of the goose.

Mean feather-element values for each breeding colony are summarized in Table 26. Although these mean values obscure intracolony variations, especially those related to geological or mineralogical diversity within colony areas, they probably do provide a perspective adequate to the present discussions. (The reader is referred back to the data on colony feather mineral patterns for a more detailed understanding of intracolony variation.)

The credibility and validity of the feather mineral technique in determining the birthplaces of geese of unknown origins relate primarily to the geological diversity of colony areas. They also rest on the assumption that experimental error and such intracolony variations as those related to sex and age as well as mineralogical diversity within the feeding range of a colony do not exceed or negate intercolony differences in mineral patterns. Although it is a more remote possibility, we acknowledge that some slight differences may exist between populations in genotypic control of the absorption and excretion of these elements. The likelihood of some differences related to genetic control between the two species we are concerned with here is also admitted. If present, such disparity would likely relate to differences in amino acid composition of the primaries. Values for the coefficient of variability (V) constitute a summing up of intracolony variability factors (Table 27).

The number of significant associations among elements in the primary feathers appears to provide a rough measure of the geological diversity of the feeding area of a colony, reflecting the mineralogy of the ecosystem as well as the absorption of these minerals by food plants and geese. The significant correlation values (r) between elements for colonies from which the reference samples were of ad-

equate size are summarized in Table 28. The number of paired samples of uncensored data on which these correlations are based are given in Table 29. The number of significant correlations are summarized by element and colony in Table 30.

The frequency with which each element was correlated with another element in the series is shown in Table 31. The meanings of these data are more readily apparent when they are reordered in decreasing frequency and expressed as percentages. Also pertinent to the interpretation of the mineral data are the number of colonies in which various pairs of elements were significantly associated (Table 32).

ECOLOGICAL, PHYSIOLOGICAL, AND BEHAVIORAL FACTORS

The ecological, physiological, and behavioral factors related to mineral intake in all their facets are complex, as illustrated, for example, by the vast literature dealing with the physiology of the homeostasis of minerals. Yet if even a rudimentary grasp of the quantitative aspects of feather minerals as reflectors of the mineral environment is to be gained, an attempt must be made to winnow out a few salient principles. If the result is oversimplification, limitations of space and lack of competence outside of our fields of specialization can be blamed.

It is probably superfluous to mention in passing that the metabolic responses of birds differ in some respects from those of mammals used in most experimental studies. Probably less appreciated is the fact that waterfowl, particularly geese, differ from most other birds in the nature of their mineral intake and the relative importance of their various modes of mineral excretion. Except for early gosling stages, geese consume only plant foods. However, some coastal populations of Canada geese consume appreciable amounts of invertebrates, particularly snails. The food plants of geese occupy, of course, the pivotal position in the nutrient chain of the ecosystem. As Sutcliffe (1962:5) points out: "A plant probably contains at least traces of all the elements present in the environment in which it grows." Also of crucial importance in respect to the absorbability of the elements present in plant tissues is the fact that "the bulk of each metallic element, however, exists as inorganic compounds or ions, and much of it is dis-

26. Summary of mean values and their standard errors (ppm) of minerals in reference feathers of blue and lesser snow geese and Ross' geese collected on their breeding grounds or banded there and shot elsewhere

Geographic Area and Colony	Ca		Mg		Na		K		P	
	n [a]	Mean	n	Mean	n	Mean	n	Mean	n	Mean
HUDSON BAY										
Baffin Island, N.W.T.	53	1303 ± 86	56	345 ± 23	58	313 ± 18	58	54 ± 3.4	28	200 ± 15
Southampton Island, N.W.T.	18	990 ± 140	13	415 ± 53	23	379 ± 22	23	64 ± 5.4	9	98 ± 84
Cape Henrietta Maria, Ont.	24	1721 ± 96	24	342 ± 33	24	359 ± 29	24	51 ± 4.4	2	175 ± 18
McConnell River, N.W.T.	24	814 ± 134	28	436 ± 41	38	330 ± 19	37	60 ± 4.0	25	280 ± 49
Cape Churchill, Man.	4	1700 ± 102	4	413 ± 97	4	332 ± 46	4	45 ± 6.4	0	<200
CENTRAL ARCTIC										
Karrak Lake, N.W.T. (Ross' geese)	20	2995 ± 122	20	605 ± 40	20	485 ± 14	17	30 ± 3.8	0	166 ± 4.3
WESTERN ARCTIC										
Banks Island, N.W.T.	3 [b]	2083 ± 146	1	150	3	162 ± 68	3	45 ± 3.5	3	292 ± 65
	11 [c]	1198 ± 149	9	358 ± 40	11	244 ± 13	10	80 ± 12	5	107 ± 42
Anderson River, N.W.T.	5	<500	1	150	5	194 ± 44	5	29 ± 6.7	5	320 ± 2.5
Kendall Island, N.W.T.	7	1800 ± 92	7	579 ± 55	7	396 ± 31	7	197 ± 36	1	175
Wrangel Island, U.S.S.R.	15	1400 ± 112	16	418 ± 27	17	207 ± 13	17	45 ± 7.0	16	313 ± 26

[a] n = number of uncensored data.
[b] Museum skins collected at Cape Kellett and Sachs Harbor, Banks Island.
[c] Fall migrants collected on the west side of the mouth of the Mackenzie River.

solved in the aqueous sap of cell vacuoles'' (Sutcliffe 1962:3).

Ecological Factors

A central theme of ecology is the association of animals with rather specific habitats. For the student of waterfowl, the collective term "blue and snow geese" is often associated with "coastal marsh," and the latter term may sometimes be loosely used to imply "salt marsh." Such usage, granting the impreciseness of the terms, may or may not be justified by ecological realities. McIlhenny (1932:281–82), an observer of blue and lesser snow geese on the Louisiana coast over a fifty-year period, suggested they "might be called Salt Water Geese or Salt Marsh Geese, as they are never found feeding or sleeping more than eight miles back in the marshes from the salt beaches."

Obviously, he was not aware of the incongruity of his statement. Nevertheless, we suspect that, subconsciously at least, many would associate blue and lesser snow geese with brackish marshes. The influence of sodium on plant growth and physiology and on the absorption of the essential macrominerals in birds and mammals (and, directly or indirectly, on the absorption of trace elements that may be affected by calcium concentrations in the lumen of the intestine and by calcium absorption) appears to be so profound that some understanding of sodium levels in the marshland environments used by blue and lesser snow geese during their annual cycle is essential to an understanding of their overall response to their total mineral environment, as indicated by mineral levels in their primary feathers.

Braided river mouths, deltaic islands, or stream-

Fe		Zn		Mn		Cu		B		Si		Al
Mean	n	Mean	n	Mean	n	Mean	n	Mean	n	Mean	n	Mean
224 ± 19	59	143 ± 3	24	5.7 ± 0.6	59	9.2 ± 0.4	53	3.0 ± 0.3	57	5403 ± 651	59	200 ± 15
167 ± 20	23	133 ± 3.7	12	1.7 ± 2.3	23	8.8 ± 0.6	21	3.2 ± 0.4	19	2663 ± 492	23	64 ± 7.6
246 ± 25	24	164 ± 5.8	13	3.2 ± 2.1	24	11 ± 0.9	23	4.1 ± 0.5	24	5454 ± 840	24	113 ± 16
274 ± 32	38	135 ± 3.5	29	19 ± 3.4	38	11 ± 0.5	33	5.1 ± 0.6	38	4155 ± 580	38	114 ± 15
187 ± 16	4	147 ± 5.5	0	<3	4	9.5 ± 0.3	4	2.8 ± 0.5	4	4725 ± 874	4	93 ± 13
175 ± 7.2	14	149 ± 2.7	20	28 ± 2.2	20	22 ± 1.5	20	2.9 ± 0.2	20	5535 ± 558	20	115 ± 5.6
106 ± 4.0	3	148 ± 8.1	3	28 ± 16	3	20 ± 1.7	3	5.3 ± 0.9	3	2633 ± 485	3	63 ± 12
432 ± 39	11	132 ± 2.5	11	13 ± 1.8	11	8.9 ± 0.4	10	3.4 ± 0.3	11	$12,273 \pm 1948$	11	254 ± 35
175 ± 18	5	93 ± 3.3	5	10 ± 0.8	5	9.4 ± 0.8	5	5.0 ± 0.5	3	1731 ± 1331	5	104 ± 13
135 ± 16	7	141 ± 6.3	7	7.9 ± 1.4	7	11 ± 0.7	5	2.5 ± 0.4	3	869 ± 510	7	65 ± 11
170 ± 13	17	100 ± 4.5	17	18 ± 3.5	17	8.2 ± 0.6	17	4.8 ± 0.3	13	2582 ± 464	17	96 ± 10

dissected coastal plains, as summarized earlier, characterize breeding colony areas in the Hudson Bay region, the Anderson River colony site, and the Kendall Island colony. Although coastally situated, these nesting areas are more intimately associated with fresh than with brackish waters. Furthermore, the location of these colonies can be related more significantly to their protective isolation from predators than to the relative salinity of the adjacent marsh. The Banks and Wrangel island colonies, on the other hand, are located well inland. In any event, the mineral profiles of the immediate nesting areas are immaterial to the diet of these geese during the molt, because dispersal from the nesting areas after the young have hatched characterizes the behavior of goose populations in the North. Some segments of the Banks and Wrangel island populations may gravitate toward the coastal marshes to feed, but, insofar as is known, the *tidal marsh* vegetation is not significantly used by any of these blue and lesser snow goose populations. Rather it is the adjacent low tundra or more inland areas, both subject to annual spring flooding from meltwaters, that constitute the primary feeding grounds of goslings and flightless adults (Figures *138–39*). Assuming that high, wind-driven tides cause some temporary salting of areas adjacent to the coast, the spring runoff assures annual leaching of the unfrozen soil horizons. Combined family flocks may occasionally resort to bare tidal flats at low tide, perhaps to feed on exposed algae for its mineral content, but also for the purpose of using these areas for loafing (and temporary escape from flies). The Wrangel Island geese are reported to feed on invertebrates along the coast, as cited earlier, but this proclivity may

27. Coefficients of variability (V) of elements in reference feathers from nine colonies of blue and lesser snow geese and on colony of Ross' geese

Colony	Ca	Mg	Na	K	P	Fe	Zn	Mn	Cu	B	Si	Al	Mean Per Colony[a]
EASTERN COLONIES													
Baffin Island	55	44	41	42	1038	65	16	342	9.1	75	89	70	44.3
Southampton Island	60	47	28	41	257	59	14	459	34	59	81	56	47.9
Cape Henrietta Maria	27	48	40	41	14	50	17	238	39	63	75	71	47.1
McConnell River	81	50	36	40	88	71	16	94	30	65	86	82	55.7
Cape Churchill	10	40	27	25	—[b]	18	80	—[c]	5.2	33	32	25	22.3
Sub-mean per element	46.6	45.8	34.4	37.8	349	52.6	14.2	283.2	23.5	59.0	72.6	60.8	43.5
CENTRAL AND WESTERN COLONIES													
Karrak Lake (Ross' geese)	18	29	130	57	11	18	8.1	35	31	38	45	22	39.6
Banks Island	12	(10)[d]	72	14	39	6.5	9.4	96	13	28	32	32	22.8
Anderson River	11	11	98	31	12	25	9.4	23	25	14	73	19	22.8
Kendall Island	14	25	21	48	6.1	32	12	47	17	36	101	45	35.1
Wrangel Island	33	27	26	64	34	32	18	78	30	25	74	43	36.3
Sub-mean per element	17.6	20.4	51.8	42.8	20.4	22.6	11.4	55.8	23.2	28.2	65.0	32.3	31.3
Mean per element	32.1	33.1	43.1	40.3	—[e]	37.7	12.8	—[e]	23.3	43.6	68.8	46.3	37.4

[a] Excludes phosphorus and manganese.
[b] No variation in original (censored) data (all <200).
[c] No variation in original (censored) data (all <3).
[d] Arbitrary figure; no variation in original (censored) data (all <200).
[e] Extreme variations make a mean figure here meaningless.

28. Significant correlation coefficients (r) between elements in reference feathers from various colonies of blue, lesser snow, a Ross' geese from Karrak Lake

Element and Colony	Ca	Mg	Na	K	P	Fe	Zn	Mn	Cu	B	Si	Al	Total Numb of Significa Correlatio
Ca													
Baffin Island		.70		.51		.65		.57			.65	.63	6
Southampton Island		.51		.71	.67	.66		.66			.81	.59	7
Cape Henrietta Maria		.90	.52	.69		.91	.61	.80	.80	.81	.94	.95	10
McConnell River								.66			.51	.48	3
Karrak Lake (Ross' geese)		.82	.48	.32				.72				.36	5
Kendall Island		.79		−.55		.76	.64	.88	.63		.80	.64	8
Wrangel Island		.92		.41			.71	.59	−.57			.33	6
Mg													
Baffin Island	.70			.44		.50		.72			.47	.57	6
Southampton Island	.51			.76	72	.83		.92		.60	.81	.78	8
Cape Henrietta Maria	.90		.65	.68		.74	.50	.70	.74	.71	.86	.87	10
McConnell River					.42					.65	.58	.67	4
Karrak Lake (Ross' geese)	.82		.46	.31		.34		.60		.60			6
Kendall Island	.79		.70			.97	.94	.84	.64	.76	.89		8
Wrangel Island	.92			.42			.64		−.46			.39	5
Na													
Baffin Island				.56									1
Southampton Island				.45	.66								2
Cape Henrietta Maria	.52	.65		.68			.54		.62	.64	.58	.56	8

176

Table 28. *Continued*

Element and Colony	Element												Total Number of Significant Correlations
	Ca	Mg	Na	K	P	Fe	Zn	Mn	Cu	B	Si	Al	
McConnell River					.40					.35			2
Karrak Lake (Ross' geese)	.48	.46			−.34		.39	.49	.43	.44			7
Kendall Island		.70		.87		.69	.87		.67	.80			6
Wrangel Island				.43	.53						.34		3
K													
Baffin Island	.51	.44	.56			.66		.70			.39		6
Southampton Island	.71	.76	.45			.60		.68	.70		.64	.50	8
Cape Henrietta Maria	.69	.68	.68			.79	.47	.90	.58	.68	.73	.79	10
McConnell River													0
Karrak Lake (Ross' geese)	.32	.31			.34		.52						4
Kendall Island	−.55		.87			.85	.54			.55		.67	6
Wrangel Island	.41	.42	.43							.42			4
P													
Baffin Island													0
Southampton Island	.67	.72	.66				.87	.79		.81			6
Cape Henrietta Maria													0
McConnell River		.42	.40			.52			.46	.71			5
Karrak Lake (Ross' geese)			−.34	.34						.36			3
Kendall Island							.89	.86	.60	.75	.96	.35	6
Wrangel Island			.53					−.36			.43		3
Fe													
Baffin Island	.65	.50		.66				.57			.69	.73	6
Southampton Island	.66	.83		.60				.80			.88	.88	6
Cape Henrietta Maria	.91	.74		.79			.49	.81	.72	.80	.87	.93	9
McConnell River					.52				.55		.39	.37	4
Karrak Lake (Ross' geese)		.34									.74	.65	3
Kendall Island	.76	.97	.69	.85								.81	5
Wrangel Island									−.36		.58	.88	3
Zn													
Baffin Island									.33				1
Southampton Island					.87								1
Cape Henrietta Maria	.61	.50	.54	.47		.49			.74	.68	.62	.57	9
McConnell River													0
Karrak Lake (Ross' geese)			.39	.52						.32			3
Kendall Island	.64	.94	.87	.54	.89			.78	.71	.92	.78		9
Wrangel Island	.71	.64								.38			3
Mn													
Baffin Island	.57	.72		.70		.57					.44	.58	6
Southampton Island	.66	.92		.68	.79	.80				.79	.76	.79	8
Cape Henrietta Maria	.80	.70		.90		.81			.53	.58	.72	.83	8
McConnell River	.66										.43	.43	3
Karrak Lake (Ross' geese)	.72	.60	.49									.33	4
Kendall Island	.88	.84			.86		.78		.73	.76	.92	.65	8
Wrangel Island	.59				−.36					.43		.41	4
Cu													
Baffin Island							.33						1
Southampton Island				.70						.54			2
Cape Henrietta Maria	.80	.74	.62	.58		.72	.74	.53		.85	.73	.75	10
McConnell River					.46	.55							2
Karrak Lake (Ross' geese)			.43										1

177

Table 28. *Continued*

Element and Colony	Ca	Mg	Na	K	P	Fe	Zn	Mn	Cu	B	Si	Al	Total Number of Significant Correlations
Cu													
Kendall Island	.63	.64	.67		.60		.71	.73		.79	.65		8
Wrangel Island	−.57	−.46				−.36					−.34	.38	5
B													
Baffin Island													0
Southampton Island		.60			.81			.79	.54				4
Cape Henrietta Maria	.81	.71	.64	.68		.80	.68	.58	.85		.84	.83	10
McConnell River		.65	.35		.71								3
Karrak Lake (Ross' geese)		.60	.44		.36		.32						4
Kendall Island		.76	.80	.55	.75		.92	.76	.79		.68		8
Wrangel Island			.42				.38	.43				.38	4
Si													
Baffin Island	.65	.47		.39		.69		.44					5
Southampton Island	.81	.81		.64		.88		.76				.93	6
Cape Henrietta Maria	.94	.86	.58	.73		.87	.62	.72	.73	.84		.98	10
McConnell River	.51	.58				.39		.43				.82	5
Karrak Lake (Ross' geese)						.74						.84	2
Kendall Island	.80	.89			.96		.78	.92	.65	.68		.59	8
Wrangel Island			.34	.43		.58			−.34			.59	5
Al													
Baffin Island	.63	.57				.73		.58					4
Southampton Island	.59	.78		.50		.88		.79			.93		6
Cape Henrietta Maria	.95	.87	.56	.79		.93	.57	.83	.75	.83	.98		10
McConnell River	.48	.67				.37		.43			.82		5
Karrak Lake (Ross' geese)	.36					.65		.33			.84		4
Kendall Island	.64			.67	.35	.81		.65			.59		6
Wrangel Island	.33	.39				.88		.41	.38	.38	.59		7
Total Number of Significant Correlations	45	47	29	38	23	36	26	41	29	33	41	42	430

29. Number of uncensored data used in computations of correlation coefficients (r)

Colony	Ca	Mg	Na	K	P	Fe	Zn	Mn	Cu	B	Si	Al
Baffin Island	43	40	49	48	18	49	49	29	49	35	48	49
Southampton Island	18	13	23	23	9	23	23	12	23	18	19	23
Cape Henrietta Maria	24	24	24	24	2	24	24	13	24	23	24	24
McConnell River	29	28	38	38	37	38	38	29	38	33	25	38
Karrak Lake (Ross' geese)	20	20	20	20	20	20	20	20	20	20	20	20
Kendall Island	11	11	11	11	11	11	11	11	11	11	11	11
Wrangel Island	17	17	17	17	17	17	17	17	17	17	17	17

178

30. Summary by colony and element of the number of significant correlations between elements in reference feathers

	Colony							
Element	Baffin Island	Southampton Island	Cape Henrietta Maria	McConnell River	Karrak Lake (Ross' geese)	Kendall Island	Wrangel Island	Total Number of Significant Correlations
Ca	6	7	10	3	5	8	6	45
Mg	6	8	10	4	6	8	5	47
Na	1	2	8	2	7	6	3	29
K	6	8	10	0	4	6	4	38
P	0	6	0	5	3	6	3	23
Fe	6	6	9	4	3	5	3	36
Zn	1	1	9	0	3	9	3	26
Mn	6	8	8	3	4	8	4	41
Cu	1	2	10	2	1	8	5	29
B	0	4	10	3	4	8	4	33
Si	5	6	10	5	2	8	5	41
Al	4	6	10	5	4	6	7	42
Total number of significant correlations	42	64	104	36	46	86	52	430

be for the purpose of obtaining small mollusks to compensate for calcium deficiencies in the ecosystem. In brief, we are not aware of any extensive use by blue and snow geese of saline-adapted plants (halophytes) on their breeding grounds, as Canada geese use *Puccinellia* sp., for example, along the coasts of James and Hudson bays.

The coastal marshes of the south and west coasts of Hudson and James bays are famous for their fall concentrations of migrant blue and lesser snow geese, but large numbers of these geese also forage well inland to feed on berries. The proximity of these coastal marshes to seawater insures that there

will be some small, wind-borne inputs of nutrients into the plants growing there. But as Art et al. (1974) point out, the uptake and retention of sodium by plants is low. Although the salinity of these marshes has not been determined, the salinity of the shallow, inshore waters along the west coast of James Bay is relatively low due to the enormous influx of freshwater from the numerous large rivers

31. Frequency of significant correlations per element by number and percent

Element	Number of Correlations	% of Total
Ca	45	10.5
Mg	47	10.9
Na	29	6.7
K	38	8.8
P	23	5.5
Fe	36	8.4
Zn	26	6.0
Mn	41	9.5
Cu	29	6.7
B	33	7.7
Si	41	9.5
Al	42	9.8
Total	430	100.0

32. Summary of elements significantly correlated in reference feathers from six colonies of blue and lesser snow geese and the Karrak Lake Ross' goose colony

Number of Colonies for Which Significant Correlations Were Found	Element Pairs Significantly Correlated
7	Ca-Mn, Ca-Al, Fe-Al, Mn-Al
6	Ca-Mg, Ca-K, Fe-Si, Al-Si
5	Ca-Si, Mg-K, Mg-Fe, Mg-Mn, Mg-B, Mg-Si, Mg-Al, Na-K, Mn-Si
4	Ca-Fe, Na-P, Na-B, K-Fe, P-B, Zn-B, Mn-B
3	Ca-Zn, Ca-Cu, Mg-Na, Mg-Zn, Mg-Cu, Na-Zn, Na-Cu, K-Zn, K-Mn, K-B, K-Si, K-Al, P-Mn, Fe-Mn, Zn-Cu, Cu-B, Cu-Si
2	Ca-Na, Mg-P, Na-Si, K-Cu, P-Zn, P-Cu, P-Si, Fe-Cu, Zn-Si, Mn-Cu, Cu-Al, B-Si, B-Al
1	Ca-P, Ca-B, Na-Fe, Na-Mn, Na-Al, K-P, P-Fe, P-Al, Fe-Zn, Fe-B, Zn-Mn, Zn-Al

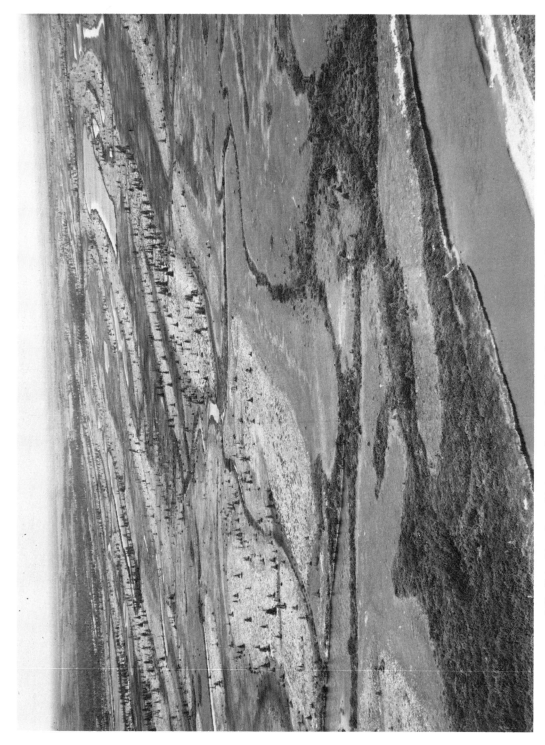

246. Raised beaches with intervening marshes along the coast of Hudson Bay. Looking westward from near the mouth of the Black Duck River (foreground) on the Ontario-Manitoba border.

that dissect the Hudson Bay lowlands. Beach ridges, formed during postglacial isostatic rebound, border the south coast of Hudson Bay and extend exceptionally well developed along the west coast of James Bay as far south as Ekwan Point. These beaches act as barriers to high tidal waters, and the intervening marshy swales act as catch basins for meltwaters and rainwaters (Figures *246–48*).

Annual precipitation along the west and south coasts of Hudson Bay varies from 14 to 18 inches (35.6 to 45.7 cm) in Manitoba and from 20 to 22 inches (50.8 to 55.9 cm) in the north to 30 inches (76.2 cm) in the south (Longley 1972; Chapman and Thomas 1968). Snow cover in northern Ontario varies from 20 inches (50.8 cm) to 5 feet (1.52 m) and maximum cover occurs along the coast of Hudson Bay in mid-March. Average snow

cover at Moosonee is about 30 inches (76.2 cm) (Potter 1965).

During breakup, coastal plain areas in low arctic and subarctic regions are subject to widespread flooding from meltwaters (Figures *138–39*). This annual flooding alone would insure that the input of wind-borne salts into soils and plants would not be cumulative. In addition, as a result of ice damming, coastal flooding often occurs during the spring breakup around the mouths of major rivers draining into Hudson and James bays. Along the southwest and south coast of James Bay innumerable small streams and drainage channels, which derive their waters from the enormous area of muskeg that extends westward from the coastal marshes (Figure *249*), "irrigate" these marshes to varying extents.

Svatkov (1958:82, 84) has commented similarly

247. Beach ridge and marsh along the coast of James Bay about one hundred miles south of Cape Henrietta Maria. The complex beach ridge systems along the west and south coasts of Hudson Bay are formed as a result of isostatic rebound following the melting and retreat of the glacial ice sheet. Rate of rebound in the Cape Henrietta Maria area has been estimated and four feet per hundred years (Webber, Richardson, and Andrews 1970). View looking south westward.

248. The coastal marsh of James Bay just south of the Kapiskau River. View looking southward.

as to why salts do not accumulate in most lowland soils on Wrangel Island:

"One of the peculiarities of the water conditions in arctic gley soils is the washing of their surface horizons by snowmelt and rainwater. The penetration of rain into the soil is facilitated by the numerous cracks caused by drying out. . . .

"No investigation was made of deep salinization of the arctic polygonal solonchak soils, since the snowmelt and rainwater carry the salts off to the rivers (or the ocean)."

The coastal marshes of Louisiana present many physiographic and hydrologic parallels to the coastal marshes of Hudson and James bays. (For an understanding of this area, we are indebted to John J. Lynch 1968; and personal communication of 27 September 1974). The eastern coastline of Louisiana, from Marsh Island to the mouth of the

249. Coastal marsh along the coast of James Bay a few miles north of the Moose River. View looking south eastward.

Mississippi, a distance of 190 miles, is ragged and deltaic in character (Figure *174*) and its barrier beaches and islands are too discontinuous to protect the marshes from daily tidal influences. Normal Gulf Coast tides are eighteen inches, but wind tides may increase normal high tide levels by several feet.

The delta of the Atchafalaya River, an early post-Pleistocene channel of the Mississippi River, was formed about 4,600 years BP (Morgan 1970). Until the normal flow of the Atchafalaya River was curtailed by what is termed an overbank control structure completed in 1959, its discharge of water at low stage was equal to 10 percent of the flow of the Mississippi River (Figure *250*) (Russell 1936). Waters formerly carried by this distributary now facilitate navigation on the lower Mississippi River, but at flood stage, the Atchafalaya's chan-

nel is used to alleviate high-water conditions downstream (Figure *257*) on the Mississippi River. According to John J. Lynch and John D. Newsom, about thirty thousand blue and lesser snow geese (particularly the blue phase) used the Lafourche and Terrebonne marshes between the Mississippi River delta and Marsh Island as late as 1955, but this number was but a remnant of earlier concentrations. At about this time, fifteen thousand geese still used the marshes of St. Bernard Parish and sixty to eighty thousand were concentrated on the Delta N.W.R. and the Pass A Loutre marshes near the outer limits of the Mississippi River delta. These latter marshes and those formed by the Lafourche distributary and the Atchafalaya River are no longer used by blue and lesser snow geese. The timing of the abandonment of these marshes by blue and lesser snow geese suggests that the cut-off

of distributary flows (the Lafourche in 1904, the
Atchafalaya in 1959) and the channelization of the
lower Mississippi River by man are chiefly responsible. The sediments of the Mississippi River, estimated annually to equal 258,000 acre-feet
(Slusher 1968), or 3.2 billion cubic yards (Morgan
1970), are now funneled directly into the Gulf of
Mexico instead of being allowed to enrich the eastern Louisiana marshes with their nutrient load.
Some of the adverse effects of channelization on
the Gulf Coast marshes, particularly ecological
changes resulting from the increased salinities of
marsh waters, have been reviewed by Chapman
(1968).

The Marsh Island-Mississippi River delta section of the Gulf Coast was evidently used more ex-

tensively and intensively by geese in earlier times
(Bent 1951). In 1910, McAtee (1911:272) concluded that the center of abundance of the blue
goose was restricted to "a narrow strip extending
along the coast of Louisiana from the Delta of the
Mississippi to a short distance west of Vermilion
Bay." However, from his account, one cannot be
certain that he observed blue geese in most of the
marsh intervening between these two areas. In
1932, McIlhenny reported that in some years a
large flock would use the marshes of Point au Fer
on the east side of Atchafalaya Bay and the
marshes bordering Morison's Pass, but he did not
consider these areas a part of their usual feeding
range. It thus appears that the blue and snow geese
have made dynamic shifts in their range in historic

250. The overbank structure of the Atchafalaya River at its origin and juncture with the Mississippi
River. The Atchafalaya River constituted the main channel of the Mississippi River in early post-
Pleistocene times. Until its flow was occluded by this structure in 1959, it was an important distributary of the Mississippi River. At times of peak severe flooding on the Mississippi River, gates permitting channel discharge are opened.

times, and the thesis can be developed that these shifts relate directly to increases in the salinity of the outer marshes and a reduction in the input of nutrients from periodic flooding (Figure *251*). And both factors can be traced to a single cause—the channelization (or is it better termed "canalization"?) of the Mississippi River south of Baton Rouge, Louisiana.

The concentrations of blue and snow geese in central and, in earlier times, eastern, Louisiana were probably significant expressions of the fertility of the soils of this sector of the Gulf Coast. Thus, Shacklette et al. (1971:D–6) in their study of the composition of surficial soils in the conterminous United States found low concentrations of many elements in soils of the coastal plains. "The general exception to this low concentration is the part of the Coastal Plain crossed by the Mississippi

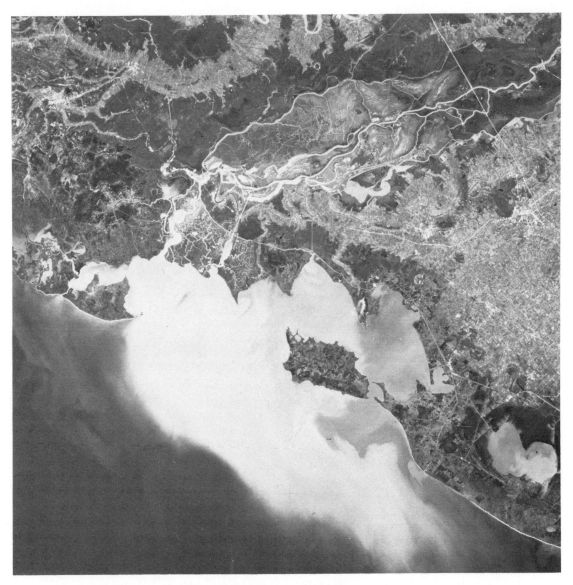

251. Sediment discharge of the Atchafalaya River at its terminus on the eastern Gulf coast of Louisiana. This photograph, taken from Skylab on January 30, 1974, provides some appreciation of the sediment loads carried by the Mississippi River at flood stage that no longer enrich and build this sector of eastern Louisiana marsh except in rare, peak-flood years.

River.'' In a similar vein, Odum (1964:72–73) notes that outflow of waters from rivers transversing low fertility soils, as exemplified by the Altamaha River of Georgia, are a nutritional handicap to coastal estuary marshes because of their clay load and the diluting effect of the low-nutrient waters. Although practically the entire coast of Louisiana has been referred to as a deltaic plain, i.e., the depositions of active deltas and the remnants of earlier deltas formed by numerous small rivers (bayous), the eastern half of the Louisiana marshes is the product of both earlier and recent delta formations of the Mississippi River and its distributaries (Russell 1936).

The major portion of the present delta of the Mississippi River and its major distributary, the Lafourche, was deposited between 1,500 and 700 years B.P. The maximum discharge of the Lafourche is estimated to have taken place prior to 700 years ago, but *flood waters and sediments were carried by the Lafourche until 1904 when it was dammed at Donaldsonville* (Slusher 1968) (our italics).

We asked John J. Lynch whether man-made levees along the lower Mississippi, by the containment of flood waters, had denied enrichment to the marshes west of the present delta, had caused their subsequent salinization, and thereby had been responsible for the abandonment of the eastern sector of Louisiana coastal marsh by blue and lesser snow geese. His reply was most stimulating:

A few days ago, while viewing the Skylab photo on pages 486–7 of October '74 *National Geographic* (Vol. 146:4) [Figure *251*], I was reminded of your letter of 9/23. In it you brought up the possibility that many Snow/Blue Geese may have quit wintering in the marshes of southeast Louisiana (Delta NWR & vicinity) because man's levees prevented discharge of Mississippi River over-flow waters into these marshes. My reply of 9/27 offered some information that would argue against this possibility. Yet the Skylab photo mentioned above *could* be interpreted as *supporting* it.

This Skylab image shows Mississippi River floodwaters (channel-discharge, not merely bank-overflow) surging out of the Atchafalaya Floodway, and bathing all the Vermilion Bay Region [Figure *251*]. So one could hypothesize as follows: The quarter-million or so Snows and Blues that now winter in these Vermilion Bay marshes are more similar in some respects to the relatively few S/B Geese still wintering at Delta NWR & vicinity (150+ air-miles east) than they are to geese wintering a mere 25 miles west of the photo-area. The Vermilion Bay geese have much the same color-ratio as Delta birds, with most White-phase Snows being in ''mixed

groups'' that also contain dark ''Blues'', rather than being in ''pure'' groups having all White birds. Also, the Vermilion Snows & Blues could be considered primarily ''wilderness birds'' that continue to have great affinity for natural coastal marshes; whereas all other Snows & Blues of southwest Louisiana and east Texas make regular feeding forays to the ricefields (and many could now be considered full-time winter residents of agricultural lands).

We could then say . . . when the old Snows & Blues of southeast Louisiana, who never had much opportunity to get acquainted with agricultural lands, decided to abandon the present active Delta of the Mississippi (Delta NWR & environs), they went west until they encountered the *next* ''wilderness area'' where Mississippi River waters flowed directly into the Gulf Coast marshes. This, of course, would be the mouth of the Atchafalaya River, from whence Mississippi River floodwaters and sediments would be projected westward to Vermilion Bay (as in the Skylab photo).

Having thus perpetrated a truly remarkable ''Post hoc, ergo propter hoc,'' we should then immediately consider the entire subject closed, and move briskly on to contemplation of other matters. But we cannot take this ''easy out.'' Not until we explain (1) *Why* did the old Delta-wintering geese ''abandon'' southeast Louisiana in the first place? And (2) The Vermilion Bay marshes *already* had a hefty wintering population of several hundred thousand Snows & Blues; would these have ''courteously moved elsewhere'' (like to the ricefields), so as to leave ample room for the geese that had ''abandoned'' southeastern Louisiana and were looking for a new ''marshy-type'' winter home? And (3) being an early-vintage ''Astronaut'' (of the ''Tree-top type'') who has viewed this part of the Gulf Coast many times from the more ''primitive satellites'' (airplanes), I can assure you the Vermilion Bay receives as much Mississippi River discharge as is portrayed in the Skylab photo, only at times of major floods; and these are quite infrequent and always occur so late in spring that geese have already left the Gulf Coast and headed North.

Being thus unable to ''explain away'' the foregoing imponderables, I'm afraid I'll have to stick with my original thesis. Those old Snows & Blues that once wintered in considerable numbers in the active Delta region of SE Louisiana simply ''faded away'' (anyway for the most part). This would be especially true after deteriorating climate in the eastern N. American Arctic (starting in 1961) made their consistent reproduction there almost impossible, and perhaps encouraged some survivors to nest elsewhere and subsequently to migrate and winter further west.

Nevertheless, if only to stimulate further field study of the problem, an attempt should be made to ''explain away'' the questions that Lynch raised. We suggested that: 1) more efficient containment of floodwaters and the dispersal of mineral-rich sediments directly into the Gulf have been the direct and indirect causes through ecological change—perhaps subtle in nature—of the progressive abandonment by blue and lesser snow geese of the outer delta; 2) the building of levees on the Mississippi River farther upstream in the delta and control structures on the Lafourche and Atchafalaya rivers were among major factors causing the gradual abandonment of this sector of the marsh by blue and lesser snow geese in recent times; and 3)—in support of the first two arguments—there has been no known abandonment of the breeding grounds of blue and lesser snow geese in the southern sector of their breeding range on Baffin Island, the source of the Delta N.W.R. flock.

In the light of the apparent cause-effect history of the abandonment of the eastern Louisiana coastal marshes by blue and lesser snow geese discussed above, it is apropos to point out here that recent proposals by the Quebec government to divert and harness for hydroelectric power the three main rivers (the Nottaway, Broadback, and Rupert) emptying into the south end of James Bay pose a comparable—probably more serious—threat to the welfare of blue and lesser snow goose populations.

The coastal marshes at the south end of James Bay constitute the richest and most extensive fall feeding grounds for migrant blue and lesser snow geese in the Hudson-James bay area. It is reasonable to suspect, particularly in years of late hatch, that the survival of large numbers of these geese in the ensuing months may relate to the fall feeding period on these marshes. It is here that these geese gain their energy stores for their direct flight to the Louisiana marshes (Bent 1951:183; Soper 1942:158). The proposed plan would block the flow of the Nottaway and Broadback rivers about one hundred miles from the coast and divert their waters into the Rupert River. The long-term effect of these diversions and the resulting loss of sediments to impoundments would be certain to increase the salinities of and to cause extensive deterioration of the present south coast marshes.

For an appraisal of the salinity of the Gulf Coast marshes used by blue and lesser snow geese and a review of the feeding habits of these geese, see Lynch 1967. Also on this subject, we again quote John J. Lynch (letter, 9 September 1974).

Perhaps we *over*-simplify matters when we divide the Gulf marshes into "Fresh," "Brackish" and "Salt". A given tract of marsh that is "brackish" today, could be quite saline a few weeks hence be-

cause of lack of rainfall and incursion of a Gulf storm tide. Subsequent heavy rains might quickly convert this marsh to a nearly-"fresh" state if its soils were organic peats that would leach readily; at another extreme continued dry weather following the storm tide would create "hyper-saline" conditions, especially where soils were less-permeable "muddy-peats" or clays.

Let's concentrate on the plants that geese prefer in these marshes. Snows and blues were (and still are) very partial to the "three-square" sedges (*Scirpus olneyi* and *S. robustus*), cattails (*Typha* sp.), and other plants that are not true halophytes. While such plants are often abundant in "brackish" marshes, they make their best growth (and flower and fruit most readily) when salinities do not exceed the "faintly-brackish" category (less than 1/10 of sea-strength). They *tolerate* periods of higher soil- and water-salinities by the simple expedient of becoming temporarily dormant. (I've not seen chemical analyses of these plants, but can say that their rhizomes never have the slightest "salty" taste when chewed.)

In the same marshes, and associated with the above preferred food plants of geese, are other plants that *are* true halophytes. The latter include several species of *Spartina* (particularly *S. patens*, also *S. alterniflora*, *S. cynosuroides* and *S. spartinae*), and *Distichlis spicata*. These halophytes continue to grow well even when salinities are quite high. Furthermore they survive sudden changes in salinity without suffering cell-plasmolysis. They maintain osmotic balance in their tissues by taking salt water into the vascular circulation, and can lose any excess via transpiration (leaving conspicuous crystals of salt around leaf-stomata in dry weather). [See Sutcliffe (1962:159–62) for a description of the histology of salt glands of halophytes.] Any goose that eats such vegetable matter is going to get quite a slug of sea-salt. Although Blues & Snows *prefer* the freshwater plants such as threesquare, they also consume some halophytes altho in small amounts as a rule. But occasionally geese will eat large quantities of salt-plants such as *Distichlis*. In so doing, they don't get very fat, and their "palatability for humans" drops to damn near zero. Yet they seem to survive in good order ingestion of such salty provender. (What is really puzzling is that geese will pull such crazy stunts when there is no apparent reason for them to do so.) But as a general rule, geese feeding in "brackish" Gulf marshes are actually feeding primarily on freshwater plants.

The remainder of the Louisiana coast in the west and the eastern Texas coast as far as Galveston is protected by well-developed barrier beaches (sea rim) that act as effective dams against most wind tides. South along the coast from Galveston to south Texas and Mexico, again quoting Lynch:

. . . average annual rainfall decreases, and seasonal droughts become more frequent. The coastal lagoons between Galveston & Corpus Christi are only partially filled with marsh, while those south of Corpus Christi into Mexico are largely open and often quite saline. Some of the winter goose-range of these regions consists of "salt-flats," where halophytes may be the only food-plants available during prolonged dry spells. But when rice culture became well-established at Lissie Prairie (Eagle Lake) and adjoining inland parts of Texas, geese from the Galveston-Corpus coastal strip soon moved in to become full-time ricefield birds. I've a strong suspicion other geese may have moved north from the old salt-flats below Corpus, so as to enjoy this assured supply of fresh water and non-saline food plants in the ricefields north of Victoria.

And geese that continue to winter in the salt-flats below Corpus are attracted by reservoirs, "cattle-tanks," and other local sources of fresh water. I think one could safely say that even these south Texas & Mexican geese, while they can tolerate saline conditions, will opt for fresher environments, given a choice.

The whole of the Gulf Coast is subject to hurricanes of varying strength, but according to Lynch, these storms are "self-limiting" in their ecological effects in that they produce their own nullifying factors. Although the southeast quadrant of the disturbance brings ashore high tides, the northeast quadrant produces torrential rains which counter and dilute the inward movement of tidal waters. The principal benefit derived from storms that do bring ashore large amounts of salt water is that they set back the succession of plants in the marsh, creating conditions favorable to species that are palatable to and readily used by geese. The organic layer in the older western marshes is only three to five feet deep and often less. The 60-inch (152 cm) annual rainfall of the region usually assures that any salinity gained from storms is soon lost. The response of blue and lesser snow goose populations to the food resource of rice fields is evidence that soil structure and moisture; the kind and accessibility of available food (roots, seeds); and the succulence, physiognomy, and physical structure of food plants are all factors that govern their feeding habits rather than dependence on salt marsh plants per se.

To sum up, although analyses indicate that the sodium content of some plants in the coastal marshes in the northern breeding grounds may be considerably higher than that of plants growing in the uplands, blue and lesser snow geese (on the evidence of their salt glands) have a salt intake well below that of many races of Canada geese. (An alternate explanation, in lieu of any observed en-

largement of the salt glands, would be that snow geese have proportionately larger and more efficient kidneys than Canada geese and hence are genetically better endowed to handle high salt loads.) But the fact remains that, on the Gulf Coast, the preferred feeding areas of blue and lesser snow geese can be related only to the freshwater or very slightly brackish marshes—chiefly the *head-of-the-tide* and *lagoon* marshes that surround the upper portions of the coastal lagoons and the *sea-rim* marshes that lie behind the barrier beaches in western Louisiana and Texas.

Extrarenal Secretion of Minerals

Skin, sweat and sebaceous glands, hair, and nails constitute significant avenues of mineral excretion in mammals in addition to the kidneys and alimentary tract (Schwartz 1960:337–86). Feathers are, of course, homologous to mammalian hair, and it is a primary thesis of this report that in birds, at least in ducks and geese, feathers constitute a significant but little realized mode of excretion.

Of special interest here is the role of feather keratin in relation to the relative roles of absorption and secretion in achieving homeostasis, i.e., as a recorder of mineral turnover. As the flight feathers in geese are all grown under the same period of stress—the molt and its many physiological ramifications (Hanson 1962)—we are not concerned here directly with the many seasonal changes involving variations in mineral metabolism linked to hormonal, vitamin, and some environmental (e.g., temperature) influences. It is generally held that a close monitoring of intestinal absorption by means of various feedback mechanisms and the function of bone over longer periods as a *put-and-take* reservoir under hormonal control are chiefly responsible for maintaining long-term homeostasis of many minerals in blood plasma (Bauer et al. 1961; Copp 1969:458). However, our data on feather minerals suggest that the importance of feathers as an excretory route may be underestimated. An understanding of this facet of the metabolism of geese would have pertinence in relation to laboratory studies of domestic geese, readily available as laboratory animals but little used despite their unique advantages, as shown in studies of lipogenesis (Benedict and Lee 1937) and osteoporosis (Hanson unpublished).

All birds possess supraorbital glands or "salt glands" as they have commonly come to be known. These glands, when active, function mainly as an extrarenal means of sodium excretion

252. Adult and gosling Canada geese off the south coast of Akimiski Island in James Bay. Due to the influx of large amounts of fresh water into James Bay and the shallowness of offshore waters along the west coast, these waters are brackish. Note attempt of adults to form a protective cordon around the goslings, a characteristic response of loosely consolidated family groups to threatening situations.

and are particularly active in those species associated with saline waters. Races of Canada geese that come in contact with salt or brackish water for extended periods are notable in this respect, as their salt glands become enlarged and extremely active. The enlargement of these glands may be seasonal, as in the case of the Canada geese on Akimiski Island, which used the offshore waters of James Bay for resting or escape (Figure 252) or, as in the case of the coastal Labrador and other populations in contact with brackish and salt water the year around, these glands may be notably and permanently enlarged (Figures 253–54) (Hanson unpublished). This enlargement has no relation to body weight (Figures 255–56). In the case of the Akimiski Island Canada geese, the progress of the wing molt can be used as a relative expression of the time that an individual has been in contact with salt water. The progressive, seasonal enlargement of the salt gland, therefore, can be directly related to the wing molt (Figures 257–58). In these geese

the interorbital portion of the skull underlying the salt glands does not become modified, but in individuals associated with salt and brackish waters throughout the year, the tremendous enlargement of these glands results in a reduction of the interorbital width and a ''seating depression'' of the underlying bone as a result of the physical pressure exerted upon the skull and also partly as a result of their hyperactive metabolic state (Figure 259). Consequently, living examples of *saltwater races* can be distinguished from specimens of *freshwater races* by the round-headed profiles of the former and the flat-headed profiles of the latter (Figures 255–56). The skull alone, depending on whether there is a reduction of the interorbital width combined with well-defined indentations that partially reflect the outline of the overlying gland, can serve to separate the skeletons of saltwater and freshwater races (Hanson unpublished). In either case, whether the salt glands undergo temporary or permanent enlargement, the stimulus for growth and

253. Adult male Canada geese (*Branta canadensis interior*) from Horseshoe Lake (1959) showing (A and B) the relative lack of development of the salt gland and (C) the flat-headed appearance of these and other large Canada geese with inland ranges.

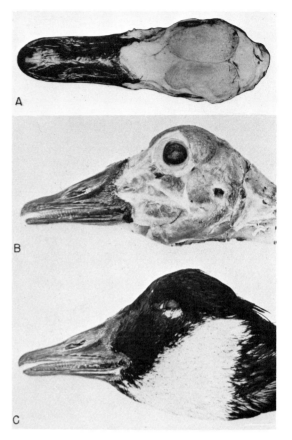

254. Adult male Canada geese from Pea Island N.W.R., North Carolina, 1959. Note extreme development of salt glands over the eyes and how, dorsally, they overlap each other over the top of the head.

maintenance of large size is increased levels of sodium in the blood (Goss 1964:306).

Molt Migrations

Unlike the males of most species of North American ducks, which undergo the molt at points distant from the females with their broods, both parents of geese remain together with their broods. Hence, the mineralogy of the ecosystems in which these components of breeding populations of geese feed should be essentially the same. In most populations of blue and lesser snow geese, the nonbreeding geese and failed breeders also remain within the general range of the breeding geese (Hanson et al. 1972); hence they too are subject to the same environmental influences as are the breeding geese. An exception to this generalization are the nonbreeding lesser snow geese from the Egg River colony on Banks Island, which undergo the molt in the Thomsen River valley at the north end of the Island. The nonbreeding component of some populations of Canada geese also undergo the molt in areas distant from the breeding segment of the population. One of the most notable ex-

amples is that of the Giant Canada goose (*Branta canadensis maxima*), which breeds in the Midwest and on the Great Plains but whose nonbreeding component molts in the central Canadian Arctic (Hanson 1965:78–82). Care must be taken to distinguish these components of Canada goose populations when evaluating their feather mineral patterns.

Timing of the Molt

In geese the timing of the molt varies somewhat between members of breeding pairs. In the case of Canada geese on Akimiski Island the molt of the primary feathers by the female precedes that of the male by seven to ten days (Hanson 1962). In blue and lesser snow geese, the interval is reported to be as short as three to four days (Palmer 1972:78). In neither case would the difference in the onset of the molt be likely to result in a significant differen-

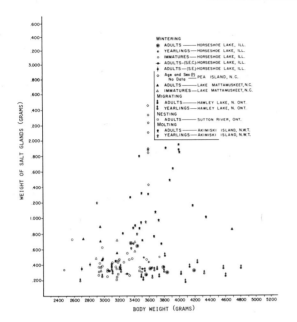

255. Weights of salt glands versus body weights in various populations of large Canada geese (males). Most of the geese now wintering at Pea Island N.W.R. breed along the coastal drainage areas of Labrador. Geese that breed in northern Ontario winter in southern Illinois. Geese that breed on Akimiski Island may winter in southern Illinois or the mid-Tennessee Valley.

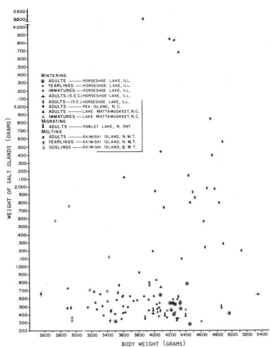

256. Weights of salt glands versus body weights in various populations of Canada geese (females).

tial in the input of minerals into the primaries due to differences in the mineralogy of the terranes to which the sexes are exposed during the growth of their flight feathers.

PHYSIOLOGICAL AND FUNCTIONAL ASPECTS OF THE MOLT

Ducks and geese constitute one of the eleven families of birds that molt and then regrow all of their flight feathers simultaneously (Stresemann and Stresemann 1966). For ducks and geese the molt—with its host of associated metabolic changes—is probably the most stressful period of their life cycles. Hence, the mineral patterns of flight feathers must be evaluated in this context. The physiological exigencies of the molt are perhaps best illustrated by the fact that the basal metabolism of molting hens (*Gallus domesticus*) averages 45 percent above that of laying hens, whereas egg laying has no appreciable effect on their basal metabolism (Mitchell 1964, 1:651).

During molt the breast and leg muscles of Canada geese and other species of geese and ducks undergo massive changes in size which are inversely

related to one another (Hanson 1962:44). The breast muscles undergo a 30- to 41-percent decrease in weight, depending on sex and age, and by midway through the molt the leg muscles have enlarged by 41 to 57 percent over their weight at the end of the migration period. Mitchell (1959:38) has emphasized that birds and mammals will draw upon the amino acids of the body tissues (proteins), particularly the methionine-cystine needed, to ensure the success of the replacement of feathers or hair, i.e., vital insulative coats. To explain partly the evolutionary significance of the loss in weight of the pectoral muscles, it was postulated by Hanson (1962:36) that a sulfur deficiency in terms of cystine and methionine was a salient fact of the molt in Canada geese and that they attempt to meet the need for these amino acids by withdrawing them from pectoral muscle. It was also suggested by Hanson (1962) that molting geese meet a hypothetical sulfur imbalance by conversions of the chondroitin sulphate and keratin sulphate (Herring 1973) of the tarsometatarsi, which becomes osteoporitic at this time, especially in adult females. This conclusion is in general agreement with Mitchell's (1964, 11:509) statement that to maintain homeostasis in the face of

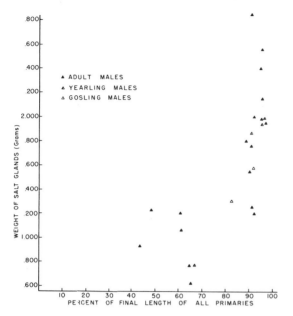

257. Relation of weights of salt glands to total length of wing primary feathers in male Canada geese, Akimiski Island, July–August, 1959.

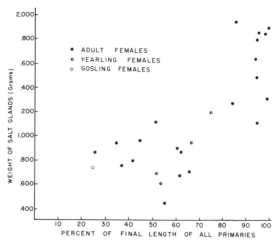

258. Relation of weights of salt glands to total length of wing primary feathers in female Canada geese, Akimiski Island, July–August, 1959.

mineral deficiencies in the food supply, an animal will withdraw required nutrients from its body tissues with a resultant catabolic wasting of the donor tissues.

The simultaneous replacement of the wing feathers in waterfowl in contrast to the leisurely successive replacement of the feathers in upland game and most other birds is an obvious enough reason why the molt period should be more stressful for geese than, for example, ring-necked pheasants (Anderson 1972). The ratios of weight of wing feathers to body weight may also explain differences in the metabolic demands molt places upon these two groups. However, there may be more subtle reasons, relative to the amino acid composition of feathers which may differ fundamentally between upland game birds and waterfowl if findings for chickens, white domestic turkeys, and domestic geese are criteria (Blackburn 1961). "It seems quite definite, however, that turkey feather barbs contain less cystine, tyrosine and phenylalanine than goose feather barbs or goose down. The sulphur contents substantiate the conclusions about differences in cystine content" (Schroeder et al. 1955:3908). The demand for these essential amino acids is less, therefore, for turkeys during the molt than for geese, and raids upon muscle tissue for these amino acids with con-

comitant wastage of other amino acids are correspondingly less.

The rapid hypertrophy of the leg muscles of geese during molt was initially explained as relating to the use-disuse phenomenon, as during the flightless period of the molt, waterfowl are entirely dependent on their legs for escape either by running or swimming (Hanson 1962). On the basis of recent studies of the free amino acids of captive Canada geese carried out in collaboration with Dr. James D. Jones of the Mayo Clinic and the dissection of lesser Canada geese in the central Arctic of Canada in 1974, it was more fully realized by Hanson that the hypertrophy of the leg muscles has well begun before the actual onset of the flightless condition—probably somewhat before the loosening of the major wing feathers (Hanson 1962:23), Hence, the timing of the onset of the hypertrophy of the leg muscles and their subsequent development must, like the changes in the size of the breast muscles, also be explained primarily on an evolutionary basis rather than on a purely functional use-disuse basis.

Amino acids for the hypertrophy of the leg muscles are, in all likelihood, partially derived from the degradation of protein in the breast muscles. The rapid hypertrophy of the leg muscles is probably also equally related to the metabolic use of essentially nonfunctioning flight muscles of the breast as a source of phosphorus (Irving 1964:257). The concurrent development of osteoporosis of the tarsometatarsi may perhaps also be explained in part by the functional use of bone

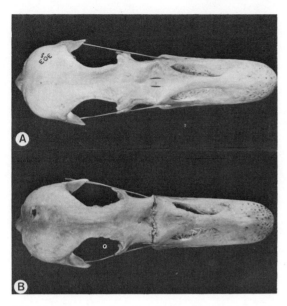

259. Dorsal view of skulls of (A) adult male Canada goose from Horseshoe Lake, Illinois, and (B) adult male from Pea Island N.W.R., North Carolina. Note reduced interorbital width of skull of goose from North Carolina (B).

tissue to meet an overall deficit of phosphorus in the body pool (see discussion of phosphorus in chapter 7). It is also pertinent to note that the sheath that encloses the growing feather is rich in phosphorus (Table 11).

Does the withdrawal by Canda geese of calcium from the leg bones also exemplify priority use of interdependent mineral storage pools? It has recently been shown that the rapid replication of tissues during growth is associated with elevated serum calcium (Perris 1971:101–26). A lowered serum calcium appears to be associated with osteoporosis of the tarsometatarsi during the molt in wild Canada geese but data for caged birds is conflicting (see Table 42). Breast muscle calcium is elevated during the molt (Hanson and Robert E. Johnson unpublished), a period during which this muscle loses and regains 30 to 40 percent of its weight. But the rapidity with which the flight feathers are regrown suggests that, on a dry-weight basis, feather keratin may be regenerated at a faster rate than other protein tissues in the body, both in the rapidly growing gosling and in the adult bird.

To sum up, egg laying and the molt occurring in quick succession place a double stress on the adult female. This stress is reflected in greater overall loss of body weight, greater loss of pectoral muscle protein, and more extensively developed osteoporosis in this age-sex class than in adult males or nonbreeding yearlings. Which elemental component of bone—calcium, phosphorus, or sulfur—is, at least in a theoretical sense, the first to be drawn upon in response to demands to meet a deficit in the body pools resulting from feather growth and reorganization of body musculature can only be conjectured at this point.

Finally, it should be pointed out that the sequence of the molt in Canada geese, and doubtless other northern waterfowl, has a significance of its own both from the standpoint of spacing metabolic demands and meeting functional needs. In brief, while the wing feathers are being replaced and survival is enhanced by the hypertrophy of the leg muscles, the body feathers remain intact. In adult females belly feathers plucked for the nest are largely regrown during incubation (Hanson 1959). Thus, the retention of the body feathers during the flightless period insures both insulation and buoyancy when water is resorted to for escape and other uses. Subsequently, when powers of flight are again achieved, which occurs when the cumulative length of all primaries attain about 85 percent of their final total growth, the molt of the body feathers is initiated. Being again capable of flight, the geese are no longer completely dependent on full aquatic competence for escape and survival. Among the last feathers to be replaced are those of the neck and, finally, the head.

STATISTICAL PARAMETERS OF FEATHER MINERAL VALUES

The number of significant correlations between minerals in reference feathers of geese appears to provide an inverse index as to the geological diversity of a colony range. The highest number of correlations was found in the reference feathers from the Cape Henrietta Maria colony (Table 30). The feeding range of this colony is easily the most restricted of the large Hudson Bay colonies, and the mineralogy of its ecosystem is affected only to a limited degree by drainage from diverse rock types (notable exception, iron; see chapter 4).

Lesser snow geese from the small Kendall Island colony have the highest number of feather mineral correlations of the western colonies. As noted earlier, these geese take their newly hatched young to the adjacent mainland to feed. It is reasonable to assume that the soil mineralogy of this deltaic arctic coast has a high degree of uniformity as a result of the commingling of sediments derived from the Mackenzie River basin thereby accounting for the high number of significant correlations found. In contrast, reference feathers from geese using feeding areas subject to diverse mineralogical input from different drainage basins, or interspersion of rock types, or both, show markedly fewer correlations, e.g., Baffin Island, Karrak Lake (Ross' geese), and Wrangel Island (Table 30).

Viewed in another way, elemental variability in the reference feathers also tends to reflect geological or soil-mineral diversity (Table 27) although the relationship is less consistent. The smallest of the Hudson Bay colonies, both in number of birds and geographical range, is the Cape Churchill colony. The mineralogy of its range relates mainly to its present depositional soils and minerals derived from coastal waters rather than to sediments from an interior drainage, and the variability of mineral

contents of reference feathers from the colony is correspondingly low.

Geese that feed across the coastal delta lands of numerous drainage systems (e.g., the McConnell River colony) emanating from the diverse rock types of the Canadian Shield have the highest mean coefficient of variability of feather minerals.

However, the overall higher variability of feather minerals in the Hudson Bay colonies as compared with the western colonies is not readily explained. The Ross' geese of Karrak Lake and the Wrangel Island geese are presumably exposed to more diverse geology and varied mineralogical patterns in the soils of their feeding ranges than are geese of the eastern colonies, yet the variability of their feather elements is lower. This low variability may reflect appreciably smaller sample size (not true in the case of the Wrangel Island sample) or more restricted feeding ranges.

THERE ARE, in effect, three general kinds of approaches to gaining insights into the mineral metabolism of animals: 1) *in vitro* studies, as exemplified by the use of excised sections of intestine or the incubation of isolated cells, portions of cell membranes or of a specific organelle; 2) laboratory studies using living, but confined, orthodox experimental animals (e.g., the domestic fowl); and 3) studies of free-ranging wild animals, an approach particularly exemplified by investigations of the mineral content of the keratin structures. Although admitting that the objectives and precision of the latter two approaches differ from each other as does also the variability of their environments, the third approach affords the opportunity to profit from the results of an almost infinite number and variety of natural experiments that a lifetime of laboratory work with caged animals could not duplicate.

How animals contend metabolically with their mineral environment is but one aspect of their continuing evolutionary adaptation and development. This aspect of evolution has long been a major topic of research on aquatic organisms. It has been a generally less recognized area for research on birds. For example, it has become apparent from our studies that most geese are contending with the problem of mineral surpluses rather than deficiencies. And it has also been evident, as will be discussed, that for most elements studied here, the regulation of excretion is more important in maintaining homeostasis than regulation of absorption.

We earlier suggested that, in identifying geographic origins of goose populations, feather mineral patterns can be thought of as analogous to fingerprints. (The term "chemograph" has been suggested by Eugene LeFebvre for the feather mineral patterns.) Similarly, in physiological terminology, feather mineral patterns are analogous to a kymograph record or, to update an instrumental analogy, to a permanent digital computer printout of nutrition on the breeding grounds and metabolic states during the molt. It must, therefore, be concluded that most feather mineral concentrations reflect, albeit distorted, rates of mineral absorption, and that homeostasis is achieved chiefly by excretion. When a mineral gradient in the nutrient chain is steep, the rate at which the keratin being synthesized binds or "blots up" the excess is accelerated, supplementing, as it were, other modes of excretion.

One would not expect a shift in mineral intake to cause plasma values to change dramatically unless such changes were tied to specialized metabolic

CHAPTER SEVEN

The Biogeochemistry of Feather Minerals

functions (e.g., egg laying). The situation may be somewhat analogous to the example given by Bauer et al. (1961:617), that if bone resorption and formation are matched, they can proceed over a wide range of rates, yet deviations from balance between calcium absorbed and excreted will be zero. Ignoring losses and gains of body tissue per se and transient osteoporosis of the tarsometatarsi, mineral balance studies of geese during the molt would show less important deviations from zero if the mineral content of the feathers were considered part of the total excretory output of minerals. Conversely, if mineral losses were related only to losses via the digestive tract, kidneys, and salt glands, then a partially inverse relationship between intake and output could be expected during molt because mineral input into the feathers would not be accounted for.

Aside from whatever other functions they may have, each of the essential elements under review here plays an important role in enzymatic reactions. Brief summaries of the roles played by monovalent and divalent metals are contained in reviews by Suelter (1970) and Williams (1967), respectively. In general, the light or monovalent metals sodium and potassium combine with the substrates and enzymes to form functional ternary compounds, presumably acting as a bridge structure between the latter two (Suelter 1970), a concept now well established (Jones and Hix 1973). In contrast to the relatively mobile cations (a category which also includes calcium and magnesium), the heavy trace metals—iron, zinc, manganese, and copper—(manganese excepted) are more tightly bonded and are usually found as isolated atoms or,

33. Mean contents of elements in basic and acid igneous rocks, metamorphic rocks, shales, and carbonate rocks

Rock	Elements (%)					
	Ca	Mg	Na	K	P	Fe
Basic Igneous	7.0–7.6	3.8–5.2	1.7–2.1	.58–1.0	.12–.16	7.2–9.2
Acid Igneous	1.3–1.8	.45–.66	2.6–2.9	3.2–3.4	.061–.083	1.4–2.9
Igneous (Undifferentiated)	3.63	2.09	2.83	2.59	.10	5.00
Metamorphic[a]	2.4	1.4	3.0	2.4	.07	3.3
Shale	2.2–5.2	1.3–2.0	.50–.96	1.7–2.9	.074–.20	3.2–4.7
	2.21	1.5	.96	2.66	.07	4.72
Carbonate	30.0–31.0	0.4–4.7	.04–2.0	.27–.29	.035–.04	.38–.90
	30.2	4.7	.04	0.27	.04	.38

Note: Averages are calculated from sample sizes ranging from hundreds to thousands.

[a] Data for metamorphic rocks are from Canadian Shield specimens.

[b] nd = no data.

at the most, as a few clustered atoms. The interchangeability of some metals in biological systems relates to their similar valences and sizes, but the geometry of metal protein complexes is unusually irregular and presumably relates to the incompatibility of the structural demands of the proteins and of the metal ions (Williams 1967:98). Simkiss (1974:121) has pointed out that calcium is generally associated with extracellular enzymes and magnesium with intracellular enzymes, a division in accordance with their relative distribution in the body. Significantly, metals are bound to the active sites of enzymes; inhibition of enzyme action prevents the exchange of that fraction of the metal ions concerned.

In the discussions that follow we have attempted, on the basis of the literature, to provide a few salient facts on each element regarding: 1) the prevalence in the parent rocks characteristic of the feeding range and drainage basins related to each breeding colony of geese; 2) the chemical and physical forms of water-transported elements; 3) the essentiality, use, and prevalence in plants; 4) the essentiality, absorption, and function in mammals and, particularly, birds if data are available; and 5) modes by which it is excreted.

Our findings on elemental concentrations in the flight feathers of geese are discussed from the standpoint of: 1) the magnitudes of differences among means for various colonies and populations; 2) the relation of elemental variability for each colony to characteristics of the parent rocks and soils; 3) correlations between elements and their meanings in respect to ecosystems and metabolic functions; 4) the relative roles of mineral concentrations in the ecosystem and metabolic regulation of mineral absorption and excretion in affecting elemental concentrations in the primary feathers; and 5) the relative roles of absorption and excretion in affecting homeostasis. Concerning the latter, the reader should be cautioned that our remarks on metabolic regulation of homeostasis refer only to the absorbability of various elements and the rapidity with which they are absorbed, circulated, and excreted—not to what might be called the second stage of the regulation of homeostasis, the cellular regulation of mineral metabolism by cell membranes of muscle or nervous tissues, as examples, or by organs. Whereas cellular mechanisms may rigidly control the mineral uptake of most tissues, the developing keratin obviously more nearly reflects what is absorbed and circulated. The nutritional demands of the growing feathers are so great that it is reasonable to suspect that increased vascularization of the primary feather follicles results in a greater proportion of the blood flow being shunted to the wings during the molt than at other times of the year.

CALCIUM

In Rocks and Soils. Calcium is over eight times more abundant in carbonate sedimentary rocks (limestones) than it is in igneous rocks (Mason 1966:180; also see Table 33). These ratios, however, bear little relation to calcium levels in the primaries of geese unless the calcium is in a form available to plants or can be extracted by geese from the grit and soils. In general, plants grown on calcareous soils contain high concentrations of in-

Elements (parts per million)				Elements (%)		
Zn	Mn	Cu	B	Si	Al	Reference/Table(s)
30–100	1300–2200	20–100	1–5	22.6–23.4	8.2–8.7	Rösler and Lang (1972)/93, 95
40–60	390–930	10–30	10–15	32.2–33.2	7.6–7.7	Rösler and Lang (1972)/93, 95
70	950	55	10	27.7	8.13	Mason (1966)/6.5
nd[b]	620	nd	nd	30.0	8.5	Rösler and Lang (1972)/111
80	770	57	100	23.7–27.2	8.0–8.2	Rösler and Lang (1972)/94, 95
95	850	45	100	7.3	8.0	Mason (1966)/6.5
20–35	1000–1100	4–30	20–55	2.4–3.2	.42–1.1	Rösler and Lang (1972)/95
20	1100	4	20	2.4	.42	Mason (1966)/6.5

tracellular calcium (Epstein 1972:354; and Table 8). This relationship implies a high dietary intake of calcium by grazers on such soils but not necessarily a high rate of absorption of calcium.

Role in Plants. Calcium has important functions in plants (Hewitt 1963:167–72) by: 1) forming calcium pectate in the middle lamella and thus providing for tissue integrity and affecting membrane permeability; 2) being essential to normal cell division; and 3) forming complexes with deoxyribonucleic acids and thereby providing for the integrity of genetic material. Calcium also plays a role in maintaining acid-base relationships within the cell and, in association with ionized sodium and potassium, contributes to the activity of water within the cell and thereby to cell hydration. The formation of calcium oxalate crystals within cells is thought to provide a mechanism for reducing levels of oxalic acid. With such diverse functions, it is not surprising that calcium is the most abundant of the alkali or alkaline-earth elements in vascular plants.

Absorption in Mammals. In most laboratory mammals and in the chicken, calcium absorption occurs primarily in the upper portion of the small intestine (duodenum and jejunum) (Bronner 1964:357; Hudson et al. 1971:59). Regulation of intestinal absorption is believed to be the chief mechanism for achieving calcium homeostasis in man (De Grazia 1971:151); the intestinal excretion of calcium, on the other hand, is said to be passively linked to the production of digestive juices and is not a dynamic factor in calcium homeostasis (Bauer et al. 1961:665). The amount of calcium in the diet, per se, bears little relation to the amount

of calcium absorbed (Bronner 1964:354–60). Although it has been widely held that absorption rates of calcium are an inverse function of the degree of calcium saturation of the body (Bronner 1964:366; Wasserman and Taylor 1969:390–91; Irving 1973:22), an important consideration in interpreting feather mineral data is Bronner's (1964:366) conclusion that the *intensity of calcium absorption may be related to the intensity of calcium turnover in the body* (our italics). More recently, Irving (1973:18) has postulated similarly that "it is possible that *the rate of calcium loss from the body* and not its calcium saturation is the factor that controls calcium absorption in the intestine" (our italics). Judging from our feather data, we are in agreement with the concept that calcium turnover rates bear a closer relationship to calcium absorption rates than to the degree of calcium saturation in the body.

Rates of calcium absorption in geese must be linked in part to the mineral content of the food and to a small, but undefined extent, to the mineral content of water, grit, and soil ingested, but rates of calcium absorption are more importantly related to the absolute and relative intake levels of calcium and sodium. Sodium, as will be discussed, appears to have a profound effect on calcium excretion and turnover and thus, in turn, influences the rate of calcium absorption. Thus, we are brought full circle to the thesis that, in geese, turnover rates of calcium, and those of most of the other elements under discussion, are related directly or indirectly in some measure to elemental concentrations in the environment.

Calcium Levels in Primaries of Geese. The preeminent position of calcium in feathers of wild geese is apparent (Tables 30–32). It was by far the most useful element in distinguishing geese from the various colonies of blue and lesser snow geese (Table 14), and after magnesium, it was involved most frequently in significant correlations with the other eleven minerals. In fact, it is doubtful whether the difference between calcium and magnesium is meaningful (forty-five versus forty-seven correlations, Table 31).

Mean calcium values in our reference samples of primaries from blue and lesser now geese vary at least sixfold (<500 to 3,000 ppm) (Table 26). Both lowest and highest values for blue and snow geese were found in the McConnell River series (Figures *72–83*), and comparatively high values were found in the reference feathers of Ross' geese from Karrak Lake, N.W.T. (Figure *142*). The feeding ranges of these two colonies of geese lie within outcrop areas of the Canadian Shield or adjacent to it. In contrast, mean calcium values are low or only moderately high for the primaries of blue and lesser snow geese from Baffin Island, Southampton Island, Cape Henrietta Maria, and Cape Churchill—colony areas that are underlain by Paleozoic limestones.

Mean coefficients of variability for calcium values for the Hudson Bay colonies were more than 2.5 times those recorded for the western colonies (Table 27). In succeeding elemental comparisons between geese from these two regional—and geologically different—areas, similar correlations between high values and low variability and low values and high variability will be noted.

Relationships of Content of Calcium in Feathers to Ecosystems. Off Southampton Island carbonate sediments constitute 50 percent of the clays, and off Cape Churchill and Cape Henrietta Maria, 30 to 50 percent (Leslie 1964). Similar values would hold for adjacent colony areas that have emerged from the sea in the last thousand years (Hanson et al. 1972). Thus, it is difficult to project purely on the basis of calcium levels in the substrate (rock, soils, and sediments) an explanation for the low calcium values in the feathers of the geese from these colonies. But it is noteworthy that the reference collection of feathers representing the McConnell River colony contained some samples that were low in calcium and others high in calcium (Figures *77–78, 82, 86, 88*). As the inshore sediments of the sector of coast used by this colony average less than 10 percent calcium, it might be presumed that the feeding grounds of these geese, which extend inland from the coast for at least five miles, include soils both high and low in calcium

content. However, for reasons that will be discussed below, variations in calcium levels found in the feathers of geese from this colony more likely reflect the variations in the areal distribution of sodium and variations in sodium intake.

Our data *seemingly* exhibit an inverse relationship between calcium in the rocks and soils of the various ecosystems and absorption as indicated by the amounts of calcium laid down in the feather keratin. In this respect our data are somewhat in conflict with some findings on the relationship of levels of dietary calcium and the absorption of calcium: "Observations in both man and animals indicate that increased calcium intake in the diet results in increased calcium absorption by the intestine. Yet the amount of calcium absorbed is not a simple function of the amount ingested" (DeGrazia 1971:164). However, we hypothesize that high calcium levels in the feathers reflect high calcium intake and high turnover as a result of high sodium intake.

Relation of Magnesium to Calcium Absorption. High levels of calcium in soils or plants are generally reported to depress absorption rates of other elements in plants and animals. Magnesium, however, is a notable exception; its antagonistic role toward calcium absorption in plants is well established (Epstein 1972). A reciprocal relationship between calcium and magnesium has been indicated for small laboratory animals (Wacker and Vallee 1964:500–501), but other studies have shown that magnesium can increase or decrease calcium absorption, depending on experimental conditions (Wasserman and Taylor 1969:344; Tansy 1971:196). We found no inverse relationship between the prevalence of these two elements in the ecosystem and the amounts present in the keratin of the vanes of the primary feathers (Table 26); rather, concentrations in the feather vanes appeared to be, in general, reflections of probable calcium-magnesium ratios in the carbonate rocks.

In some populations of wild geese, we found relatively high calcium levels accompanied by high magnesium levels—e.g., extreme values for Canada geese from Akimiski and the Belcher islands (Figure *252* and see Figure *266*)—findings which seem to support the results of *in vivo* experimental studies by Clark and his associates (Bray and Clark 1971:313–14). Nevertheless, there is no consensus about the reciprocal effects of these two macroelements. Because we have only fragmentary knowledge of the mineral content of the diet of blue and lesser snow geese, we can only make inferences regarding the reciprocal effects of these two ions. Nevertheless, it is of interest to note that in data for lesser snow geese collected in the Imperial Valley of California, the calcium-magnesium ratio in the feathers of these geese was as high as 8–1 (Figure *240*), and that in the case of lesser snow geese from Jenny Lind Island and Sherman Gulf, the ratios were 14–1 and 16–1 respectively (Figures *95–96*). In these geese high magnesium intake was not requisite for high calcium absorption. The reverse of this situation—extremely high magnesium values—was found in the Akimiski Island population of Canada geese (Figure *252*). The feathers of these geese were also very high in calcium; hence, it cannot be argued that if magnesium intake were sufficiently high it would effectively suppress calcium absorption. Because large quantities of either element or both may be in the flight feathers, the evidence favors excretion as the major mode of achieving calcium homeostasis, whether by a passive mechanism, such as the digestive fluids, or by an active one, such as the kidney, in which rates of calcium excretion are, in part, metabolically tied to sodium intake.

Relation of Sodium to Calcium Absorption. One of the highest levels of calcium in primary feathers was found in Canada geese shot by Indians off the coast of Akimiski Island in James Bay (Figures *252* and *260*). We surmised early in the preparation of this report that a high intake of sodium was somehow metabolically related to high calcium levels, but no consistent relationship between these elements was at first apparent for other Canada goose populations or for blue, lesser snow, and Ross' geese. The reasons became clear. Blue, snow, and Ross' geese shun salt and brackish waters; some populations of Canada geese do not avoid such waters, and because they have highly active salt glands they are able to cope with relatively high rates of salt intake. The blue and lesser snow geese of Southampton Island, although exposed to a soil and mineral substrate of limestone, absorb only nominal amounts of calcium, as indicated by levels in their primary feathers (Figure *64*). The Canada geese of Akimiski Island, on the other hand, thrive in an ecosystem based on a limestone substrata but feed in tidal marshes on various sedges, particularly *Puccinellia*. As mentioned earlier, to accommodate a shift to salt water during the brooding season, the salt glands undergo marked enlargement, that coincides with the rapid growth stage of the goslings and the use of tidal marshes and tidal waters for feeding, resting, and escape (Figure *253*). The feathers of this population, as noted above, have the highest mean levels

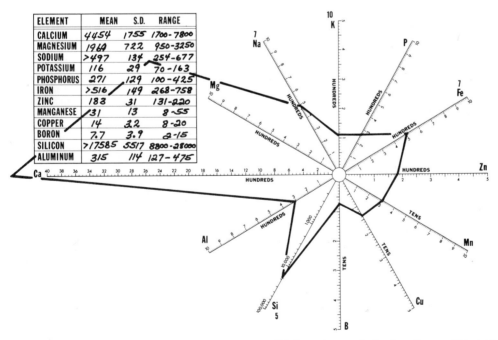

ELEMENT	MEAN	S.D.	RANGE
CALCIUM	4454	1755	1700-7800
MAGNESIUM	1969	722	950-3250
SODIUM	>497	137	254-677
POTASSIUM	116	29	70-163
PHOSPHORUS	271	129	100-425
IRON	>516	149	268-758
ZINC	183	31	131-220
MANGANESE	31	13	8-55
COPPER	14	2.2	8-20
BORON	7.7	3.9	2-15
SILICON	>17585	5517	8800-28000
ALUMINUM	315	114	127-475

260. Feather mineral pattern of Canada geese (n = 13) from the population breeding on Akimiski Island. Note extreme values for calcium, magnesium, and sodium.

of calcium of any geese studied; significantly, average sodium levels exceeded the upper limits of the sodium curve for which the optical spectrograph had been calibrated. With the exception of the Belcher Islands population (see appendix 2) comparable elemental responses could not be found in the feathers of other populations of Canada geese breeding on coastal areas and having varying degrees of contact with salt water (e.g., populations of coastal Newfoundland-Labrador; Newfoundland; Copper River delta, Alaska; coastal British Columbia; and the Queen Charlotte Islands), because the geological substrates of their feeding areas are of igneous rather than carbonate rocks. Under the discussion of the feather mineral patterns of Ross' geese, we pointed out the apparent relationship of birds with low sodium values to areas of basic rocks and birds with relatively high sodium values to areas of sodium-bearing marine clays.

The sodium-calcium relationship poses intriguing questions relating to the direct and indirect effects of sodium concentrations on the absorption and metabolism of other minerals. Feather mineral values indicate that sodium apparently stimulates increased absorption of calcium from foods as well as high excretion rates, possibly triggered in part initially by calcium loss from body pools. To what extend these events occur, in what sequence, and how they may affect the welfare of goose populations must be conjectured at this point. Obviously, the Ross' goose population that breeds in the Queen Maud Gulf lowlands is thriving, as evidenced by the increase in numbers from possibly as low as three thousand in 1949 to thirty thousand in 1965. Whether this population suffers a calcium deficiency stress during the breeding cycle might be indicated by a comparison of the relative osteoporosis, during the molt, of the tarsometatarsi of these geese with those of Canada geese collected on Akimiski Island. Coastal Canada goose populations on low-calcium soils probably compensate for calcium-depleting levels of sodium intake by the ingestion of snails and small bivalves (see discussion of phosphorus).

The physiological relationships between calcium and sodium appear to be profound and interdependent, affecting the rates of absorption and excretion. The sodium ion is required by almost all physiological transport systems; it is pertinent, therefore, that sodium appears to facilitate or is essential for the transport of calcium and phosphate ions and that levels of calcium absorption are directly affected by levels of sodium intake. High so-

dium levels are reported to inhibit calcium absorption, competition for a common binding site being one possible explanation (Wasserman and Taylor 1969:341–43) although, we suspect, an unlikely one. Conversely, calcium is essential for maintaining normal intestinal permeability. Studies of the rat intestine showed that the presence of the ion was requisite for the absorption of sodium, but that calcium in excess amounts inhibited the absorption of sodium (Wasserman and Taylor 1969:325–26). Thriving populations of geese probably seldom ingest diets as extreme in mineral content as are some media used in *in vitro* experiments. Mineral concentrations used in some experiments are so much in excess of levels animals have experienced in their life history as well as their evolutionary history, it is difficult to appreciate the relevance of the experiments. The level of sodium in brackish water ingested by Canada geese is probably effectively reduced before reaching the intestine by the rapid absorption of a large portion of the sodium in the lower portions of the esophagus and in the proventriculus and by sodium elimination via the salt glands.

Calcium-Sodium Relationships in the Ecosystems of Pheasants. In 1931 Leopold suggested that soil calcium might be the factor limiting the southward distribution of the ring-necked pheasant (*Phasianus colchicus*) in the United States (Leopold 1931:125). Since that time numerous field and pen studies have been carried out to resolve the question, but the findings have fallen short of being conclusive. In Illinois, as elsewhere, pheasants have thrived on the calcium-rich prairie soils of Wisconsinan drift origin, but pheasants introduced appreciably south of this drift have invariably failed to establish themselves even though production of chicks occurred in the first spring following the release of adult stocks. Despite intensive research on the failures of these adults and their progeny to survive, reasons for the failures have remained obscure. In our initial study reporting the relation of feather mineral levels in blue and snow geese to mineral levels in the ecosystem, we expressed the belief that "because the amount of a mineral in feathers appears to be an expression of the extent to which it occurs in excess of metabolic needs in the nutrient chain, feather calcium or its ratio to some other element or elements may provide new insight into the question of pheasant range limitations" (Hanson and Jones 1968:6). Our reevaluation of the data in Labisky et al. (1964), Jones et al. (1968), and Anderson and Stewart (1969, 1970, 1973) has confirmed our confidence that this approach would indicate some of the basic—and perhaps the most important—factors limiting pheasant range. Because these data on pheasants and mineral levels in their environments provide crucial support to our evaluation of the primary role played by sodium in ecosystems in relation to mineral metabolism in geese, we believe it is apropos to review here similar relationships in pheasants and the relation of mineral levels in the ecosystem to limitations of pheasant range.

Studies of the ring-necked pheasant in Illinois have been largely concentrated on two areas: 1) the Sibley area, an area in Ford and McLean counties on the Wisconsinan drift capable of sustaining high pheasant populations when crops and farming practices permit, and 2) the Neoga area, Cumberland County, an area on the Illinoian drift on which repeated efforts have been made to establish pheasant populations by means of releases of birds raised on state game farms.

A superficial understanding of the soils of these two areas is fundamental to an appreciation of how their ages and histories relate to their mineral contents and quantitative differences in the nutrient chains in their ecosystems. If the ecological and physiological interrelationships suggested have validity for pheasant populations, they in turn, lend support to our interpretation of ecological and physiological factors determining mineral levels in goose primary feathers.

"These two areas differ markedly in geomorphology, age, and nature of soils. The Sibley Area lies within that region of rolling topography formed by the Normal and Cropsey morainic system; about two-thirds of it lies on a gently undulating morainic ridge and the remainder on a broad, flat outwash apron. The till is calcareous; the Cropsey moraine, in La Salle County, has a *mean calcium carbonate content of 23 percent* [our italics] (Jones *et al.* 1966:366). The soils, which reflect the fine-textured nature of the glacial till, are silt loams or silty clay loams. The major soil series found on the Sibley Area are Elliott and Saybrook, both classified as Brunizem or Prairie soils, and Drummer, classified as a Humic-Glei soil. Elliott and Drummer occur on the moraine and Saybrook on the outwash apron. All of these soils are near neutral in pH in their surface horizons and have high natural productivity; most of the soils are tile drained. The age of the soils of the Sibley Area is probably about 16,000 years. However, a small amount of younger loess is incorporated in the surface horizons (Jones and Beavers 1963:439–440).

"The Neoga Area lies immediately south of the outwash apron bordering the Shelbyville moraine, the terminal moraine of Wisconsinan glaciation. . . . Topographic relief is slight at Neoga. The soils of the study area, being of pre-Wisconsinan origin, have had a longer and more complicated history of development than those of the Sibley Area. The till is calcareous only at depths greater than 3.4 meters. Two predominant soil series, Cisne and Ebert, are found at Neoga. Cisne, a Planosol, is characterized by a shallow clay layer that impedes drainage. Ebert, an intergrade soil having properties between a Planosol and Humic-Glei soil, is more poorly drained than Cisne. Both soils are moderately acid; the pH in the surface horizons ranges from 5.3 to 6.0 in Cisne soils and from 5.6 to 6.5 in Ebert soils. These soils developed in 1.2 meters of loess (Fehrenbacher et al. 1965:568) overlaying Illinoian till that has an ancient soil or paleosol in it. This paleosol is quite impermeable and restricts internal drainage. Thus, these soils, which have low natural productivity, are mostly surface drained. The Neoga soils may be 100,000 or more years old" (Jones et al. 1968).

Harper and Labisky (1964) and Labisky et al. (1964), on the basis of analyses of aggregate soil samples conducted by a commercial laboratory, concluded that Neoga soils were not inferior to Sibley soils in calcium content. In fact, values for the latter soils were lower (0.15 percent) than the average for Neoga soils (0.23 percent). They also concluded that the consumption of calcareous grit in the two pheasant populations was probably not significantly different. Nor were any differences in the sodium contents of the soils of these two areas reported. However, analyses of soils of these areas at the University of Illinois, which we believe to be more definitive, revealed that they did indeed differ significantly in respect to key elements. Levels of calcium, magnesium, and potassium in the Neoga soils were notably lower than they were in the Sibley soils, and sodium levels in the Neoga soils were significantly higher than they were at Sibley (Table 34). Calcium-sodium ratios underline the extent of the basic differences between these areas: Neoga, 11.5–1; Sibley, 18.5–1.

The question arose as to whether the food plants of the pheasants in these areas reflected soil differences. Jones, Labisky, and Anderson (1968) evaluated the question from analyses of foxtail grass (*Setaria faberi*) and corn (*Zea mays*), but their evaluation of element ratios considered only the classical couplets, calcium-magnesium and sodium-potassium. (We strongly suspect that prog-

34. Prevalence of selected elements (ppm) in soils of pheasant study areas in the vicinity of Neoga and Sibley, Illinois

Element	Neoga		Sibley	
	Mean	Range	Mean	Range
Ca	6,640 ± 598	3,600–9,900	8,910 ± 619	6,600–13,064
Mg	860 ± 72	540–1,240	1,874 ± 139	1,400–2,460
Na	8,150 ± 100	7,400–8,400	6,780 ± 140	5,900–7,600
K	16,210 ± 349	14,000–17,300	21,500 ± 531	19,500–24,700

Reference: Jones et al. (1968).

ress in mineral metabolism research has been slowed by failures to break with these traditionally studied element pairs.) Although the differences were not statistically significant, probably because of the variations in the soil types studied, sodium and potassium levels were higher and magnesium was significantly higher in Neoga corn (the major food of pheasants) than they were in the Sibley corn samples although potassium and magnesium were significantly lower in Neoga soils than they were in Sibley soils (Table 33). If sodium-calcium ratios in the corn are considered, however, differences between the two areas are emphasized: Neoga, 71/31 = 2.3; Sibley, 60/38 = 1.6. Mineral levels in foxtail differed significantly only in magnesium, being higher in the Neoga samples.

Significantly, Harper and Labisky (1964:726) reported that clover and grass leaves were found more frequently in the crops of the Sibley pheasants than in those from Neoga pheasants and that the calcium content of the Sibley forage was three times that of the clover and grasses ingested by the Neoga pheasants. Earthworms consumed by Sibley pheasants may also have contributed importantly to the total calcium intake of these birds.

It was suggested that the higher ratio of sodium to potassium found in the soil, foxtail, and feathers from the Neoga areas as compared with those of the Sibley area could imply nutritional imbalances (Jones et al. 1968). However, we believe that if a nutritional imbalance can explain the failure of pheasants to become established on the Neoga area, it is probably related to sodium-calcium ratios in the ecosystem, more specifically in the nutrient chain. For example, it is known that increases in exhangeable sodium in soils results in decreases in the calcium and potassium content of wheat (Pearson and Bernstein 1958; Schrenk 1964).

It can be inferred from our findings that levels of elements in primary feathers (Table 35) more effectively monitor the mineralogical status of the pheasants in the Sibley and Neoga areas than do other organs or tissues or more subtle and sophisticated physiological measurements. The effectiveness with which mineral levels of feather keratin indicate proportional changes in mineral absorption rates and small but significant shifts in mineral levels in the blood can be inferred from the extremely high correlation coefficients found between calcium, magnesium, sodium, and potassium in the primaries of pheasants from Neoga and Sibley (Table 35). The apparent inverse relation of sodium levels in the ecosystem and nutrient chain to the absorption of calcium, magnesium, and potassium is indicated by plottings of feather mineral values in Figures *261, 262,* and *263*. The ex-

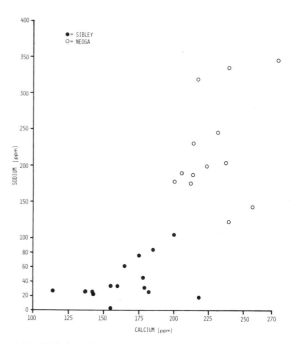

261. Relation of calcium to sodium levels in vane portions of primary feathers in populations of ring-necked pheasants from Sibley and Neoga, Illinois.

35. Ash weight and mineral content of femurs and primary feathers (in percent of ash) of ring-necked pheasants from Neoga and Sibley, Illinois

Element	Tissue	Age						Reference
		Four Months		Seven Months		Adult		
		Neoga	Sibley	Neoga	Sibley	Neoga	Sibley	
Ash Weight	Feathers			0.37	0.32			Jones et al. (1968)
	Feathers	1.28	1.45	1.34	1.48	1.28	1.28	Anderson and Stewart (1970)
	Femur			64.90	65.37	79.57	78.68	Anderson and Stewart (1973)
Ca	Feathers			5.59	5.25			Jones et al. (1968)
	Feathers	5.00	5.00	6.00	3.75	7.00	5.25	Anderson and Stewart (1970)
	Femur	27.25	35.50	29.00	33.00	33.75	30.00	Anderson and Stewart (1970)
	Femur			24.36	23.97	28.91	28.24	Anderson and Stewart (1973)
Mg	Feathers			3.00	2.59			Jones et al. (1968)
	Feathers	4.00	3.13	3.38	3.38	3.13	2.13	Anderson and Stewart (1970)
	Femur	0.50	0.45	0.55	0.45	0.50	0.50	Anderson and Stewart (1970)
	Femur			0.39	0.41	0.33	0.32	Anderson and Stewart (1973)
Na	Feathers			5.97	1.41			Jones et al. (1968)
	Feathers	6.00	6.50	5.13	4.50	6.25	5.38	Anderson and Stewart (1970)
	Femur	0.35	0.44	0.40	0.48	0.46	0.53	Anderson and Stewart (1970)
	Femur			0.60	0.53	0.60	0.61	Anderson and Stewart (1973)
K	Feathers			1.94	0.88			Jones et al. (1968)
	Feathers	3.50	4.50	2.50	2.50	2.63	3.50	Anderson and Stewart (1970)
	Femur	0.13	0.19	0.14	0.15	0.09	0.10	Anderson and Stewart (1970)
	Femur			0.21	0.19	0.16	0.13	Anderson and Stewart (1973)

tremely close correlation between sodium and potassium (Table 36, Figure 254) can be related to their close reciprocal relationship in cellular metabolism.

Because sodium levels in the diet appear to influence the absorption and excretion rates of calcium and magnesium, dietary intake and absorption rates of sodium must be taken into consideration in evaluating absorption and excretion rates of the other two minerals. The difference in average sodium levels in the feathers of the Neoga and Sibley pheasant populations is startling: Neoga 221 ppm versus Sibley 45 ppm, a nearly fivefold difference. This difference reflects significant differences in the soils and somewhat higher levels in Neoga corn (Jones et al. 1968).

Relation of Calcium and Sodium to Bone Composition. The relation of calcium levels in femurs of Illinois pheasants to levels in the soils is less consistent than in the case of feathers. Data published by Anderson and Stewart (1970) indicate that immature pheasants from Neoga have less calcium in the femurs than have the Sibley birds; and in the adult age class, the Neoga birds are higher in calcium (Table 34). More recent data (Anderson and Stewart 1973) indicate higher levels of calcium in both age classes of birds from Neoga than

are present in Sibley-area pheasants. Thus, despite lower levels of calcium in the soil of the Neoga area, both the feathers and femurs of the Neoga birds are higher in calcium than are those of the Sibley pheasants. But we believe these relationships may be deceptive. Anderson (1969:985) concluded from data on condition factors and organ weights that immature pheasants on poor and fair quality ranges were stressed less than were immature birds living on high-quality range. However true this may have been for pheasants that had reached physical maturity and for the times of observations, this assessment must be superficial in respect to actual mortality factors, as pheasant populations are short-lived in the Neoga area. Anderson (1969:986) suggested that Neoga pheasants on the Illinoian drift absorb less calcium than do Sibley birds. Yet an earlier study (Jones et al. 1968) showed a higher level of calcium in the feathers of the Neoga pheasants than in the Sibley birds, and similar results were obtained by the pooled averages of Anderson and Stewart (1970) (Table 35). As femur calcium in the Neoga birds was equal to or possibly higher than femur calcium in the Sibley birds, it must be concluded that calcium absorption per se was not lower in the Neoga birds; their systems merely worked harder to

36. Correlation coefficients for four elements in primary feathers of hen pheasants collected from poor (Neoga) and from favorable (Sibley) ranges in Illinois

	Ca	*Mg*	*Na*	*K*
Ca		.78	.79	.73
Mg			.85	.86
Na				.94

Notes: The correlations are based on twenty-eight observations; all are significantly different from zero at the 0.001 level. Calcium and magnesium were analyzed by atomic absorption; sodium and potassium levels were determined by flame emission spectroscopy.

achieve parity with birds on good range. Indicative of the latter is the finding that the parathyroids of the Neoga pheasants average 25 percent larger than those of the pheasants from Sibley (Anderson 1969). It thus appears that the Neoga birds are indeed stressed in relation to calcium deficiencies in the environment, their response being comparable to the parathyroid enlargement found in laying hens on a calcium-deficient diet (Bloom et al. 1960). And we strongly suspect that high sodium

levels in the Neoga ecosystem underlie or aggravate this stress.

The only investigator of the physiology of bone that we are aware of who has seriously concerned himself with sodium-calcium relationships is Dequeker (1972). His report supports our appraisal of the role of sodium in calcium metabolism: "The following hypothetical sequence of events could be put forward: sodium excretion with the concomitant calcium excretion leads to negative calcium balance and decreased serum calcium, as has been reported experimentally by other workers (Brickman et al., 1971); lowered serum calcium leads to increased parathyroid hormone secretion, with enhanced bone resorption and release of hydroxyproline. *These observations indicate that the proper interpretation of calcium metabolism changes cannot be made without knowledge of the simultaneous excretion of sodium. The study of the latter may be more important to explain calcium*

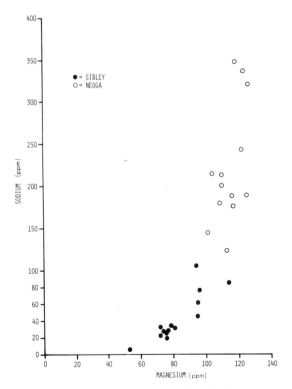

262. Relation of magnesium to sodium levels in vane portions of primary feathers in two populations of ring-necked pheasants from Sibley and Neoga, Illinois.

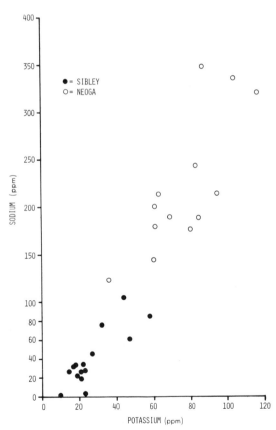

263. Relation of potassium to sodium levels in vane portions of primary feathers in populations of ring-necked pheasants from Sibley and Neoga, Illinois.

excretion variations than the intake and absorption of calcium'' (Dequeker 1972:187). (Our italics.)

Relation of Sodium to Calcium Metabolism in Anseriformes and Galliformes. The families Anatidae (ducks, geese, and swans) and Phasianidae (quails, pheasants, and peacocks) undoubtedly differ profoundly in ability to handle high levels of mineral intake, particularly sodium, and important species and subspecies differences also exist. The salt or supraorbital glands are probably not functional in Phasianidae (in any important sense), whereas, in the family Anatidae, species and races associated with salt or brackish water have greatly enlarged glands. For example, the Greenland race of the mallard (*Anas platyrhynchos conboschas*) has very large salt glands; the glands of the inland, continental races are not enlarged.

Krista et al. (1961) made a study of the effects of saline drinking water on chicks, laying hens, turkey poults, and ducklings, the results of which demonstrate differences in tolerance of salt water between families as well as an age-sex related differential in salt-water tolerance and mortality. White Leghorn chicks maintained on water containing 4,000 ppm of sodium increased their water intake, produced watery feces, and exhibited loss of appetite; at 12,000 ppm, the chicks experienced a marked rise in mortality, although some (10 to 15 percent) survived levels of up to 12,000 ppm. White Leghorn pullets increased their water consumption with increased salt levels in their water (4,000 ppm, 7,000 ppm, and 10,000 ppm). Egg production dropped significantly at 10,000 ppm, but body weight and appearance were not affected. Day-old turkey poults experienced a significant increase in mortality at a sodium level of 4,000 ppm in their water and a decrease in the rate of weight gain. Abnormalities occurred mainly at the 4,000 ppm, 7,000 ppm, and 8,000 ppm levels. In contrast, Rouen ducklings held on water containing less than 10,000 ppm of sodium suffered little or no mortality; at the latter level and above, all died within the first week. At all levels, however, there was a marked graded response in growth. Ducklings on high levels of salt (12,000 ppm) exhibited various nervous disorders—staggering gait and backward head bobbing.

We interpret the above results in this way. The high mortality of chicks using water containing 7,000 ppm of salt is indicative of the relative intolerance of Phasianidae to sodium; laying hens survived the highest sodium levels presumably only because calcitic grit (oyster shell?) was available to them and the laying mash was high in calcium.

Turkey poults were the least tolerant of sodium chloride added to their drinking water. Ducklings tolerated relatively high levels of sodium, presumably because they have at least partially functioning salt glands. They failed to grow normally because of a sodium-stimulated turnover of ingested calcium; erratic nervous movements at high levels of sodium intake are attributed to lack of sufficient calcium for normal functioning of the myoneural junctions.

Dr. Donald J. Bray of the University of Illinois, College of Agriculture, has related to us the case of a turkey grower west of Peoria, Illinois, who of necessity one year had to provide his flock with water containing relatively high amounts of sodium chloride. To compensate, he fed them a salt-free diet. In winter, they did well on this regime, but with the advent of hot weather, mortality was high—sudden deaths occurred from salt stress due to increased water consumption, and delayed mortality occurred due to the latent effects of disease.

Some of the immature pheasants released at Neoga are known to have survived until the following spring and produced clutches of normal size and shell thickness. The history of the pheasant releases at Neoga mirrors the experiments reviewed above in which some chicks survived high salt intake and the pullets produced eggs. The conclusion seems clear: the failure of pheasant releases to succeed at Neoga was due primarily to poor growth and high chick mortality related to relatively high levels of sodium in the ecosystem.

Renal Secretion of Calcium. "A clear-cut relationship between salt intake and renal excretion of calcium in man has been known since 1905" (Walser 1969:255). (Our italics.) If dietary calcium is held constant, calcium excretion parallels changes in sodium excretion, the clearance of these ions being linear in both the dog and man. It appears that when the excretion of sodium is high, the excretion of calcium is also high (Walser 1969:255). Conversely, in man, a greater reduction in calcium output via the kidney is achieved by restricting salt intake than by lowering calcium intake (Walser 1969:290). These relationships, a consideration in the treatment of kidney stones, again point out the effect of sodium intake on calcium turnover and parallel our findings on the relationship of the content of these elements in feathers to their prevalance in the ecosystem. The parathyroid gland may be involved, as an intravenous injection of hypertonic solution of saline causes greater excretion of calcium in hyperparathyroid patients than it causes in normal patients (Walser

1969:255). Also, hyperparathyroidism, per se, increases urinary calcium excretion and is associated with a negative calcium balance in man (Spencer et al. 1973:715). The ring-necked pheasant populations studied at Neoga, Illinois, as discussed above, are, because of their enlarged parathyroid glands, in a metabolic state analogous to the hyperparathyroid patient. Perhaps the most salient feature of this relationship between ions is that the calcium secreted via the kidney is endogenous calcium (Jaffe 1972:112).

Present knowledge of renal mineral secretion in birds is scanty; Sykes (1971:270–72) advises the use of basic textbooks and reviews to find detailed information on principles. However, our knowledge of mineral excretion is not as scanty for birds in general, as indicated in the recent review by Shoemaker (1972), but even this relatively detailed review is concerned mainly with the mechanisms of secretion and resorption of sodium, potassium, and chlorides. The salt glands are also known to secrete potassium and relatively minute amounts of calcium and magnesium, but data on renal calcium secretion are lacking (Shoemaker 1972:557). In man when no calcium is present in the diet, 100 to 200 mg of calcium are secreted daily in the urine and 300 to 400 mg in the feces. Thus, renal secretion of calcium, in a "basal metabolic" state with respect to calcium intake, presumably ranges between one-fourth and two-thirds of the quantity lost via the digestive tract.

In the domestic fowl notable increases in the excretion levels of calcium by the kidneys (amounts not stated) occur following the laying of each egg (Sykes 1971:270). This loss, presumably the result of a time lag in homeostatic control mechanisms, must be regarded as wastage. An extreme example of the effects of high calcium demands is seen in laying chickens placed on a calcium-deficient diet; in two weeks as much as 40 percent of the calcium may be depleted from the skeleton (Hazelwood 1972:496). As the wild ring-necked pheasant produces an average of twenty eggs in a season (Labisky 1968), the seasonal calcium requirements to meet egg shell needs and post-ovulation losses must be considerable. Pheasants are obviously well able to meet this demand when living on high-quality soils, but for pheasants situated on soils with low calcium values *combined* with sodium concentrations in ecosystems well above metabolic needs, calcium depletion per se may well be a very real survival factor. This depletion probably has no immediate effect on the survival of the individual in the wild but is apparently gradual and insidious in its effect on pheasant populations, particularly for the chicks hatched in marginal ecosystems in respect to calcium-sodium ratios. To sum up, we believe that sodium plays an important and heretofore overlooked role in calcium, and probably magnesium, metabolism in birds and that sodium levels, as they relate to calcium in soils and in the nutrient chain, may explain in part the failure of pheasant populations to become established on Illinoian drift areas.

Calcium-Barium Relationships. Anderson and Stewart (1973) were concerned that calcium levels at Neoga were insufficient, in effect, to suppress the uptake of toxic levels of barium, the one trace element that was found to characterize calcitic grit particles in the Neoga soils. Our findings for geese provide little indirect support for a simple inverse relationship between calcium levels and the absorption of trace metals. It is more likely that barium levels in pheasant tissues directly reflect barium levels in the nutrient chain. Anderson and Stewart (1973) did not speculate as to the metabolic pathways by which barium may achieve toxicity, but it is known that ingested barium becomes concentrated almost exclusively in bones and is "affected by the same conditions which affect calcium metabolism" (Bauer et al. 1961:638–639). It is cleared from the plasma through skeletal accretion at twice the rate of calcium, and its excretory clearance rate is considerably higher than that of calcium (Bauer et al. 1961). These findings are in agreement with our findings on the metabolic fate of other nonessential trace elements: they are easily absorbed and rapidly excreted (see chapter 8). However, the fact that barium can be a proxy for calcium at the neuromuscular junction and in the neurohypophyseal secretion of vasopressin (Rodan 1973:192–94) is indicative that barium can enter into a variety of complex physiological events.

Osteoporosis, Calcium Balance, and Sodium Intake. The development of a temporary osteoporosis in the tarsometatarsi of birds during molt was established by Meister (1955). In Canada geese it is a particularly well developed phenomenon in adult females as compared to adult males and nonbreeding yearlings (Hanson 1958; Hanson unpublished). The development of osteoporosis in the tarsometatarsi during molt suggests that geese are in a state of negative calcium balance at that time. Unfortunately, no studies of calcium balance in geese during this period have been carried out, but because of mineral deposition in feathers associated with their growth, studies of mineral balance in geese would be complex and difficult. In view of this gap in our knowledge and the fact that osteoporosis is a phenomenon common to both geese and humans, the literature on human osteoporosis is of particular interest.

Defects in calcium absorption, reduced calcium intake, and increased resorption of bone are among the many factors frequently cited as contributing to this complex, ill-defined disease regarded as an exaggeration of a process occurring normally after middle age (Irving 1973:125). Although clinicians have reported (personal communications; Urist 1969:441) that patients with slowly developing osteoporosis can compensate for sodium-induced losses of calcium by increased absorption of calcium and that they do not go into a state of negative calcium balance, it should be noted that the osteoporotic patient loses approximately 300 mg of calcium per day in the urine versus 150 mg for the normal individual (Hoffman 1970:609). Also, other investigators have stated on the basis of large-scale radiographic studies of bone loss that, regardless of calcium intake, 30 mg of calcium are lost per day after the age of forty (Garn et al. 1967). In view of the higher rates of renal calcium excretion established for osteoporotic patients, the well-documented stimulatory effect of sodium on the renal secretion of calcium, and the apparent existence of a negative calcium balance after the age of forty (scarcely detectable by short-term balance studies using presently available techniques), it is surprising that the effect on calcium losses in man of the long-term use of excess salt in the diet perhaps has not yet been satisfactorily explored. Sodium-calcium relationships are not discussed in major reviews of bone physiology and osteoporosis, nor is a low or conservative salt intake suggested as a possible calcium-conserving measure for osteoporotic patients (Wolstenholme and O'Connor 1956; McLean and Urist 1961; Frost 1966; Collins 1966; Jackson 1967; Jowsey and Gordan 1970; Fourman and Royer 1968; Barzel 1970; Harris and Heaney 1970; Vaughan 1970; Nichols and Wasserman 1971; Hopps and Cannon 1972; Jaffe 1972; Irving 1973:125–28; Sognnaes 1960; Zipkin 1973).

There is, however, one notable exception to the lack of concern with sodium-calcium relationships in human osteoporosis in the literature of bone physiology. Dequeker (1972) expresses precisely the same inferences that we have drawn from the findings on feather minerals about the effect of sodium on bone resorption: "Calcium deficiency can also be due to calcium loss in the urine: primary or

secondary hypercalciuria. Dietary factors such as increased salt intake or carbohydrate may cause a secondary hypercalciuria. Such a possible cause of calcium deficiency resulting in lowered serum calcium, increased parathyroid excretion and enhanced bone resorption *has never received much attention. This mechanism might prove to be more important than calcium lack in the diet* [our italics]. In support of this hypothesis is the observation that negro populations [presumably in Africa] who consume less salt have a low incidence of symptomatic osteoporosis and that in alcoholics, an increased rate of osteoporosis has been found (Saville 1965; Dent and Watson 1966" (Dequeker 1972).

We believe that the remarkably similar conclusions reached by Dequeker and by us from unrelated points of inquiry suggest the possible long-term benefits of moderation in salt consumption by persons over fifty years of age—particularly by women, in whom the rate of diagnosed osteoporosis exceeds that in men by at least the ratio of three or four to one. However, Collins (1966:114) has emphasized that a sex differential in the prevalence of this bone disorder may not, in fact, exist. The need for intensive research on sodium-calcium relationships in many disciplines seems manifest.

MAGNESIUM

The origins of most of the magnesium in the ecosystems to which the blue and snow geese of the Hudson Bay area are exposed are seawater, carbonate rock sources, and areas of basic rocks (Table 33). The availability of magnesium, occurring in the illite and chlorite clays of the glacial and littoral sediments, to the geese ingesting these clays is indeterminate at this time. (Earlier, in chapter 4, we hypothesized that the magnesium in the bentonite to which the lesser snow geese at the Anderson River delta are exposed accounts for the high magnesium observed in the feathers of these geese.) Magnesium occurs in seawater to the extent of 1.27 percent by weight, dolomites contain 13.1 percent magnesium, and magnesium in the clays ranges from barely detectable levels to about 8 percent.

Magnesium is the only metal in the chlorophyll molecule of plants, amounting to 2.7 percent by weight but accounting for about 10 percent of leaf magnesium (Nason and McElroy 1963:511). Magnesium also occurs in bound and ionized conditions in plant cell protoplasm, and there is a striking organization of the element to seed during seed formation and filling stages. A number of enzyme reactions (Nason and McElroy 1963:512) are activated by magnesium, particularly those involved in group transfer and specifically those incorporating phosphate.

Studies indicating antagonistic action between calcium and magnesium have been briefly discussed in the subsection on calcium. A similar reciprocal relationship of absorption rates between magnesium and potassium is reported to occur in plants (Epstein 1972:299–300).

Absorption areas of the intestines vary somewhat with species (Tansy 1971:194). No studies indicating the primary segment of the intestine involved in magnesium absorption appear to have been carried out for birds, but, as in the case of calcium, the principal area of magnesium absorption can reasonably be assumed to be the jejunum.

Inferential evidence suggestive that feather concentrations of calcium and magnesium reflect both their independent absorption on different sites of the mucosal cells of the intestine and thus, to a degree, the relative prevalence of these ions in the environment is buttressed by observations that the active uptake of magnesium by plant cells occurs at sites on the plasma membrane apparently distinct from those involved in the uptake of calcium, barium, and strontium (Sutcliffe 1962:60; also see Chapman 1960:302). The mainly extra- and intracellular functions of calcium and magnesium, respectively, would also seem to be a priori reasons why these elements might be independently absorbed. Hughes (1974:94–95) has pointed out that the catalytic powers of these elements are quite different and that "it is possible to explain the differing catalytic efficiencies of Mg^{2+} and Ca^{2+} purely in terms of their preferences for different binding sites (i.e. binding groups and site symmetry). This seems a more reasonable explanation."

Knowledge of the regulation of the absorption of magnesium in mammals is still incomplete, and the results of much research are regarded as equivocal (Tansy 1971:193–94). "For the most part, it is still uncertain whether the absorption of magnesium is regulated by the needs of the body or absorbed independently of body requirements" (Tansy 1971:194). We believe that magnesium excretion, and in turn, absorption rates, are affected similarly to calcium excretion rates by variations of sodium intake, but according to Walser (1969:255) the relationship between these ele-

ments is not clear. Texter et al. (1971:392) state that "the major regulator of absorption is the quantity of magnesium in the intestinal lumen rather than the nutritional requirements of the animal." But these authors also note that "sodium uptake by the intestinal segment increases the passage of magnesium into the bowel" (1971:392). *In vitro* studies of rat small intestine have shown that magnesium uptake was dependent on sodium concentrations in the bath and it has been suggested that grass tetany in cattle may be related to sodium intake (Tansy 1971:195).

In attempts to deal with the problem of reciprocal relationships between the absorption of calcium and other elements, studies of colostomized hens revealed that an increase of calcium absorption on shell-forming days was accompanied by increases in the rates of absorption of magnesium, sodium, potassium, and phosphorus. The above findings are in accord with our concept that under conditions of *normal* dietary intake, the rates of absorption of metal ions are positively interrelated, reflecting their abundance in the nutrient chain of the ecosystem (see discussion of iron in relation to the uptake of related ions) rather than inversely related to each other. In general, our findings do not support the idea of a common, and therefore competitive, mechanism for calcium and magnesium transport, as has been suggested for the rat but refuted by other data (Tansy 1971:196). With reservations as to the role of sodium, our findings do not conflict with Wacker and Vallee's (1964:494) statement that "no factor is known which affects the absorption of magnesium as vitamin D does that of calcium."

Magnesium values varied fourfold in the feathers of geese from the various breeding colonies, but average values were notably higher for the western colonies (Table 26). Coefficients of variability were significantly lower for values from the western colonies (Table 27). Only zinc and copper exhibited appreciably lower variation (Table 27).

Magnesium was more frequently correlated in primary feathers with other elements than was any of the other minerals (Table 28), particularly with calcium, potassium, iron, manganese, boron, silicon, and aluminum (Table 32). The close association of magnesium levels with calcium levels may reflect in part a consistency of ratios of these elements in plants in areas of similar geologies, as is the case for the eastern colony sites of blue and lesser snow geese, rather than physiological control of absorption in the intestine. In the rat, magnesium is reported, under normal dietary intake, to promote calcium absorption (Tansy 1971:196).

Calcium and magnesium ratios in primary feathers of geese, considered this time from the standpoint of magnesium, offer constructive insights—at least during the physiologically demanding period of feather growth—into the question as to whether significant reciprocal relationships exist between the absorptive rates of these elements. Generally, the ratio of calcium to magnesium in the feathers of blue, lesser snow, and Ross' geese varied between 1.5 and 3.5 to 1, but in the case of snow geese collected on Banks Island or bagged in Utah and in the Imperial Valley of California, presumably all representing the same population, this ratio was as high as 14 to 1, 24 to 1, and 8 to 1, respectively (Figures *100, 207, 240*). But more important and enlightening, as well as exhibiting ''the reverse of the coin,'' are data for Canada geese, notably the large race breeding in central and southern British Columbia (Hanson unpublished). Two samples of primaries of geese from this area had extremely high magnesium values (average 1,500 ppm). For most of

these geese, the calcium-magnesium ratio averaged 1 to 1, but for some individuals magnesium values exceeded calcium values by ratios of between 2.1 and 4.0 to 1 (Figure *264*). Presumably, soils and clays derived in part from dolomites and dolomitic rocks of the region (Douglas 1970) explain the extreme levels of magnesium found in the flight feathers of these geese. These data again point to the extreme gradients in the mineral environments with which geese must contend, further emphasizing the role of excretion in achieving homeostasis.

The Canada geese on Akimiski Island (where the clay soils are underlain by Paleozoic limestones) have available to them in the nutrient chain large enough concentrations of calcium and magnesium to exceed excretory losses stimulated by high sodium intake. As a result, the high gradients of these three elements with which these geese must contend is reflected in high rates of turnover of these elements in the circulatory system.

In accordance with our data for geese, which suggested that high levels of sodium intake may tend to increase the rate of magnesium absorption, the Neoga, Illinois, pheasants also exhibited sig-

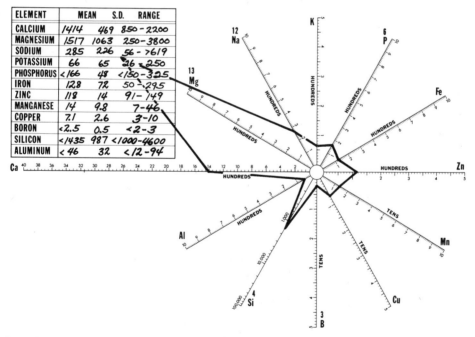

ELEMENT	MEAN	S.D.	RANGE
CALCIUM	1414	469	850-2200
MAGNESIUM	1517	1063	250-3800
SODIUM	285	226	56->619
POTASSIUM	66	65	26-250
PHOSPHORUS	<166	48	<150-325
IRON	128	72	50->295
ZINC	118	14	91-149
MANGANESE	14	9.8	7-46
COPPER	7.1	2.6	3-10
BORON	<2.5	0.5	<2-3
SILICON	<1435	987	<1000-4600
ALUMINUM	<46	32	<12-94

264. Feather mineral pattern for Canada geese (n = 14) from the population breeding in central British Columbia. Note extremely high magnesium value and low calcium value. This extreme skewing of the calcium/magnesium ratio in the feathers of these geese is believed to represent a response to dolomitic strata in the region.

nificantly higher levels of magnesium in their primary feathers than did the Sibley pheasants (Jones et al. 1968). This finding points to a higher sodium-related rate of turnover of magnesium in the Neoga pheasants than in Sibley pheasants (Figure 262 and Table 36), as in the case of calcium, because magnesium levels in the soils of the two areas do not differ significantly (Table 34; Jones et al. 1968).

In summary, therefore, it can be reasonably concluded that environmental factors, especially sodium intake, rather than physiological controls are determining factors influencing the magnesium content of the primary feathers and that regulation of excretion rather than absorption is important in aiding the tissues to achieve their rigorously controlled homeostasis. These conclusions are supported by Tansy's (1971:197) conclusion, drawn from his appraisal of the literature, that "the apparent lack of endogenous regulation of magnesium absorption under normal dietary conditions is emphasized."

SODIUM

Sodium has been rated as the third or fourth most abundant element in crustal rocks. Reflecting its ionic potential, it is the element most readily leached from soils—this characteristic contributing to the salinity of the oceans. Sodium is particularly abundant in acid igneous rocks, ranging from 2.6 to 2.9 percent in sodium content, as compared with basic rocks, which range from 1.7 to 2.1 percent in sodium content (Table 33). Shales tend to be less than 1.0 percent sodium, and carbonate rocks range from 400 ppm up to about 2.0 percent, partially in response to clay content. Metamorphic rocks of the Canadian Shield reflect the contents of their igneous rock precursors. Seventeen samples of glacial clays in Illinois from the Wisconsin Pleistocene deposits, ultimately derived from Paleozoic sediments, averaged 0.98 percent sodium (White 1959).

Sodium is present in varying degrees in most species of plants; it is requisite for growth in some and accumulated by others (Gauch 1972:242; Epstein 1972:57). Values for plants collected in the course of field work for this study are given in Table 8. There is some evidence that sodium is essential to halophytes, e.g., salt bush (*Atriplex vesicaria*) and *Halogeton glomeratus*. Under circumstances of low potassium supply, sodium assumes a supplementary role to the biochemical functions of potassium, and positive dry-matter responses

have been observed in experiments where sodium has been substituted for potassium, especially in root crops and small grains (Nason and McElroy 1963:319). Increases in succulence have been noted where sodium has replaced potassium (Hewitt 1963: 183–84), but the effects of chloride ions or the ratio of absorbed/free ions in plant cells may also affect succulence (Chapman 1960: 300–302).

Sodium differs from other cations by the extreme rapidity with which it is absorbed. In the chicken, selective absorption and active transport from the crop has been demonstrated (Hill 1971:43); in man some absorption takes place in the stomach (Oser 1965:348). Homeostatic mechanisms for this ion are necessarily highly efficient. Excretion of sodium in the chicken is via the kidney, but, as discussed extensively above, in birds exposed to a saline aquatic environment—sea ducks, brant, and Canada geese—large amounts of sodium are excreted by way of the supraorbital or salt glands.

Variations in average sodium values in the feathers studied were threefold, but values for the Hudson Bay colonies, which averaged higher than the western Arctic colonies, did not differ significantly from each other (Table 26). In the western colonies, sodium levels differed significantly in four intercolony comparisons (Table 31). Highest mean sodium levels were found in the Ross' geese from Karrak Lake (Table 26). We indicated earlier that the high intake of sodium by these geese was probably related to their feeding in an area of sodium-enriched marine clays.

All of the nesting areas of the Hudson Bay colonies of blue and lesser snow geese are coastal, and the immediately surrounding vegetation is subject to occasional flooding from wind-driven tides. However, ice cover in Hudson Bay during the nesting season would, presumably, protect the coast against wind tides during the actual nesting period. The plants adapted to the salinity of these coastal salt marshes are, in varying degrees, high in sodium (Table 8), but the tendency of most blue and lesser snow geese to feed away from the coast reduces their consumption of plants exposed to brackish conditions. On the other hand, Canada geese on Akimiski Island in James Bay, even after achieving the power of flight following the molt, seek out stands of *Puccinellia,* which are subject to daily tidal flooding. Thus, it is not surprising that the primaries of some individuals of this population exceed 600 ppm of sodium and that the mean for thirteen birds is somewhat greater than 497 ppm. Values of a few primary feathers exceeded values for analytical calibration (Figure *260*). (Also see appendix 1 in regard to high sodium levels in the primaries of Canada geese from the Belcher Islands, N.W.T.)

Sodium correlations with other minerals occurred notably fewer times than did other intermineral correlations (Table 30), but it was significantly paired with potassium in five of the seven colonies for which data approached adequacy (Table 32). The frequent association of these elements may reflect a relative constancy of the ratio of these elements in marshes, regardless of their relative degrees of salinity.

Critical regulation of circulating levels of sodium and potassium is requisite. Potassium and sodium levels in the feathers of blue and lesser snow and Ross' geese varied only twofold and threefold, respectively; these minimal variations indicate that ecosystem levels of these minerals play a secondary role in determining their levels in feathers. Presumably, the salt glands play some auxillary part in the excretion of sodium in blue and snow geese, but the fact that these glands undergo great seasonal enlargement in Canada geese exposed in summer to brackish water (Figures *252–54*), whereas in blue, lesser snow, and Ross' geese they do not, suggests that there may be important species and genera differences in the relative role of the salt glands in sodium metabolism.

POTASSIUM

Potassium, as an element, constitutes from 0.5 to 3.0 percent of the igneous crustal rocks, 1.7 to 2.9 percent of shales, and 0.27 to 0.29 percent of limestones (Table 33). A significant feature of the composition of limestones is the dominance of potassium over sodium (Mason 1966:154). Potassium has a lower ionic potential than sodium has, and thus it is more readily adsorbed by clays than is sodium, which tends to remain in solution (Mason 1966:162, 179). Potassium from seawater is incorporated in certain clay mineral species carried to the sea by rivers; these clays are illitic (micaceous) and are important in the geochemical cycling of potassium.

The specific biochemical roles of potassium in plants are uncertain (Nason and McElroy 1963:512–15) although potassium is definitely associated with carbohydrate and protein synthesis. Relatively large amounts of potassium are required by most higher plants, apparently because relatively high concentrations are required within the

plant cell to activate numerous enzyme systems (Gauch 1972:213). There are, nevertheless, moderately wide variations in the potassium contents of domestic plant foods (Altman and Dittmer 1968) and in native plants in the northern parts of North America (Table 8). This element is readily translocated and tends to be concentrated in meristematic tissues. Potassium probably forms coordinated complexes with some enzymes and is important in providing for conformational characteristics (Evans and Sorger 1966), a substantial amount of potassium being needed in stoichiometric terms as compared with the alkaline earths or the transition metals. The ability of sodium to stimulate growth at low potassium levels in culture mediums has been referred to above.

In man, potassium absorption occurs in the small intestine (Jackson and Smyth 1971); presumably, the same section of the gut is chiefly involved in birds. Excretion of potassium by the gut takes place primarily in the colon.

Although potassium levels in the primary feathers of geese varied from 30 to 197 ppm, means for the various colonies fell within a remarkably limited range of from 30 to 65 ppm (Table 26). It is possible that the moderately high value for the Kendall Island colony is not representative of free-flying birds from this population, as the primaries analyzed were from flightless birds collected on the nearby mainland. As pointed out in the introductory discussions, the sheath material surrounding the growing feather is high in potassium; clippings of the "unfurled" feather tips apparently retain some of this material despite vigorous washing. However, the high potassium values found for plants collected near Churchill, Manitoba (Table 8), indicate that extreme values for clippings of partly grown primaries of immature and adult blue and lesser snow geese from Churchill (Figure *93*) are based on ecosystem realities. High values for feathers of immatures—but not adults—were also found in the McConnell River colony. In contrast, potassium values for the partially grown primaries of flightless geese from Cape Henrietta Maria and Anderson River are within the range of values for fully fledged geese.

In geese, potassium was, quantitatively, most frequently correlated with calcium, magnesium, and sodium (Table 28). These findings are counterparts of the remarkably high correlations between these elements in pheasant primaries cited in the discussion of calcium (Table 36). The low variability of potassium between colonies is in keeping with its crucial role in cellular metabolism and nerve function and is evidence of fairly stringent homeostatic controls—particularly absorption.

It might be assumed that the correlation of potassium values with those of calcium and magnesium in feathers is primarily an expression of the regional abundance of potassium in the native clays; however, the physiological roles of this element are so vital that it is more reasonable to suspect that some degree of "proportional homeostasis" may exist in its relation to these other elements. We believe that sodium is the dominant ion affecting the levels of absorption and excretion of the other three.

PHOSPHORUS

Igneous rocks contain from 1,200 to 1,600 ppm of phosphorus, shales 740 to 2,000 ppm, sandstones 170 ppm, carbonates 350 to 400 ppm, and glacial clays 960 ppm (Table 33) (Mason 1966:43, 180). The role of this element in the ecosystem is related to its availability to plants and animals. Aluminum and iron compounds of phosphorus in acid soils are not readily available to plants; hence, soils characterized by these forms of phosphorus may be relatively infertile for grazing animals. Soils of Arctic and Subarctic regions—from which blue and snow and most races of Canada geese originate—are generally low in available phosphates, a deficiency reflected in generally low phosphorus concentrations in primaries of geese. Mammals grazing on soils deficient in phosphates eat the soils themselves (Irving 1964:250); this habit at least raises the question as to whether the significance of soil ingestion, or geophagy, by geese is understood. The dietary intake of phosphorus for most geese in the North must be adequate, as evidenced by flourishing populations, but one can also reasonably conjecture that in most northern areas the abundance of phosphorus in soils is marginal.

Phosphorus is absorbed by plants mainly as the orthophosphate ion, PO_4^{3-}. Northern plants may have evolved the ability to overcome low levels of phosphorus in the soils by increasing the efficiency of their root systems in extracting available phosphates (Chapin 1974; also see Scaife 1974 and rebuttal by Chapin 1974 [*Science* 186:847]). Plant cytoplasm levels of phosphorus are extraordinarily high compared with those in soil solution and reflect the active and efficient nature of roots in exploring the soil for this element. In vascular plants phosphorus occurs as the inorganic form in large quantities and as phytic acid, phospholipids, and phosphorylated sugars (Nason and McElroy

1963:508–9), and, of course, phosphorus is related to every synthetic process in the cell through adenosine triphosphate (ATP). The element is rapidly translocated within the plant to meristematic tissue, where it occurs in abundance and is associated with energy transfer, respiration, and photosynthesis.

Phosphorus is notably accumulated in the bodies of animals, where it is requisite for the most basic metabolic and structural functions. Bone constitutes a major reservoir of phosphorus, but soft tissues must also be regarded as a functional reserve, one that is more labile than bone (Irving 1964:257). The homeostasis of phosphorus must, therefore, be achieved in relation to pools in both kinds of tissues.

Although phosphorus is not an important component of egg shells (Romanoff and Romanoff 1949:355–56), high turnover of calcium, particularly the transient medulary bone of the tarsometatarsi laid down during egg laying, results in wastage of phosphorus via the kidney (Sykes 1971:273). Phosphorus is the primary mineral of egg yolk in which it is complexed as a lipophospho-protein, but it is a relatively insignificant element of the albumin (Gilbert 1971:1388–92). The phosphorus demands of egg laying, soon followed by the dramatic reorganization of muscle tissue during the molt with its accompanying osteoporosis of the cortex of the tarsometatarsus, would suggest a state of phosphorus deficiency during this stressful period.

Absorption of phosphorus in mammals occurs at numerous locations in the intestine, primarily the upper portions. Homeostasis is achieved mainly by regulation of excretion via the kidney, the parathyroid exerting a major control in this respect (Irving 1964:257).

Our evaluation of the phosphorus content of feathers was handicapped by incomplete analyses, i.e., the inability to obtain determinations below the lower limit of the operational curve or settings of the optical spectrograph (150 to 200 ppm, see chapter 2). Thus, values of less than 200 ppm in Table 26 have limited meaning. However, it is reasonable to conclude that actual values probably varied no more than fourfold and very likely not over threefold.

Phosphorus values for feathers of geese from western colonies were higher than those from the Hudson Bay colonies, a finding in accord with low phosphorus levels in carbonate rock. However, some of these differences can presumably be attributed to the enriching influence of river sediments, both organic and glacial clays derived from igneous rocks of the Cretaceous interior (e.g., Anderson River and Kendall Island colonies), and to the consumption of invertebrates by geese.

Coefficients of variability for phosphorus were higher for the Hudson Bay colonies than they were for those in the western Arctic (Table 27). This relationship occurs in part because many values for the eastern colonies were below the detectable limits. However, relatively high values for the Baffin Island stocks of geese can probably be related to the high phosphorus content of some plants on the Bluegoose Prairie (Table 8). Of the macroelements, phosphorus levels were least frequently correlated with those of other elements (Tables 28, 31), a finding perhaps indicative that phosphorus absorption is largely independent of the absorption of other ions and closely regulated by metabolic needs. There is a vast literature on calcium-phosphorus relationships' effect on absorption of imbalances in diets, but our findings are more in accord, as indicated above, with Irving's (1973:44) statement: "It may be concluded on the balance of the evidence that quite large increases in ingestion, either as phosphoric acid or neutral phosphate, do not affect calcium metabolism in man to any significant extent."

Phosphorus values for most populations of Canada geese were similar to or below those for blue, lesser snow, and Ross' geese, but were relatively high for the populations breeding and wintering along the Pacific coasts of British Columbia and Alaska south of Yakutat Bay. Phosphorus levels in sixteen individuals from this area ranged from 350 to 750 ppm and averaged 522 ppm (Figure 265). This maritime population of Canada geese is particularly known for its inclusion of littoral animals in its diet (e.g., the snail, *Littorina sitkana*) (Delacour 1954:170; and personal communications to Hanson by local residents). Hence these coastal geese can probably be distinguished from other inland populations of western Canada geese on the basis of high levels of phosphorus alone; presumably, the high content of invertebrates in the diets of the coastal geese is sufficient to account for the high phosphorus levels in their primaries.

It is of passing interest that, in conjunction with the consumption of invertebrates by coastal populations of Canada geese, lipids may also serve to

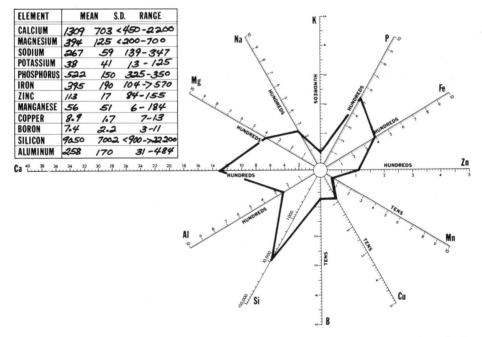

ELEMENT	MEAN	S.D.	RANGE
CALCIUM	1309	703	<450-2200
MAGNESIUM	394	125	<200-700
SODIUM	267	59	139-347
POTASSIUM	38	41	13-125
PHOSPHORUS	522	150	325-350
IRON	395	190	104→570
ZINC	113	17	84-155
MANGANESE	56	51	6-184
COPPER	8.9	1.7	7-13
BORON	7.4	2.2	3-11
SILICON	9050	7002	<900->22200
ALUMINUM	258	170	31-484

265. Feather mineral pattern of Canada geese (n = 16) from the population breeding along the Pacific coast of the Alaska panhandle. Note high phosphorus level that is believed to be a response related in part to a relatively high consumption of small marine mollusks and other marine invertebrates.

identify these populations. In the course of tax-onomic studies on Canada geese it was noted that most specimens of these large coastal geese had a gummy exudate over the surface of their tarsome-tatarsi and feet, obviously of a fatty nature. The most likely explanation for the presence of this mobile fat is that it is composed of unsaturated fatty acids derived from the littoral invertebrates which are included in the diet of these geese. Fats of marine invertebrates are characterized by a wide range of unsaturated fatty acids, and "the deposi-tion of the marine type of fat by marine birds is fur-ther evidence of the influence of the dietary fat" (Shortland 1961:695, 697). Fatty exudates are sel-dom seen on museum skins of inland races of Can-ada geese that feed on leaves and seeds of wild and domestic plants although their tarsometatarsi and other bones may vary in color from yellowish to yellowish-brown (after degreasing and drying) as a result of the amounts and kinds of fat incorporated in the marrow cavity and in the bone tissue.

If accurate determinations were made of phos-phorus levels, significant age differences would probably be found, higher levels occurring in im-matures (six to nine months old) than in older geese. Earlier studies of inorganic phosphorus in the plasma of Canada geese wintering in southern Illinois showed that the phosphorus level was sig-nificantly higher in young birds (five to eight months old) than in yearlings (seventeen to twenty months old) or in adults (Hanson 1958; Hanson un-published).

In summary, a subjective appraisal of phos-phorus metabolism in geese is that levels in feathers generally reflect strict metabolic control of homeostasis via excretory routes. Only in popula-tions of geese having high phosphorus intake due to the consumption of the invertebrates or in those feeding on the relatively rich deltaic soils (Ander-son River and Kendall Island populations) can eco-logical factors be shown to effect relatively high levels of phosphorus in the primary feathers.

IRON

Iron ranks third in abundance of the elements in crustal rocks (Mason 1966:42–43). The two oxida-tive states of iron constitute 5.0 percent of crustal rocks ranging from 1.4 to 2.9 percent in acid rocks to 7.2 to 9.2 percent in basic rocks (Table 33). Carbonate rocks contain much less, from 0.38 to 0.90 percent, and shales contain from 3.2 to 4.7 percent.

In evaluating the ranges of wild geese in respect to the mineral content of their primaries, the enriched states of glacial and marine clays, in con-trast to most limestones, should be noted, but in-sofar as rock types are concerned, a basic principle stated by Mason (1966:179) should be borne in mind: "The data indicate that the common sedi-mentary rocks seldom show marked enrichment in minor or trace elements over the amounts in ig-neous rocks—in fact sandstones and carbonates are usually depleted in these elements." Conversely, the corollary of this general relationship for macroelements is: "The chemical composition of sedimentary rocks is exceedingly variable, more so than that of igneous rocks, since sedimentation generally leads to a further diversification" (Mason 1966:152). Intimations of the latter rela-tionship can be seen in various intercolony and in-tracolony variations in feather minerals.

Assuming that findings for the Yukon and Ama-zon rivers (Gibbs 1973) offer acceptable models for the transport of iron by river systems in the Arctic, iron carried downstream to the coastal plain of Hudson Bay by rivers emanating from Sut-ton Hills south of the Cape Henrietta Maria area (Figure 48) and by the McConnell River (Figure 71) would be carried in three forms: 1) as precipi-tated metallic coating, roughly 40 to 47 percent; 2) as crystalline particles in the transported sedi-ments, 45 to 48 percent; and 3) the remaining iron incorporated in organic compounds.

Among the few soils analyzed, iron values were particularly high from Cape Henrietta Maria, from the McConnell River delta, and from a site along the Soper River, southeastern Baffin Island (Table 7). Transport of iron via rivers and glacial action from known iron-rich sedimentary rocks readily account for the high iron values in the first two of these soils. The occurrence of soils rich in iron on southern Baffin Island is not surprising in view of the Precambrian rock terrane.

Iron absorption by plants is facilitated by the excretion by the roots of organic acids of low mo-lecular weights, such as malonic and citric acid (Christopher et al. 1974:134). On calcareous soils an iron deficiency may arise in plants from the in-terference with iron absorption of carbonates and bicarbonates (Epstein 1972:300–301). Iron is present in a number of enzymes common to plants and animals. In plants, it is requisite for chloro-phyll synthesis and is important in the cytochromes and is therefore related to the transfer of electrons to oxygen peroxidases, and catalases are closely related iron-bearing enzymes associated with the control of the level of hydrogen peroxide in the cell

(Nason and McElroy 1963:478–680). The chlorotic symptoms associated with a deficiency of iron are thought to be related to a lack of iron for the production of chloroplast protein, and Granick (in Nason and McElroy 1963:483) has suggested that the protoporphyrin 9 precursor goes either to chlorophyll by the incorporation of magnesium into the precursor molecule or to hematin by the incorporation of iron.

Among the small collection of plants analyzed, the highest iron levels were observed for plants collected on Cape Henrietta Maria and the McConnell River delta (Table 8), two localities influenced by iron deposits.

In the animal kingdom the role of iron in enzymes involved in oxygen use by tissues and its role in oxygen transport (as the key component in hemoglobin) are too familiar to require further comment. It is of particular interest here, however, that in chickens the demand for iron during egg laying is great as compared to the demand during nonlaying periods. "With the onset of laying, the efficiency of iron absorption is increased and serum levels are markedly raised. There is usually a small fall in blood hemoglobin levels, but no significant reduction in storage iron is apparent. It seems, therefore, that the increased iron requirements imposed by egg laying is met by these means" (Underwood 1971:46). These relationships suggest that a sex differential in iron levels in the primaries of geese may be found in some populations of waterfowl as a result of higher residual stores in females—the most probable explanation for the sex differential in the Ross' geese from Karrak Lake, although a sex differential in iron was not found in the feathers of that population. As in the case of copper, increased absorption of iron in laying hens may be related to increased binding rates mediated by estrogen. Thus, plasma levels of iron in birds do vary significantly and can readily explain variations in feather levels.

Iron is generally reported to be poorly absorbed in man. Only 5 to 10 or 15 percent of the available dietary sources are used, and iron is absorbed only as Fe^{++}, involving two iron-binding proteins in an active transport system. The most active sites of iron absorption lie in the upper portions of the small intestine—the duodenum and jejunum. The regulation of iron absorption is believed to be the primary mode of achieving homeostasis, as most authorities relegate excretion to a minor role (Forth and Rummel 1971:171; Underwood 1971:28; Forth 1974:199). However, other experimental data are not in agreement with the concept of a tightly regulated control of iron absorption. Thus Christopher et al. (1974:135) report that "many experimental facts are not consistent with such a proposal [the 'mucosal block' hypothesis]. It was not possible to show saturation of iron absorption over a 3000-fold range in iron dose. Studies under conditions of maximal ferritin concentrations in the mucosa of the small intestine revealed that iron was still absorbed. Also absorption has been shown to increase in patients with hemochromatosis [excess tissue levels of iron] even though there are high iron levels in the mucosa." Our own data on levels of iron in primaries of wild geese are in accord with the latter investigators' findings.

In the chicken, increased dietary phosphorus is known to decrease iron absorption (Hudson et al. 1971:62). Interactions between iron and zinc and between copper and manganese have been well established for mammals, notably at the absorption level (Underwood 1971:28; Grassman and Kirchgessner 1974), but it is appropriate to note here that amounts of all four metals are very high in the primary feathers of Canada geese from the Belcher Islands, N.W.T., reflecting high environmental levels (appendix 1, Figure 266). If ion antagonism for binding sites exists between these metals in geese, levels in feathers are obviously a poor measure of this pehomenon. However, findings for the Belcher Island geese parallel the results of studies of rats by Forth and Rummel (1971:179–86), who found that in iron-deficient rats increased iron absorption was accompanied by increased absorption of manganese and zinc but not of copper. They concluded that, because iron-deficient animals also absorb more metals related to iron, these metals are mistaken for iron by the absorption mechanism and thus compete for mucosal sites. The highest mean iron value for the reference feathers was 2.6 times greater than the lowest, but no negative correlations between calcium and iron were recorded for intracolony or intercolony values (Tables 28, 30). In feathers from four of seven colonies, calcium and iron levels were significantly positively correlated. Feather patterns notably high in iron were recorded for the McConnell River colony (Figures 78, 86), west of which is a known deposit of silicious iron ore (Figure 71). A similar relationship has been described for the Cape Henrietta Maria colony. The variability of iron in the feathers of geese from the eastern colonies is over twice that found in goose primaries from the western colonies (Table 27).

Iron and aluminum and iron and silica were significantly positively associated in seven and six colonies, respectively (Table 32). The correlation between iron and silicon may be related initially to the association of these elements in both igneous and sedimentary iron ores (Park and MacDiarmid 1964:16, 360), but in terms of the ecosystems to which blue and snow geese are exposed, these elements more likely originated from glacial and river sediments or from coastal marine clays relatively high in iron.

Among all geese studied, highest iron values were found in feathers of Canada geese, notably the Belcher Islands population, which averaged over 692 ppm (Figure 266). Iron values for two of eight individuals studied were 1,127 and 1,130 ppm (Figure 266). These findings are in accord with the iron-rich geology of these islands; one island, Innetalling, has been estimated to have nearly one billion tons of magnetite iron ore containing 27 percent iron (Buck and Dubnie 1968:967).

It is generally held that animals have little capacity to excrete iron and that man does not appear to have an active regulatory mechanism for the excretion of iron (Moore and Dubach 1962:299, 333). "Once absorbed iron differs from other metals by virtue of its very slow rate of excretion" (Forth and Rummel 1971:173). In view of the high levels of iron found in primary feathers of the Canada geese from Belcher Island, and in other populations of geese, it is difficult to hold to the concept that homeostasis for this metal in geese is achieved almost wholly by the regulation of absorption; conversely, excretion as a regulatory means probably cannot be relegated to a minor role in geese (under conditions of high intake) as it is in man (Moore and Dubach 1962:333; Forth and Rummel 1971:171).

ZINC

The average zinc concentration of the earth's crust is near 60 or 70 ppm, depending on the ratio of acidic to basic rocks comprising the crust assumed in the calculation (Wedepohl 1972:30-E-20). Common minerals in the rocks of the Canadian Shield contain zinc as an atom substituted for ferrous iron or magnesium. Biotite and magnetite are important in this regard (Wedepohl 1972) and represent the major sources of zinc in these igneous and metamorphic rocks, which contain about 70 ppm of zinc (Table 33). Zinc in shales is also associated with the micaceous and chlorite components; in addition, the incorporation of zinc in iron oxide minerals carried into the basin of deposition

is to be expected (Wedepohl 1972). Shales have an average zinc concentration dependent to some extent on organic matter content. Wedepohl (1972:30-K-12) cites an average of 95 ppm of zinc for nonbituminous shales, whereas those containing any substantial amounts of organic matter average about 200 ppm of zinc. Limestones are much lower in zinc, containing, on the average, 20 ppm (Wedepohl 1972:30-K-12) (Table 32). Bayrock and Pawluk (1967) found an average of 75 ppm of zinc among 475 samples of Keewatin till from Alberta.

Zinc is an essential element for plants and animals, being a component of a number of metalloenzymes. In plants, it is particularly associated with auxin, functioning either through the protection against oxidation or through the control of the production of tryptophan, which is necessary for the synthesis of indole-3-acetic acid, or both (Nason and McElroy 1963:465–70). As might be expected, zinc is concentrated in the apical portions of the plant. Alcohol dehydrogenase contains 0.2 percent zinc, and carbonic anhydrase contains 0.3 percent, although the latter extracted from plants has not been found to contain zinc (Nason and McElroy 1963:470). A deficiency of zinc leads to impaired flower setting and seed production and to increased levels of calcium, phosphorus, and magnesium (Nason and McElroy 1963:471); as might be expected from its association with auxins, zinc deficiency is often expressed by dwarfed and twisted leaves (Hewitt 1963:229–37) and by deranged amino acid ratios and amide contents (Hewitt 1963:299–318).

According to Underwood (1971:241), "there is ample evidence that the zinc content of plants can be influenced by soil type." Horsetail (*Equisetum*), or at least one species, appears to be notably tolerant of high levels of zinc in the soil substrate, and tissue contents as high as 5,400 ppm have been recorded (Hemphill 1972:57). A favored food of Canada geese, this plant may account in large measure for the transfer of zinc from zinc-rich formations in the Belcher Islands to the endemic race of Canada geese (Figure 266). It is therefore likely that variations in feather zinc levels reflect the zinc content of the food plants within the metabolic confines of its limited variation in blue and lesser snow geese.

In chickens, zinc is required for growth, feather, and skeletal development. Zinc deficiency in poultry produces poor feathering, resulting from the effects of hyperkeratosis and its degenerative effect on the feather follicle.

In humans, skin, nails, and hair are relatively rich in zinc. Mean zinc values for human hair range from 120 to 170 ppm (Underwood 1971). These values are similar to the range of mean values for our reference feathers (Table 26).

Zinc is absorbed in chickens partly in the proventriculus, but resecretion into the duodenum and progressive absorption through all but the last segment of the intestine make uncertain the relative role of the proventriculus. The mechanisms of absorption are little known, and rates of absorption are variable, depending upon biological and nutritional states (Becker and Hoekstra 1971:233–34, 236). Plasma levels of zinc in the laying hen are two to three times higher than they are in nonlaying hens and in immature chicks of both sexes (Panić et al. 1974:636). For these reasons, zinc levels in the primaries may prove to be useful in distinguishing adult females from the other age-sex classes of geese. Zinc-deficient hens produce eggs with lower rates of hatchability and higher rates of embryonic abnormalities. Chickens excrete zinc primarily via the feces, by way of the pancreatic secretions (Orten 1966:41; Underwood 1971:317).

Zinc in trace amounts is generally a ubiquitous element, but it is reasonable to assume that Ross' geese from Queen Maud Gulf derive the zinc in their feathers indirectly from the igneous rocks of the region, whereas the zinc in the nutrient chain of the Hudson Bay colonies is presumably derived from the marine clays.

We were repeatedly impressed during the course of this study by the relatively small variations in levels of zinc from sample to sample of the primaries of various races of Canada geese as well as of blue, snow, and Ross' geese, an impression supported by the data in Table 27. Mean values in reference feathers ranged from 93 to 164 ppm, the highest value only 1.8 times greater than the lowest. However, zinc levels in the environments of blue, lesser snow, and Ross' geese may vary too little to effect wide variations in zinc levels in the feathers. In contrast, mean zinc levels in the feathers of Canada geese on the Belcher Islands were 306 ppm (appendix 1). Thus, zinc levels in the ecosystem may override metabolic controls in some cases, but on balance, the absorption of zinc appears to be more closely regulated in geese than is the absorption of any of the other eleven elements studied. Although Cotzias and Foradori (1968:116) state that both the absorption and excretion of zinc are subject to biological regulation and that the two functions are coupled, our findings are in closer agreement with the statement by

Becker and Hoekstra (1971:229): "Zinc metabolism is controlled by homeostatic mechanisms operating at the sites of intestinal secretion, such that the zinc content of most body tissues is maintained at a very constant level despite fluctuations in zinc intake or even long periods of inadequate dietary intake."

MANGANESE

Manganese is one of the more evenly distributed trace minerals in the various rock types. According to Mason's (1966:180) compilation, igneous rocks average 950 ppm, slates 850 ppm, sandstones 50 ppm (Bowen 1966:191), and carbonates, 1,100 ppm of manganese; however, it is most abundant in basic rocks (Table 33). Manganese follows the distribution of iron among rock types, there being a substantial removal of manganese from magmas during early crystallization and formation of basic rocks (1,300 to 2,200 ppm) as compared with the removal of manganese during the formation of rocks in the acid sequence (390 to 930 ppm).

In general, limestones and sandstones are deficient in trace elements, and most such elements present in limestones are in the clay fraction (Fleischer 1972:12, 15). Manganese occurs in all rocks primarily in the divalent form as an octahedrally coordinated atom (Peacor 1972:25-A-1). Limestones and, especially, dolomite contain more manganese than do argillaceous (clayey) rocks, a response to the proxying of the element in its divalent form for calcium and magnesium in octahedral sites.

"Compared to other rocks in Finland, the enrichment of soils originating from silicic rocks is remarkable" (Lounamaa 1967:295). Manganese is probably leached from these rocks as a bicarbonate (Armands 1967:144) and is probably transported in most northern rivers in three major ways: 1) chiefly as metallic coatings (50 percent), 2) in crystalline sediments (27 percent), and 3) in solution and organic complexes (17 percent) (Gibbs 1973:72). Levels of manganese of less than 2,000 ppm in common rock types (Table 33) correspond to amounts of 100 to 5,000 ppm reported in soils and to the anomalous 11.6 percent of manganese reported in Hawaiian soils derived from basaltic parent materials (Sauchelli 1969:61). Vinogradov (1959:115) reports manganese levels in the tundra soils of the U.S.S.R. in the range of 1,000 to 2,000 ppm in which the uppermost, organic-rich layers contain more manganese than do underlying horizons. Manganese appears to concentrate in the surface horizons of soils as a result of its extraction from deeper horizons by plant roots and its translocation to aerial plant parts that eventually die and are subsequently humified. In Norway maximum concentrations of manganese (and copper) are reported to be present at the tops and bottoms of bogs (Usik 1969:6). Divalent manganese is the plant-available form and becomes the stable species below about pH 6.5. Therefore, acid soils favor adequate or, in some special cases, even toxic concentrations of the element. The presence of organic matter and anaerobiosis, conditions common to the tundra, and attendant reducing conditions also favor manganese availability. Iron and manganese are controlling factors, in the absence of large quantities of organic matter, in the oxidation potentials of soil systems.

Manganese is essential for many plant and animal enzymes. In vascular plants, it is accumulated in the leaves. The element activates a number of enzymes important in oxidation-reduction and decarboxylation reactions, being especially important in these regards in the citric acid cycle and activating enzymes involved in hydrolysis and group transfer (Nason and McElroy 1963:496). Therefore it is not surprising, because of the role that manganese plays in activating enzymes, that plants deficient in the element become chlorotic largely because of the part played by manganese in chloroplast formation. Additionally, magnesium may proxy to a slight extent for manganese.

The manganese content of plants may vary substantially, depending on the soils on which they grow. In Finland needles and twigs of conifers varied in manganese content from 1,000 to 30,000 ppm. Variations in deciduous trees and shrubs were similar (Lounamaa 1967:297–98).

Manganese is essential for normal growth, reproduction (normal egg shell formation), and bone formation in birds and mammals. Its requirement for normal bone formation may relate to its role in metal-enzyme complexes in the synthesis of mucopolysaccharides (Leach 1974). It is associated with melanin pigments in feathers and hair, but this factor is not a variable in the present study, as the same feathers—the primaries—were analyzed for each goose, and melanin type and contents were not judged to be sufficiently variable. The manganese content of feathers and wool has been shown to vary with levels in the diet (Underwood 1971:181).

Manganese is poorly absorbed from the gut, but of the portion that is absorbed, the largest fraction reaches the liver, and lesser amounts, the pancreas. From these sites it is excreted back into the intestine where it becomes available as a mobile pool. A second pool exists in the mitochondria of the liver itself, and both pools are held in a mobile state. A more stable, trivalent, form is bound to a specific protein in the plasma (Oser 1965:564). Manganese levels in the plasma of chicks increase with age up to nineteen weeks, and values for mature laying pullets are always greater than they are in young laying pullets (Hill 1974:633). Excess dietary intakes of calcium and phosphorus are known to decrease the availability of manganese in the diet (Underwood 1971:197).

Mean manganese values in the reference feathers varied tenfold to fifteenfold (Table 26). Low values were associated with colonies located on limestones of the Paleozoic era, and high values with colonies whose ecosystems included igneous rocks of the Canadian Shield or whose soils were affected by sediments from adjacent Canadian Shield areas, e.g., Queen Maud Gulf area (Ross' geese), McConnell River colony area, and Anderson River. In brief, manganese levels in feathers bore a direct relationship to the abundance of this element in the primary rock types. A simple antagonistic (inverse) relationship between calcium and manganese is not indicated by the levels of these elements in feathers, as calcium levels in the primaries of western lesser snow geese equaled or exceeded levels in the feathers of eastern blue and lesser snow geese, and correlations between these elements are positive (Table 28). It is conceivable, however, in view of the data presented in the discussion of calcium, that whatever suppressive effects calcium concentrations may have on the absorption of other elements, the level of sodium ingested may be the ultimate, indirect, controlling factor.

Coefficients of variability were markedly high for the Baffin Island, Southampton Island, and Cape Henrietta Maria colonies, findings that are further suggestive of a "pockety" distribution of manganese in these ecosystems (Table 27). The variability of manganese in the primaries of geese of the western colonies was only one-fifth as great as that for the eastern colonies, a finding in accordance with an assumption of higher and more evenly distributed levels of manganese in ecosystems on, or that receive mineral input from, Canadian Shield areas.

The extremely variable nature of manganese concentrations in the reference feathers (Table 27) points to highly localized manganese concentrations in the feeding ranges of the colonies. The variation of manganese levels in plants from the McConnell River area (Table 8) exemplifies the patchy occurrence of manganese in the ecosystem; this mode of occurrence is reflected, in turn, by the high levels of manganese found in some of the reference feathers of geese from the McConnell River colony (Figures *83, 86, 88*). Feeding areas with plants high in manganese are likely to be found on acid, boggy sites in the tundra where manganese would be most readily available to plants. Perhaps the extremely variable manganese patterns in feathers from the majority of the colonies reflects the tendency for the distribution of heavy metals in both parent rocks and some soils to be log-normal in nature (Kauranne 1967:189).

Manganese ranked as one of the minerals most frequently correlated with other minerals studied (Table 32). In all seven colonies manganese was positively correlated with calcium and aluminum. These findings suggest that the prevalence of these elements in the primaries of blue and lesser snow geese is a relatively direct expression of the manganese levels of the soils in the areas in which these geese feed. The correlation of manganese in five of seven colonies with silicon and magnesium may be further expressive of these relationships. The relationship between manganese and magnesium may be indicative of the close biochemical affinities of these elements, alluded to above, but the correlation of manganese and calcium is less clear in view of the negative effects of calcium on manganese absorption in the gut cited earlier (Underwood 1971:197) and the possible intermediary effect of sodium levels.

Our findings on the relationship between manganese levels in feathers and the relative prevalence of manganese in various rock types is in agreement with findings from experiments with manganese deficiency that "the amount of metal absorbed is proportional to that presented for absorption" (Cotzias 1962:417). Furthermore, in a manganese-deficient state, manganese is not conserved. "These findings argued against the existence of a mechanism which rapidly shut off manganese excretion in the presence of deficiency" (Britton and Cotzias 1966:206) Cotzias and Foradori (1968:116) report that "unpublished results indicate that bilary excretion can be readily saturated by administering orally large amounts of manganese, in which event intestinal excretion becomes significant. Yet manganese is not effectively regulated (Britton & Cotzias 1966)."

These conclusions are at variance with the statement that Cotzias and associates have "established the existence of specific and efficient homeostatic control mechanisms" for manganese in animals and man (Underwood 1971:178). Our findings are more in accord with those of Cotzias and Foradori (1968) and Lassiter et al. (1974:557), as it is apparent that absorption rates of manganese in geese must vary greatly—as evidenced by its variability in feathers—and that whatever semblance of manganese homeostasis exists in these birds is achieved by excretory routes, of which the feathers must be regarded as one.

COPPER

The prevalence of copper among rock types is given by Mason (1966:180): shales 45 ppm, carbonates 4 ppm, and igneous rocks 55 ppm. For sandstones Bowen (1966:182) provides the figure of 5 ppm. These data are consistent with those of Rösler and Lange (1972:229) (Table 33). The comparatively high levels of copper in shales have been attributed to the copper-concentrating ability of marine organisms that subsequently become incorporated in the shale. Under oxidizing conditions copper coprecipitates with iron. Incorporation of copper with sulphides in a reduced environment is also a mechanism for copper concentration.

Anomalous copper values in the soils of the nesting grounds of Ross' geese south of Queen Maud Gulf are inferred from feather contents (Figures *137, 142*). These occurrences are most likely associated with basic rocks interspersed with the granitic terrane of the region. Vinogradov (1959) gives a value of 200 ppm as an average copper content in basic igneous rocks in contrast to 10 to 100 ppm in acid igneous rocks. He also points out the ability of copper to diffuse from ore bodies into overlying soils although this characteristic would not appear important enough to account for the moderately high values in the ecosystem at Karrak Lake. It seems more likely that a rich local copper source at or very near the surface has been eroded by glacial ice and admixed in the till of the principal nesting area that is otherwise a reflection of glacial erosion of granitic rocks. If a generalization can be made from data for sediments carried by the Yukon and Amazon rivers, most of the copper transported by rivers is carried by crystalline particles of the sediments (Gibbs 1973:180). In view of the very high stability constant for copper exhibited by humic and fulvic acids, it is unlikely that any copper of consequence moves in the ionized form.

The chemistry of copper in soil is intimately associated with organic matter (Stevenson and Ardakani 1972). Copper has been consistently found to have the highest stability constant among the transition-series elements, a commonly determined sequence being copper, nickel, cobalt, zinc, iron, and manganese. Copper-deficiency symptoms in plants occur most often on peat and muck soils, leached sandy soils, calcareous sands, and leached acid soils and soils heavily fertilized with nitrogen (Hemphill 1972:54). Insoluble copper-organic matter complexes tend to accumulate in the surface horizon, giving the organic horizon the largest copper content of any of the soil's horizons. This high affinity of organic matter for copper can decrease the soil's capacity to supply this element to the extent that the lack of copper can limit plant growth. Many northern areas where agricultural exploitation of peat or muck soils has been attempted need supplemental applications of copper to be made productive. These brief remarks are given to suggest that if cupriferous rocks occur in the Karrak Lake area, the overlying soils, largely organic in nature, would have high levels of copper in their organic horizons. Although considerable competition for this copper would exist between the plant roots and soil organic ligands, somewhat elevated levels of copper would be expected in the sedge and grass flora consumed by geese; however, Cannon (1960) in her review of biogeochemical prospecting recorded that relationships between the copper content of vegetation and copper-bearing ore bodies were not consistent.

In plants copper "occurs as neutral or anionic complexes that are absorbed more readily than ionic copper" (Van Campen 1971), being associated primarily with several enzyme systems (Nason and McElroy 1963:471–78; Epstein 1972:291). Polyphenol oxidase and ascorbic acid oxidase represent two copper-bearing enzymes, the former involved in the formation of quinones which may be related through the reduction of a quinonelike substance by hydrogen during photosynthesis (Anonymous 1969:477). The copper present in cupriferous enzymes is believed to catalyze the oxidation of the substrate by gaseous oxygen.

The copper content of the aerial parts of grasses is fairly constant but may vary thirtyfold (1–30 μg/gm), differences being attributed in part to the content and availability of copper in soils (Adelstein and Vallee 1962:374). Bowen (1966) cites 14 ppm as average for land plants. Arctic plants that we analyzed ranged from 5 to 30 ppm in copper content (Table 8). Surprisingly, the copper levels in plants from areas with a limestone substrate were significantly higher than were those in plants associated with igneous-metamorphic terranes, but the number of plants analyzed from sites on or adjacent to the Canadian Shield are too few to be representative of it.

In domestic fowl copper becomes bound to a protein in the duodenal mucosa, the rate of binding there being five times faster than in the proventriculus (Hudson et al. 1971). Similarly, in mammals binding rates are more rapid in the duodenal section of the intestine (Van Campen 1971:221).

According to Van Campen (1971:221), the mechanism that regulates copper absorption is not known: "There is little evidence that copper absorption from the intestine is regulated according to need as in the case of iron." Van Campen believes that absorption involves at least two mechanisms. "When copper concentrations in the gastrointestinal tract are low, absorption exceeds what should be expected on the basis of concentration; however, as the intestinal copper level is increased, absorption appears to become proportional to concentration." From this, it must be concluded that excretion is the chief means of maintaining some semblance of homeostasis for animals on a high dietary intake of copper. Although the liver and bile are probably the primary storage depot and excretion route, respectively, for excess copper in molting geese, it appears that the feather keratin constitutes an additional disposal route for excess levels of dietary copper.

We have found a twenty-eightfold variation in the copper content of the primary feathers of blue and lesser snow geese (8 to 10 ppm to 225 ppm), a range similar to the variation cited for grasses. However, the variation in copper levels within most populations of wild geese was seldom greater than fourfold and was comparable to levels found in plants (Table 8).

Mean copper concentrations in the primaries of blue and lesser snow geese from eight of nine colonies ranged only from 8.2 to 11.0 ppm, a finding which suggests that in geese this element may be closely regulated, particularly in regard to physiological monitoring of absorption. Lounamaa (1967:303) similarly noted that "the range of copper content in the conifers studied was on the whole narrower than that of other trace elements." But in opposition to the concept of fairly strict metabolic monitoring of copper levels are data for some geese in the Banks Island populations with an average value of 20 ppm and the Ross' geese from Karrak Lake in the Queen Maud Gulf lowlands with an average of 28 ppm in their primary feathers (Figures *100, 142*). Most noteworthy of all were the primaries of the one species of wild goose that contained 225 ppm of copper. Copper concentrations in the nutrient chain of its feeding area in the North can only be surmised.

Copper values in the primary feathers of most Canada geese fell within the 8 to 14 ppm range, but some notable exceptions occurred. Canada geese from Belcher Island averaged 20 ppm; one from the Mackenzie Valley drainage had 106 ppm; two geese from Utah had 93 and 99 ppm in their flight feathers; and one from central British Columbia had 101 ppm. In general, concentrations of copper in the primaries of most blue and lesser snow geese populations closely reflect the relatively copper-poor limestones and clays that largely characterize the substrates of most of the breeding grounds. The areas from which geese with high copper levels in their feathers originated are all known or suspected to contain economic concentrations of metals and have been or are being actively prospected. Hence, again the conclusion must be reached that it is the mineral gradient in the ecosystem that determines the quantity of a given mineral in the primary feathers—in this case copper.

Next to those of zinc, average coefficients of variability were easily the lowest for copper among the twelve elements studied (Table 27). As these low coefficients were associated with low concentrations in the feathers, they suggest low intracolony variations in the ecosystem or close metabolic controls within the framework of the copper available in the environment.

Significant associations between copper and other elements were relatively limited (Tables 28, 31, 32), a circumstance to be expected with so little range in copper content. The data for Karrak Lake Ross' geese, relatively high in copper values, might best be scrutinized for insights into significant relationships, but only sodium and copper were significantly paired. Sodium values for Ross' geese from Karrak Lake exceeded those of all the other colonies studied and were high by comparison with those of most other populations of geese. Presumably, as pointed out earlier, the high sodium values found for Ross' geese from Karrak Lake versus the low values found in samples assumed to represent the Lake Arlone area can be attributed to the higher sodium content of areas of saline marine clays versus the low sodium content of terranes influenced by basic rocks (Table 33).

Evidence for a sex differential in copper metabolism was presented in Hanson and Jones (1974) and in the introductory discussions of intracolony variability (Figure *14;* Table 37). The sex differential in mean copper levels of the Karrak Lake birds is highly significant ($P < 0.001$), and it is significant at the 5-percent level for the California sample ($P < 0.05$). The exact origin of the random sample of Ross' geese shot in California is unknown; however, it is highly probable that they summered in the Queen Maud Gulf area. Inasmuch as this breeding ground varies considerably geologically (Figure *137*), it is notable that this California sample also exhibits a significant sex differential in the copper content of the wing feathers (Table 36).

Arthur (1965) found that copper in guinea pig (*Cavia porcellus*) hair was indicative of copper levels in the diet. Petering et al. (1971) have recently observed distinctly higher contents of copper in the hair of human females as compared with that of males in age classes above thirty years. They found in males that the hair's copper content increases annually between two and ten years of age but that beyond about twelve years of age copper declines in male hair until it is equal (20 ppm) to that in female hair at about thirty years of age. However, in the hair of females a small steady increase in copper content occurs with age. Petering et al. (1971) speculated regarding differential copper needs between sexes, the availability of copper to each sex, and the apparent influence of puberty in the case of males, but no biochemical mechanism was identified.

Underwood (1971:68) states that "There are no significant sex differences in whole blood or plasma copper in most species, but plasma copper is higher in human females than in males." This difference can, presumably, be attributed to the ability of estrogens to increase the copper-binding capacity of the plasma albumins. In this conjunction scarcely significant amounts of estrogen (less than 5 pg per cc) are found in plasma samples from intact male white leghorn chickens as compared to less than 25 pg per cc in the incubating hen and to over 700 pg per cc in some laying hens (Jean Graber unpublished; Peterson and Common 1972; Hill 1974). Although gonadal tissues of geese of both sexes atrophy rapidly during the incubation period, the differentially higher levels of copper in the feathers of females may reflect significantly higher levels of residual estrogens in adult females during the post-breeding molt as compared with estrogen levels during other periods of the year (the first half of the ovulatory cycle excepted). An

37. Copper content of the vanes of primary feathers of adult Ross' geese collected at Karrak Lake, N.W.T., and at Tule Lake N.W.R., California

Source	Sex	Number in Sample	Mean ±S.E. (ppm)	Range	Probability
Karrak Lake	M	10	17.8 ± 0.7	13–20	<.001
Karrak Lake	F	10	27.1 ± 2.1	18–39	
Tule Lake	M	5	10.2 ± 1.2	7–13	<.05
Tule Lake	F	6	14.2 ± 1.3	11–18	

alternate and more likely explanation would be the existence of higher residual levels of copper metalloproteins (Mills 1974) in various tissues of females as the result of differentially higher rates of binding during the period of high estrogen levels. Thus, differentially higher storage levels of copper in the livers of females at the onset of molt may also play a role in its availability during molt. In humans the copper level in plasma increases during pregnancy. As might be expected, the administration of estrogen markedly increases serum copper and ceruloplasmin concentrations (Adelstein and Vallee 1962; Underwood 1971:69; Van Campen 1971:214).

A more specific functional reason for higher copper levels in female Ross' geese is that copper has been implicated in shell membrane formation in fowl "by evidence that the mucosa of the isthmus contains much more copper than any other portion of the oviduct (Moo-Young et al. 1970). Such a heavy endowment of Cu^{2+} reflects the activity of the isthmus mucosa in forming the keratinlike components of the shell membranes. Disulfide linkages are characteristic of keratin and the oxidative closure of sulfhydryl groups is regulated by cuproenzymes, much as elastin synthesis may be mediated by amino oxidase" (Hazelwood 1972:509). We conclude, therefore, that the sex differential observed in copper contents of Ross' geese feathers results from unusually high levels of copper in the environment and the direct mediation of estrogens in increasing the copper binding of the plasma. This pool of copper is incorporated, to some degree, in feather keratin.

BORON

Boron concentrations in crustal rocks average about 10 ppm, varying from 10 to 15 ppm in granitic rocks to less than 5 ppm in basic rocks (Table 33). Boron values for marine limestones tend to be somewhat higher than those of limestones formed

in freshwater: six samples of marine limestone ranged from 4 to 80 ppm; six samples of freshwater limestone ranged from 20 to 70 ppm (Graf 1960:45). Marine clays tend to be enriched directly in proportion to the salinity of the environment of deposition, and boron content is often used in estimating paleogeographic conditions. Bowen (1966:174–75) has given the following values for boron: igneous rocks, 10 ppm; shales, 100 ppm; sandstones, 35 ppm; and limestones, 20 ppm. It is apparent that argillaceous sedimentary rocks are richer in boron than are igneous rocks—largely as a result of the extraction of boron from seawater; the latter contains an average of 5 ppm of the element. Saline and alkaline soils are highest in boron content, the borate and kindred anions accumulating in the surface soil as a result of high rates of evapotranspiration.

Boron has been regarded as a requisite micronutrient for plants, being absorbed either as $B(OH_4)^-$ or as a monomeric boric acid (H_3BO_4), the proton-donor nature of root membrane favoring the latter (Shorrocks 1974). Although no unequivocal role for boron has been determined, maintenance of health, resistance to certain diseases, cell maturation and differentiation, as well as involvement in sugar metabolism have been cited as among the roles it plays (McMurtrey and Robinson 1938:814–16; Epstein 1972:303–4; Underwood 1971:433). Boron has been identified as forming strong sugar-borate complexes, implying the importance of boron in sugar transport (Nason and McElroy 1963:500; Hewitt 1963:257). The especially variable boron levels in apical tissues in response to the supply of the element (Nason and McElroy 1963:500–501) have led to hypotheses regarding cell differentiation. High calcium content in the growth medium reduces plant levels of boron (Nason and McElroy 1963:501). Rajaratnam et al. (1971) obtained evidence indicating that boron is required for flavonoid synthesis in higher plants. Flavonoids are sap-soluble pigments of living plant tissues which occur mainly in combination with sugar residues (Sherratt 1961:1028).

Plants show considerable sensitivity to boron in water; it may cause a decline in growth or even be quite toxic to many plants exposed to water with more than 1 ppm of boron. Highest levels of boron in plants are generally found in legumes, which normally contain 30 to 80 ppm of boron (Shorrocks 1974) and display considerable tolerance for high levels of this element in the soil—hence their use in irrigation agriculture where the boron content of the waters may be high. Grasses (leaves) normally contain 5 to 20 ppm of boron (Shorrocks 1974).

In man boron is "rapidly and almost completely absorbed and excreted, largely in the urine" (Underwood 1971:434). Safe levels for human ingestion have not been set except for an 8-ppm value established for antifungal residues of boric acid and borax residues on citrus fruits (Anonymous 1969).

In geese, boron concentrations in the reference feathers were generally low and the ratio between high and low values was only 2.1 to 1 (Table 26). Lowest values were associated with the colonies of the Hudson Bay region situated on Paleozoic limestones; highest values tended to be associated with colonies located on Precambrian rocks or soils derived from them (McConnell River) or from a range of sedimentary rocks (Banks Island). This latter situation may be associated with Polar desert soils on Banks Island.

In calling attention to correlations between feather-mineral levels, only those pairs that were correlated in five or more of the seven colonies having adequate data were cited (Table 32). This line of distinction is arbitrary, but the significance of pairs of mineral concentrations that were correlated only a few times is at least questionable. In the case of the Cape Henrietta Maria and Kendall Island colonies, ten and eight, respectively, of the twelve minerals were involved in correlated pairs. Such instances might suggest a relatively homogenous substrate and basically similar responses by the individuals to a similar mineral intake. In relation to boron only magnesium falls within this category. If boron has, in fact, no metabolic function in animals, then reasons for its correlation with magnesium must be sought in the ecosystem. In view of the involvement of magnesium in numerous plant enzyme systems and the apparent need of plants for closely controlled amounts of boron for normal metabolism, it is per-

haps not unreasonable to look to the food plants of geese for the basis of correlations of feather-mineral levels of boron and magnesium. (See chapter 8.)

Sodium, phosphorus, zinc, and manganese levels were correlated with boron levels in feathers of geese from four of the seven colonies (Tables 30, 32). A better basis for judging the significance of these correlations will be available when our continent-wide data on Canada geese have been evaluated.

SILICON

Silicon constitutes an estimated 28.2 to 29.5 percent by weight of the earth's crust; only oxygen exceeds it in abundance (Rösler and Lange 1972:229). Silicon occurs in the oxidized form under all conditions of the lithosphere, appearing in diverse compositions among minerals of divergent types. Igneous rocks are divided into basic rock types that are thought of as having low SiO_2 content—less than 55 percent—when compared with acid rock types containing up to 71 percent of SiO_2 (Table 33). Shales and noncalcareous sandstones contain substantial amounts of silicon, whereas limestones and dolomites contain the element in small proportions, reflecting silicious sand-, silt- and clay-size minerals in accessory amounts. Among minerals the silicate clay species are the major candidates for providing silicon to geophagous animals, not only because of the high silicon contents of such clays, but also because of the large surface areas of their particles and their attendant high reactivity. Clay-rich sediments, then, are thought to represent the most likely source of silicon to grazing geese. Quartz, neglecting insignificant impurities, is pure silica (SiO_2) and is ubiquitous on the Canadian Shield terrane, but this mineral is resistant to abrasion in the gizzard and very inert chemically; therefore, despite its common occurrence, it should not represent a ready source of silicon to geese.

Although silicon is ubiquitous in plants, it is not a requisite element for most species of vascular plants. However, it is a notably prominent constituent of horsetails (Equisetaceae), sedges (Cyperaceae), and grasses (Gramineae). In these families silicon is deposited in epidermal walls as a gel (Gauch 1972:294). Silica may constitute as much as 16 percent of the dry weight of horsetails (Gauch 1972:294) and 50 to 71 percent of their ash weight (Stiles and Cocking 1969:290). These plants are a highly favored food of Canada geese and constitute an important, although possibly a secondary, source of absorbed silicon in blue, snow, and Ross' geese. Silicon is also concentrated by diatoms, in which it plays an important role by providing the structural framework of the frustule.

Kovda (1956) has drawn some interesting comparisons among the major elements contributing to the ash contents of aerial parts of natural herbaceous vegetation from the northern prairies, the prairie-steppe, the dry steppe, and halophytes of the dry regions of Russia. Kovda found that ash contents were inversely related to rainfall and that only in the prairie-steppe was silica the element that contributed the most to ash content. Northern plants were highest in calcium and potassium with the anion-forming elements sulfur and phosphorus next in importance. As might be expected, the ash of halophytes contained sodium, chlorine, and sulfur in that order of abundance. Ash amounted to only 2 to 4 percent of the air-dried weight of plants from the northern prairies.

Lovering and Engle (1967) assessed silicon uptake from basalt and rhyolite by horsetail (*Equisetum hymale*) and three grasses. They concluded from infrared spectra of freeze-dried sap that silicon was bound to condensed aromatic compounds, finding levels of 312 ppm in the sap as compared with 110 ppm found in neutral water solutions in equilibrium with silica gel. From these findings, Lovering and Engle (1967:B14–B15) hypothesized that polyaromatic compounds resembling humic acid play an important role in solubilizing silicon from silicate minerals in soils, maintaining the element at high levels in solubilized condition, and providing a carrier for crossing root membranes.

Silicon accumulates with time in the plant, coating cell walls and filling the lumen of the cell—particularly in the epidermis—with opal, an amorphous hydrated form of silica (Geis and Jones, 1971). The solid opal particles are known as phytoliths. When these particles are ingested with the plant, they represent a finite level of silicon availability to an animal. Baker et al. (1961) have suggested that wear in sheep's teeth is attributable in part to the abrasive action of these opal particles, and the same workers (Baker et al. 1961) have identified opal phytoliths and diatom frustules within opaline uroliths from a ram, suggesting remarkable migration of these particles across the gut wall and through the vascular system.

Laboratory studies indicating that silicon is an essential element for birds and mammals date back

to 1964. It has been found to be requisite for normal bone growth, being primarily associated with the acid mucopolysaccharides, and, as a cross-linking agent, silicon is related to the structure and integrity of cartilage and connective tissue. It has been associated with the preosseous state in the calcification of bone and the synthesis of mucopolysaccharides as well as being an essential component of the latter (Schwarz 1974; Carlisle 1974). In man most ingested silicon passes through the gut, but of the absorbed fraction, appreciable amounts are excreted in the urine (Oser 1965:560).

Mean values for silicon in the reference feathers of blue, snow, and Ross' geese varied from 869 to 5,971 ppm (Table 26). Excepting manganese concentrations in the reference feathers from Hudson Bay, silicon was the most variable of the twelve elements studied. Despite the variability of its occurrence in the reference feathers, levels of silicon were fourth (with manganese levels) most frequently correlated with abundances of other elements (Tables 28, 30–32). Probably its frequent correlation with aluminum and iron can be best related to the association of these elements in the ecosystem, particularly in the clays and sands of the soils, rather than to metabolically related events.

As was the case for most elements, silicon was found to be highly correlated with the other elements in the reference feathers from the Cape Henrietta Maria colony (Table 30). The high degree of quantitative associations of the feather elements in this colony sample suggests that absorption is in some measure proportional to intake. Highest levels of silicon were found in Canada geese. Most notable examples were the Canada geese breeding on Akimiski Island in James Bay (Figure *252, 260*). The silicon content of the primary vanes of one of these geese exceeded 28,000 ppm.

ALUMINUM

Aluminum ranks second in abundance among elements of the earth's crust, oxygen excluded, constituting approximately 8 percent of the lithosphere (Rösler and Lange, 1972:229). Its properties with regard to distribution among rock types and availability are similar to those of silicon for which it occasionally substitutes in tetrahedral coordination. Aluminum comprises about 8 percent of igneous and metamorphic rocks and shales (Table 33). Carbonate rocks commonly contain less than

1 percent of aluminum, the amount depending, in large part, on aluminous clay- and sand-size minerals incorporated in the rock. Among the minerals of igneous rocks, most aluminum probably comes from feldspars of the granites, basic igneous rocks, and gneisses and schists of the Canadian Shield that contain from 18 to 37 percent Al_2O_3, the larger amount occurring in the calcium-bearing feldspar anorthite. In sediments, silicate clay minerals should represent some low but perceptible source of aluminum as they pass through the gut. Perhaps it is best noted at this point that the surfaces of large areas of Southampton Island consist of limestone detritus or felsenmeer and a "soil" of ground limestone. Aluminum is, appropriately, notably low in the feathers of Southampton geese.

Relatively little is known about the metabolism of aluminum, and little if any biological significance has been attributed to it either in plants or animals. However, as Gauch (1971:226) has also pointed out, "almost all of the naturally occurring elements have been detected in plants." Concentrations of aluminum as high as 3,000 to 4,000 ppm have been found in trees, and grasses are reported to contain 10 to 50 ppm of aluminum on a dry weight basis (Underwood 1971:425). Aluminum has an antagonistic effect on the copper uptake of citrus cuttings (Gauch 1971), but a similar influence in Arctic plants is not indicated by the relationship of these elements in primary feathers of geese. Although the essentiality of aluminum for plants has not been demonstrated, there is some suggestion that the element may be beneficial to cereals (Hewitt 1963:326). Tea (*Camellia sinensis*) is an accumulator of aluminum (4 percent of dry weight), and the stimulating effects of aluminum on the formation of blue pigments in the shrub hydrangea (*Hydrangea* sp.) is a phenomenon observed by many (Hewitt 1963:328). Soil pH markedly affects the nature of the aluminum molecular species in solution: the fully ionized atom—the toxic form—occurs only below pH 3.2. Therefore, significant accumulations of aluminum are noticed only in very acid soils. An appreciable proportion of the area of these soils lies along the coasts of warm temperate and tropical areas where iron pyrite-bearing marine clays have been exposed and the pyrites have been oxidized to sulfuric acid in an aerobic soil environment. A level of 1 ppm of aluminum in nutrient solution is sufficient to depress growth in corn and barley (Bould 1963:52).

In man and other mammals, "an intense disturbance of the metabolism of phosphorus as a result of intoxication by aluminum compounds" has been reported (Ondreicka et al. 1971:302). "In general the low solubility of aluminum compounds in the digestive tract, and in the ability of these compounds to form complexes within the digestive tract—results in low absorption of aluminum" (Ondreicka et al. 1971:295). Nevertheless, the concentration of aluminum in the body is believed to be a function of the amount of aluminum in the food (Ondreicka et al. 1971:303). Reasonable increases in aluminum intake in humans are accommodated by increased alimentary elimination and, to some extent, urinary elimination.

Mean aluminum values for reference feathers varied only fourfold (Table 26). Intracolony variations for this element in primary feathers of geese from the Hudson Bay colonies were twice those found for feathers from the western colonies. It is possible that the higher variations of the Hudson Bay geese reflect the postnesting dispersal of these populations, some segments going to areas where the substrate for plants is primarily carbonates and others going to more inland or to coastal areas where clay sediments dominate the mineral aspects of their nutrition.

Despite the metabolically inert nature of aluminum, the levels of aluminum in feathers were second only to those of calcium and magnesium in the number of times that aluminum content was correlated with abundance of other elements (Table 30). The explanation of this finding, in view of the apparent relative biological inertness of aluminum, appears to relate directly to the mineralogy of the ecosystems of the respective colony areas rather than to metabolic events. For example, aluminum levels in feathers were significantly correlated with silicon levels in six of the seven colony samples (Table 32). Aluminum and silicon constitute the principal structural cations of clay minerals. It is assumed that kaolinite $[Al_4Si_4O_{10} (OH)_8]$, chlorite, and illite are the most common clays of these Arctic areas. Blue and lesser snow geese are well known for their grubbing for plant roots, and they must consume significant amounts of clay, a probable partial explanation of the correlation of aluminum and silicon levels in the feathers of most colonies.

Aluminum levels were correlated with those of calcium, iron, and manganese in all seven colonies studied (Table 32); with magnesium levels in five of the seven; and with those of potassium in three of the seven. In the cases of calcium, iron, and magnesium, it is reasonable to relate their correlated occurrence with aluminum to marine clays in which these elements are prominent secondary constituents (Mason 1966:44).

Unfortunately, the primaries of blue and lesser snow geese do not, as has been pointed out, indicate the full scope of elemental variations in northern ecosystems and the environmental gradients with which wild geese must contend. Thus average aluminum values for feathers of Canada goose populations may vary as much as twentyfold, ranging from as low as 46 ppm for some central British Columbia populations to 422 ppm for individuals of an intermediate-sized race collected in southern Yukon and as high as 960 ppm for a Canada goose from Akimiski Island (Figure 266). Once again it seems that excretion must be favored over regulation of absorption as the primary stage in achievement of homeostasis.

LIKE ALL living organisms, wild geese are products of their environments and heredities—the former molding the latter through the selective processes of evolution. Some species are conservative and seemingly lack the genetic potential that may enable them to exploit a wide variety of habitats. Other species are plastic, i.e., they possess more variable gene pools and have been able to adapt to a much wider variety of habitats. The blue and snow geese, consisting of only two races (presently recognized) and being nearly circumpolar in distribution until recent times, are an example of a genetically conservative species as a result of strong selective pressures or low mutation rates or both. On the other hand, no other North American bird so well exemplifies a plastic species as does the Canada goose. It has occupied the major part of the continent north of Mexico and has become adapted to a wide variety of habitats, in the course of which adaptation the species has evolved into over thirty geographic races (Hanson unpublished).

An ability to exploit a wide variety of food sources and the possession of readily activated salt glands explain, to an important degree, the extensive range of the Canada goose. The lack of these same characteristics has confined the breeding range of blue and lesser snow geese to marshy tundra areas and freshwater or slightly brackish coastal marshes.

In chapters 4 and 5 the dominant theme was the practical aspects of feather mineralization—the use of feather mineral analyses to determine the birthplaces or summering grounds of wild geese. A discussion limited only to reconciling mineral levels in feathers with mineral levels in the nonliving components of the ecosystem could have been sufficient to terminate this report. Although such an appraisal of our responsibilities might have satisfied practical needs, it would have left a host of fascinating basic questions undeliberated.

In this report we have attempted to account for the origins of the principal minerals in the nutrient chain of blue and snow geese and trace the sequence of events—with a limited number of paired samples—that takes these minerals from the parent rocks to overlying soils and accounts for their uptake by the plants and their absorption and excretion by the geese. We have also made subjective judgments regarding the apparent relative roles of absorption and excretion in maintaining homeostasis, recognizing at the same time that some elements are tolerated within moderately wide limits and that these levels may vary some-

CHAPTER EIGHT

Wild Geese and Their Mineral Environment

what from organ to organ, the liver in particular being capable of storage (Widdowson and Dickerson 1964:146–47) although to a relatively limited extent in the case of manganese (Underwood 1971:180). This section reemphasizes some of the earlier conclusions and attempts to set them in an overall perspective.

SOIL-PLANT RELATIONSHIPS

Although our samples of plants from the breeding colony areas available for analyses were restricted both as to number and geographic coverage (Table 8), they nevertheless seemed to underscore several basic relationships. Plants from colony areas on Paleozoic limestones—carbonate terranes—have consistently and significantly high levels of calcium, magnesium, sodium, potassium, and phosphorus. Levels of two key trace elements, iron and manganese, were significantly higher in plants from areas of igneous or metamorphic rocks. Zinc levels in the plant samples, as in feathers, were relatively stable and did not differ significantly between the two rock categories. As all collections were made about mid-July, values for nitrogen are probably at least roughly comparable in the phenological sense from area to area. The notably higher levels of nitrogen, phosphorus, and the other nutritive macroelements in plants from carbonate terranes must account in part for the attractiveness of these areas to geese as nesting and feeding areas and their ability to support dense breeding populations.

A dominant role has commonly been ascribed to calcium in relation to the absorption of trace elements in plants and animals. "Roughly, it looks as if an increasing pH and Ca ion concentration renders the plants' uptake of manganese, copper, zinc, and boron more difficult" (Låg 1967:93). It has also been stated that "high levels of phosphorus in the soil will antagonize both copper and zinc uptake or use" (Schütte 1964:43). Sodium levels, however, must be considered an important third dimension of the mineralogy of soil-plant-animal relationships.

Earlier, under calcium, the physiological effects of varying ratios of calcium-to-sodium intake were treated at length, as these factors appear to be of dominating importance in the mineral metabolism of vertebrates. This theme can be carried over with profit in a discussion of soil-plant relationships, a better understanding of which brings us full circle in gaining insights into the attractiveness of environments and their capacities to sustain goose populations.

It is recognized that, at low levels, sodium is a beneficial element in the nutrition of many plants and essential for a few. It can replace potassium in pyruvic kinase of higher plants. In cotton plants under conditions of calcium deficiency and low potassium, sodium can proxy for the latter. The primary effect of sodium at low levels is to stimulate the translocation of calcium and potassium from the leaves and stems to the bolls (Hewitt 1963:320). Responses of plants to sodium vary among plant species, but are related, as might be expected, particularly to the potassium level. In barley, sodium can meet up to seven-eighths of the need for potassium (Hewitt 1963:318–21). The water content of plants is a measure of their succulence, a factor that, aside from other nutritional aspects, appears to affect the palatability of browse for geese. It is noteworthy, therefore, that the correlation between the sodium content of the nutrients supplied experimentally grown barley and the succulence of the barley was found to be very high (Hewitt 1963:184). However, the chloride ion independently or the ratio of adsorbed/free ions in the cells may also affect the succulence of plants (Chapman 1960: 300–302).

Despite the fact that low levels of sodium in soils may have beneficial influences with regard to agricultural management, it is recognized that "of all the inhibitory substances that plants may encounter in their natural chemical environment, none impairs or prevents their growth on so large a scale as salt" (LaHaye and Epstein 1969:395). Although calcium is known to play a crucial role in the selective transport or exclusion of sodium and other minerals by plant cell membranes, LaHaye and Epstein (1969) state that there appears to be "little recognition of the role of Ca in affecting the responses of plants to saline conditions." [For further discussion of this paper, see Bernstein (1970:1398) and Shear (1970:1388) and the rebuttal by LaHaye and Epstein (1970:1388).] LaHaye and Epstein found that in bean plants (*Phaseolus vulgaris*), an extremely salt-sensitive species, $CaSO_4$ in the growth nutrient prevented the mass intrusion of sodium into the plant leaf, a response believed to characterize the adverse effects of salinity. However, it should be noted that the level of sodium in their medium was only one-tenth that of seawater. Conversely, we pointed out earlier that sodium reduces the calcium content of the wheat grain.

How sodium and calcium relate metabolically to each other in the food plants of the breeding grounds and of the fall feeding grounds in the Paleozoic Basin of Hudson Bay and how these two elements affect the mineral composition of food plants on soils of the wintering grounds on the Gulf Coast have yet to be determined. In this connection it should again be noted that Gulf Coast soils associated with the formation of the Mississippi River delta contain high levels of minerals when compared with adjacent coastal soils which, like the South Atlantic coastal plain, are notably low in nutrients. The calcium content of the soils of the slightly brackish marshes which are used as wintering grounds by blue and snow geese may be an important factor that keeps the food plants of these geese palatable and within the sodium tolerances of this species.

THE PRIMARY ROLE OF SODIUM IN MINERAL METABOLISM

Can the sodium ion be regarded, in a somewhat vernacular sense, as the "master ion"—albeit one that may have only a passive or indirect influence on the uptake of other ions? Aside from indirect metabolic effects relating to the level of sodium intake, there is a teleological basis for this conjecture. The sodium ion is the primary ion which invertebrates and vertebrates have or had to contend with in the seas and in their evolutionary ascent from salt water. It is the only ion for which anatomical structures have been developed, become enlarged, or undergone seasonal hypertrophy to handle excesses of the ion or to extract it from the aquatic environment, e.g., the salt-transport organelle in brine shrimp (Copeland 1966), a rectal gland in sharks (Burger and Hess 1960), the pharyngeal villi of soft-shelled turtles (Dunson and Weymouth 1965:67), salt glands in marine reptiles (Schmidt-Nielsen and Fange 1958; Ellis and Abel 1964; Dunson et al. 1971), and supraorbital or salt glands in birds (Shoemaker 1972). Sodium is the primary component of blood involved in the maintenance of normal osmotic pressure. Lastly, its presence is requisite for most metabolic transfer systems. As Steinbach (1962:714) has stated, "Sodium provides the ubiquitous metallic component of the environment against which protoplasm apparently likes to work."

Davis (1972:66) in a review of the evidence for competition among mineral elements in their absorption from the gut concludes that competition for absorption sites is minimal and that most direct competition between elements depends on whether or not "chemical complexes may be found rendering the elements insoluble and therefore unabsorbable." Davis believes that competition for absorption sites within the body tissues is a more important factor in trace element uptake (1972:67). This opinion lends support for our conclusion that the initial uptake of minerals in excess of needs and the relative levels of minerals in feathers can partly be explained on the basis of the law of mass action.

If the absorption and turnover of sodium directly influence the absorption and turnover of calcium, potassium, and magnesium—as our data and interpretations for pheasant feathers seem clearly to suggest—and if calcium levels in the gut in turn have at least some direct or indirect effect on the level of trace-element absorption, the levels of sodium in the diet must surely influence the rates at which various regulatory mechanisms must operate to maintain homeostasis for both the bulk and trace elements. If these generalizations hold true, then sodium levels, both in experimental diets and in ecosystems, must be considered before the absorption and turnover of other elements in the animal under study can be fully evaluated.

INTERELEMENTAL RELATIONSHIPS AND FEATHER MINERALS

The problems, or perhaps questions, of the degrees of the complex interelemental influences in plant and animal nutrition are of major concern. These questions apply to interactions between macroelements and trace elements as well as to interelemental influences within these two categories. The question of interaction between calcium and

magnesium in mammals has been touched upon, and evidence for noninhibiting relationships between elements in their absorption by geese, as indicated by levels of these elements in the primary feathers, was presented.

The principal reason for the high number of interelemental correlations in feathers from the Cape Henrietta Maria population is probably related to the relatively small feeding range of the colony and the likelihood that the soil mineralogy of most of this colony area is relatively uniform. However, as noted earlier, at least one portion of the feeding area is rich in iron. Fewer interelemental correlations were found in feathers of geese that ranged over large areas having soils with more diversified mineral content.

In Table 38 feather mineral data for the Hudson Bay and Wrangel Island colonies of blue and lesser snow geese and Ross' geese were combined for tests of correlative significance. A similar set of analyses was carried out for the greater snow geese (Table 39) and for fifteen races of Canada geese (Table 40). All significant mineral correlations found were positive. Gordus (1973) found many similar positive correlations among the eighteen elements that he studied in human hair (Table 38), and he cited similar findings by Coleman in a study of human male and female subjects in Great Britain. In the latter study all correlations were positive except that between calcium and iodine and that between copper and iodine.

Increased levels of calcium in the feathers tended to be accompanied by increased levels of the other elements with the exception of potassium in blue, lesser snow, and Ross' geese. Thus, most of the feather minerals studied, to the extent that they were valid reflectors of absorption rates, failed to demonstrate clearly competitive or inhibitory influences between elements at the mineral ingestion rate of these geese.

Mineral relationships of the kind cited in an earlier discussion are reaffirmed by the data in Tables 38–40. Aluminum and silicon were found strongly associated with one another and potassium with both, reflecting the presence of the latter in micaceous clays; iron and manganese correlations reflect, at least in part, their association in soils and ore bodies; correlations of iron with silicon and aluminum were highly significant; and significant copper and zinc correlations mirror the common association of these elements in metalliferous areas. Sodium, iron, manganese, and aluminum were most frequently correlated with the other elements in Tables 38–40.

38. Coefficients of correlations between elements in vanes of primary feathers of combined samples of blue and lesser snow geese from five Hudson Bay colonies, Wrangel Island, U.S.S.R., and Ross' geese from Karrak Lake, N.W.T.

	Macroelements				Trace Elements						
	Ca	Mg	Na	P	Fe	Zn	Mn	Cu	B	Si	Al
MACROELEMENTS											
Ca		.561*	.312	.307	.175*	.349*	.459*	.485*		.432	.367
Mg	.561*		.392	.323	.298*		.489*	.367	.411	.412	.496
Na	.312	.392				.347		.327	.243		
P	.307	.323			.344	.320	.229		.619		.230
TRACE ELEMENTS											
Fe	.175*	.298		.344			.209	.146	.294	.589	.617
Zn	.349*		.347	.320				.283*		.286	
Mn	.459*	.489*		.229	.209			.335*	.335	.228	.405
Cu	.485*	.367	.327		.146	.283*	.335*			.225	.200*
B		.411	.243	.619	.294		.335			.169	.312
Si	.432	.412			.589	.286	.228	.225	.169		.829
Al	.367	.496		.230	.617		.405	.200*	.312	.829	
Number of correlations	9	9	5	7	8	5	8	8	7	8	8
Mean value of correlations	.383	.417	.324	.339	.334	.317	.336	.296	.340	.396	.432

Notes: Significance levels for samples with 183 degrees of freedom—$P < .05 = .144$; $P < .01 = .189$; $P < .001 = .240$. Correlations marked by an asterisk were found to be significantly related ($P < .05$) in human hair by Gordus (1973:Table VI). Gordus did not report sodium, potassium, or silicon.

39. Coefficients of correlations between elements in the primary feathers of nineteen greater snow geese

	Macroelements				Trace Elements						
	Ca	Mg	Na	P	Fe	Zn	Mn	Cu	B	Si	Al
MACROELEMENTS											
Ca						.505					
Mg			.620			.808	.607				
Na		.620		.559	.743	.580		.563	.608		.637
P			.559		.643		.568				.610
TRACE ELEMENTS											
Fe			.743	.643		.652				.652	.910
Zn	.505	.808	.580								
Mn		.607		.568	.652					.832	.691
Cu			.563						.886		
B			.608					.886			
Si					.645	.832					.816
Al			.637	.610	.910	.691				.816	
Number of correlations	1	3	7	4	5	4	4	2	2	3	5
Mean value of correlations	.505	.714	.616	.595	.719	.636	.675	.725	.747	.764	.733

Note: Significant levels for samples with 17 degrees of freedom: $P < .05 = .456$; $P < .01 = .575$; $P < .001 = .693$.

40. Coefficients of correlations between elements in combined samples of fifteen populations of Canada geese comprising 364 individuals

	Macroelements				Trace Elements						
	Ca	Mg	Na	P	Fe	Zn	Mn	Cu	B	Si	Al
MACROELEMENTS											
Ca		.267	.298		.422	.512	.480	.200	.109	.548	.540
Mg	.267		.319				.167				
Na	.298	.319		.111	.291	.312	.295		.362	.340	.269
P			.111		.125		.151	.139	.379	.148	.168
TRACE ELEMENTS											
Fe	.422		.291	.125		.372	.594		.286	.840	.864
Zn	.512		.312		.372		.149	.144	.234	.452	.443
Mn	.480	.167	.295	.151	.594	.149			.231	.563	.568
Cu	.200			.139		.144					
B	.109		.362	.379	.286	.234	.231			.343	.249
Si	.548		.340	.148	.840	.452	.563	.343			.829
Al	.540		.269	.168	.864	.443	.586	.249	.829		
Number of correlations	9	3	9	7	8	8	9	3	8	8	8
Mean value of correlations	.375	.251	.289	.174	.474	.327	.357	.161	.274	.508	.491

Notes: Significance levels for samples with 362 degrees of freedom—$P < .05 = .103$; $P < .01 = .135$; $P < .001 = .172$. The breeding ranges of these populations extend from coastal Labrador to central British Columbia and the Yukon-Kuskokwim River delta, Alaska.

A concept of a "hand-in-hand" absorption of minerals in some rough proportional relationship to each other and to concentrations in the ecosystem is exemplified by many of the feather mineral patterns of geese from the various colonies that represent expanded and contracted versions of the same basic pattern. In brief, feather keratin has provided a remarkably sensitive record of the environmental and metabolic relationships of minerals.

ECOSYSTEM MINERAL LEVELS VERSUS METABOLIC CONTROLS

A summation of our judgments as to the relative roles of absorption and excretion in maintaining homeostasis in geese is given in Table 41. If sodium levels in the environment were constant, absorption of calcium, potassium and phosphorus could be judged to be importantly regulated by metabolic controls. However, as sodium levels in the ecosystems to which wild geese are exposed during the molt vary widely from area to area and because mineral absorption in these birds is related to their total response to the entire complex of minerals in the nutrient chain, it must be concluded that the levels of macroelements in the primary feathers must be related first to the total complex of environmental minerals and secondarily to the metabolic regulation of absorption of elements individually. Although the absorption of magnesium is also enhanced by sodium, its absorption appears to be largely independent of calcium and more directly related to the geologies of their breeding and/or their molting areas—presumably mineral levels in the nutrient chain (e.g., Figures 95–96, 100, 260, 264, 266).

The minerals whose levels in feathers appeared to be more directly related to their levels in the ecosystem and under a lesser degree of metabolic regulation of absorption than appears to be the case for the macroelements—iron, manganese, copper, silicon, boron, and aluminum—are either trace elements or, in the case of the latter two nonessential minerals, physiological artifacts, if one will. There is, however, an exception to this grouping; we found strong evidence that zinc, which plays a vital role in some enzyme systems but which can be toxic in the presence of copper deficiencies, is the most closely regulated of the minerals we studied from the standpoint of both absorption and excretion.

We should recognize here the association found between unusually low zinc levels in hair and nutritionally deprived people in the Middle East

41. Estimation of the relative roles of elemental concentrations in the ecosystem and physiological factors in determining elemental levels in feather keratin

Element	Ecosystem Influence	Physiological Regulation	Physiological Regulation	
			Absorption	Excretion
MACROELEMENTS				
Ca	1	2	2	1
Mg	1	2	2	1
Na	1	2	2	1
K	1	2	2	1
P	2	1	1	2
TRACE ELEMENTS				
Fe	1	2	2	1
Zn	2	1	1	2
Mn	1	2	2	1
Cu	1	2	2	1
B	1	2	2	1
Si	1	2	2	1
Al	1	2	2	1

Note: Estimations of the relative roles of the two major physiological mechanisms in maintaining homeostasis of the blood plasma are also presented. Numerals indicate rank of importance.

(Strain et al. 1966). However, a later attempt to correlate paired plasma and hair samples from pre-pubescent children was unsuccessful (McBean et al. 1971).

These elemental divisions—admittedly judged from physiologically skewed data and an unorthodox point of view—are readily rationalized. The bulk minerals are major constituents of the body. Closer metabolic regulation of the absorption of these elements is manifestly requisite for normal development and body maintenance. These elements are generally available in the ecosystem in adequate amounts—if they were not, a given animal would not be there in the first place—in accordance with Liebig's law of the minimum.

Essential trace elements, on the other hand, are requisite for good health and survival, but with the exception of iron, they generally occur in trace amounts in goose feathers and in the ecosystem as well. Because these trace elements are required but are in short supply in the ecosystem, physiological mechanisms do not seem to have been evolved or to have been needed to limit their absorption closely. In a somewhat similar vein Frieden (1974:3) has commented: "What has not been fully appreciated is that the nature of the trace element requirement, particularly the metal ions, ap-

pears to have been limited by their availability in the early oceans when primitive life was evolving."

Iron, a nearly ubiquitous element, is largely bound in the soils in forms unavailable to plants, and most of the iron complexes that reach the digestive tracts of animals are poorly absorbed. Boron, silicon, and aluminum are apparently, in varying degrees, readily absorbed. They appear to be just as freely excreted. Boron, although toxic, is normally present in the nutrient chain in trace amounts; silicon and aluminum, however, are ingested by geese in relatively large amounts in clays, and silicon is the most abundant inorganic element in the grasses and sedges on which geese feed. Because these elements are largely physiologically inert, physiological barriers at the absorption level of homeostasis, particularly in the case of silicon, do not appear to have been evolved, at least in geese. Sahagian et al. (1971:323) state: "Little is known about physiological mechanisms which regulate tissue concentrations of trace metals. That there may be some control mechanisms operative, at least for the 'essential' metals, is suggested by the smaller variability observed in their tissue concentrations as compared with the 'non-essential' metals."

HOMEOSTASIS—A MULTIPHASE PROCESS

Most questions involving mineral metabolism, directly or indirectly, are ultimately concerned with the problem of homeostasis. However, in completing this report we realized that the literature has not adequately stressed the fact that homeostasis is a multiphase or multistage process. For example, on a temporal basis, the mitochondria play a more important role in maintaining cellular homeostasis of calcium than do transepithelial exchanges of calcium (Simkiss 1974:120; Wasserman 1972:355). Some studies in the past have seemingly overstressed the importance of the absorption stage, e.g., by stressing a "mucosal block" of iron. The desirability of underlining the multistage aspect of homeostasis became apparent when it was emphasized to us that magnesium is tenaciously controlled in the plasma and within cells. Yet we had concluded on the basis of feather mineral data that magnesium is not closely metabolically controlled. However, this was a judgment based on the fact that magnesium is apparently relatively freely absorbed by geese if its level is high in the nutrient chain. Our findings are, therefore, in accord with the statement in Oser (1965:556): "As is the case with serum potassium, plasma magnesium concentrations are not necessarily indicative of intracellular levels, reflecting cellular losses, intakes and urinary excretion." Thus, any statement as to the relative degree of metabolic control of elemental concentration in animals must be related to the stage of homeostasis and the tissue involved.

In Table 41 we have also summarized our subjective appraisals as to the relative importance of the role of absorption and excretion in maintaining homeostasis. If wide fluctuations in the concentrations of a mineral in goose feathers occurred on the high end of the scale, we concluded that the regulation of absorption had to be minimal and that the regulatory mechanism for excretion had to be correspondingly more important. However, it is reasonable to conclude in the cases of silicon and the elements boron and aluminum (considered nonessential) that possibly no mechanism for the regulation of their absorption exists—and that these elements are excreted as freely as they are absorbed. In our concern with the relative roles of absorption and excretion in maintaining homeostasis, we are not unaware of the complex feedback mechanism involved in maintaining fairly stable mineral levels in the plasma. Nor are we unaware of the roles of hormones, vitamin D, and bone and tissue reservoirs and of the importance of the kidneys, cell membranes, and cell organelles in maintaining plasma and tissue homeostasis. Rather, our data seemingly relegate to an inferior position the major metabolic role that absorption is sometimes believed to play in maintaining homeostasis. Or stated another way, feedback mechanisms controlling excretion and secondary absorption seem to be more sensitive than those controlling initial absorption. In large measure our independent appraisals, based on feather mineral data, on the relative importance of homeostatic mechanisms have been in agreement with those presented in the most recent literature.

GROWTH RATES OF PRIMARY FEATHERS

As discussed earlier, during the molt the metabolic capacities of the goose appear to be directed primarily to the most rapid growth of the wing feathers to enable it again to have the powers and safety of flight in the shortest period of time. Measurements of primary growth of captive geese (Table 42) revealed that the average rate of the ten primaries varied remarkably little among species and subspecies—probably not significantly so—varying between 6.8 and 8.2 and averaging 7.4 mm per day for the six species and subspecies studied. The average rate of growth of the primary feathers was 8.0 mm for an adult male blue goose and 7.7 for an adult female. Undoubtedly on the breeding grounds, the average rate of growth is somewhat faster than the rates reported here for caged geese. The rates of deposition of minerals in the feathers would, of course, tend to vary inversely with the rates of feather growth, and variations in rates of growth could account in part for expanded and contracted versions of similar feather mineral patterns from any one area. It should also be pointed out that growth rates slow up as the primaries near their terminal growth (Hanson 1965:114).

NUTRIENT PULSES IN THE PLASMA

Evidence has been presented indicating that the primary source of feather keratin minerals is the diet. Variations in mineral content in the nutrient chain must be reflected in blood levels although *relatively* small variations in mineral levels in plasma can probably account for *relatively* large variations in the feather keratin. However, to detect changes in mineral levels in plasma, at least in the case of geese, it is mandatory to sample their

42. Mean growth rates of primary feathers of six species and subspecies of captive wild geese

Species and Subspecies	Sex	Number in Sample	Average Daily Growth Rate (mm)	Range
Anser c. caerulescens	M	1	8.0	
Anser c. caerulescens	F	1	7.7	
Anser rossii	M	1	7.9	
Branta canadensis maxima	M	1	6.8	
Branta canadensis interior	M	8	7.2	6.8–7.8
Branta canadensis interior [a]	M	5	8.2	7.6–8.7
Branta canadensis interior	F	4	6.8	
Branta canadensis interior [a]	F	6	7.6	7.1–8.5
Branta canadensis hutchinsii [b]	M	1	7.7	
Branta canadensis hutchinsii	F	1	7.5	
Branta canadensis minima	M	1	7.5	
Branta canadensis minima	F	1	6.4	

[a] These geese were held in a large grassy pen, a factor that probably accounts for the faster rate of growth of their primaries as compared to the others which were held in wire-floored cages and maintained only on mixed grains and duck pellets.

[b] The geese referred to here are the small Baffin Island race.

blood while they are in an absorptive state. In the course of studies to ascertain the normal blood chemistry of Canada geese (Hanson 1958; Hanson unpublished), hundreds of blood samples were taken, but these samples were always taken from geese that had been held overnight and part of the day without food. (Despite use of heparin, the plasma of geese in the absorptive state was cloudy and often jelled, making it unusable for routine analytical procedures.)

Plasma mineral values for Canada geese in the postabsorptive state exhibit no significant ecosystem-related variations (Table 43). For example, data for 1) sodium and potassium of Canada geese wintering in southern Illinois; 2) maintained in pens at Havana, Illinois, on a diet of mixed grains and commercial poultry pellets; or 3) flightless Canada geese caught in funnel-drive traps on Akimiski Island (where they consume coastal marsh plants and resort to offshore waters [Figure 252] during diurnal high tides) did not differ significantly in respect to plasma values although large and significant differences were found for these elements in the respective feather samples. It must be concluded that feather mineral levels relate to the "nutrient pulses" of the absorptive state; hence deviations of plasma mineral levels from the homeostasis of the fasting state that are related to mineral levels in the ecosystem and in the feathers

43. Mean values of elements in plasma and feather keratin of wild and caged adult Canada geese (*Branta canadensis interior*) from the Mississippi Valley Flyway and from Akimiski Island, N.W.T., arranged according to phenology of life history

		Wintering (December–January)		
		Plasma [b] (Horseshoe Lake, Ill.)		Feathers [c] (Union County, Ill.)
Element	Sex [a]	n	Mean	n = 15
Ca	M	17	11.7 ± 0.3	1387 ± 59
	F	16	11.9 ± 0.4	
Mg				263 ± 25
Na	M	12	156.9 ± 3.2	247 ± 12
	F	7	158.4 ± 5.0	
K	M	13	3.3 ± 0.8	49 ± 3.0
	F	5	3.9 ± 0.1	
P	M	14	3.9 ± 0.2	147 ± 5.2
	F	12	3.5 ± 0.3	

Note: Mean values with standard errors given are in the following units: *feathers*—parts per million; *plasma*—calcium, mg per 100 cc; magnesium, mg per 100 cc; sodium and potassium, milliequivalents; and phosphorus, mg per 100 cc.

are likely to be detected most readily during absorptive periods.

It therefore seems likely that minerals are incorporated into feathers as invisible "nutrient bars" that correspond with feeding and feather growth during the day as opposed to the evening periods of fasting and rest. If barring of the mineral content of the feather vanes exists, the average spacing of the maxima of the nutrient bars in large Canada geese should be 7.2 mm, the average daily rate of growth of their primaries (Table 42). Again, if nutrient barring exists, in the primaries of the smaller Arctic geese it would be less readily detectable, as these geese probably do some feeding during eighteen to twenty hours of the twenty-four-hour period of light.

Faint pigment barring can be seen in the primaries and remiges, or tail feathers, of many birds (e.g., the American robin *Turdus migratorius*), but they are particularly noticeable in the feathers of passerine birds. Known as growth bars (Wood 1950), they reflect the periods of intensive daytime feeding and, coupled with the alternate unbarred portions, represent twenty-four-hour growth periods.

Inferential support for the hypothesis that "nutritional barring" exists in the primary feathers of geese is provided by findings from studies of the circadian rhythm of nutrients in the blood plasma of man (Christie 1972). Circadian rhythms have been shown for many amino acids (particularly methionine and cysteine, found in relatively high amounts in feathers) in the blood and for sodium, phosphorus, iron, and copper. No clear rhythm has been established for levels of calcium, magnesium, and zinc in the plasma. Although not necessarily of endogenous origin, circadian rhythms in the excretion of calcium, magnesium, sodium, potassium, and phosphorus have also been established.

TRANSFER STATES OF SOME FEATHER MINERAL ELEMENTS

We have not previously discussed the possible forms in which the various elements studied reach the feather follicles. Are they attached to amino acids (Bronner 1964:363) that filter through the capillary walls, or do they occur principally in ionic form and become attached to the keratin *in*

Prenesting (Late April)	Molting (July and August)						Premolting (May)
Feathers[c] (Kinoje River, Ont.)	Plasma[d] (Akimiski Island, N.W.T.)		Feathers[d] (Akimiski Island, N.W.T.)	Plasma[e] (Caged geese, Havana, Ill.)		Feathers[f] (Caged geese, Havana, Ill.)	Plasma (Caged geese, Rochester, Minn.)	Plasma (Caged geese, Rochester, Minn.)
n = 7	n	Mean	n = 13	n	Mean	n = 8	n = 10	n = 14
2064 ± 103	9	9.4 ± 0.4	4454 ± 487	16	16.3 ± 0.4	804 ± 39	10.5 ± 0.2	9.3 ± 0.2
343 ± 8.9			1969 ± 200			100 ± 10	2.5 ± 0.1	2.4 ± 0.1
478 ± 32	9	159 ± 9.8	497 ± 37	17	150 ± 3.2	131 ± 37		
32 ± 2.7	9	2.8 ± 0.4	116 ± 8	8	3.8 ± 0.3			
157 ± 30			271 ± 36			<350	4.9 ± 0.4	3.0 ± 0.2

[a] Sexes are combined for all values except plasma levels in geese wintering at Horseshoe Lake, Illinois.

[b] Methods of plasma determinations were: calcium—Clark and Collip; magnesium—atomic absorption; sodium and potassium—flame photometry; and phosphorus—Fiske and Subbarow.

[c] Geese wintering at Union County, Illinois, and breeding in the Kinoje River area of Ontario are both representative of the Mississippi Valley Flyway population and, very likely, at least a portion of the geese wintering at Horseshoe Lake, Illinois.

[d] Plasma and feather samples for geese breeding on Akimiski Island are both from geese with partially grown primaries. However, samples were taken in different years.

[e] These geese were maintained on a diet of mixed grains and duck pellets. No supplemental grit was provided.

[f] These geese were maintained on a prepared diet containing the standard University of Illinois poultry salt mix.

situ? An attempt to outline the sequence of events that accounts for the tranfer of elements from their absorption in the lumen of the digestive tract to their incorporation into the feather anlage would be an appropriate note on which to conclude this report.

Freeman (1973) has summarized findings to date on *in vitro* studies of the functional groups of amino acids and amino acid complexes involved in metal binding. A few of the salient points of Freeman's review of this complex subject are apropos here if casual readers and generalists (like ourselves in this case) are to have at least a superficial understanding of mineral transfer systems as they *may* apply to the mineralization of feathers.

According to Freeman, if acid dissociation constants are used as criteria, the sequence of metal binding tendencies by functional groups of amino acids are carboxyl > imidazole > amino. Terminal amino groups are reported, however, to be among the most commonly involved functional groups binding metal ions.

In respect to the carboxyl groups, Freeman (1973:126) states that "the tendency of carboxyl groups to occur in chelate rings has already been noted. With a metal having suitable coordination geometry, the ability of a carboxyl group to participate in chelation depends upon the presence of a second donor atom at the correct spacing for the completion of a five- or six-membered ring. This condition is satisfied in α- and β-amino acids by the availability of the terminal amino group, thus accounting for the chelate structures of the majority of amino acid complexes." Unident coordination, however, through the carboxyl groups is known only for iron and glycine.

The binding of transition metals, particularly copper, to the hydroxyl groups of serine, threonine, and tyrosine does not appear to occur within normal physiological ranges of pH. Hence the binding of metals of these amino acids does not appear to be different than for glycine and alanine; the above three amino acids are not more important in the transport of minerals than those amino acids which do not contain the hydroxyl side-chain. "For other first-row transition metals, the stability constants of serine and threonine complexes on the one hand, and α-alanine and α-aminobutyric acid complexes on the other hand, are so similar that the same type of chelation is presumed to be common to all of them. The crystal structures of Ni(L-Ser)$_2$(OH$_2$)$_2$, Cu(L-Ser)$_2$, and Zn(L-Ser)$_2$ confirm that the coordination is simply through N(amino) and O(carboxyl)" (Freeman 1974:52).

Metals notably involved in sulfhydryl linkages "occupy a triangular area near the centre of the Periodic Table" (Freeman 1974:153). Sulfhydryl linkages to silver and mercury are, of course, particularly well known. Formation constants also suggest that the enhanced stabilities of cysteine complexes of manganese, iron, and zinc as compared with amino linkages of these metals with glycine and histidine are due to accompanying metal-sulphur bonding.

The only metal concerned with in the present study that is bound to the methionine side-chain is copper (others similarly bound—lead, platinum, silver, and mercury).

Freeman (1974:159) has cautioned, in effect, regarding too literal translation at present of models based on physio-chemical findings to biological systems. "The claim has often been made that complexes of the types which have been discussed in this chapter are 'of biological interest.' In a few cases it is true that a particular complex turns up in some biological system. In many more cases the phrase 'of biological interest' expresses the hope that the systematic study of many complexes will contribute to the discovery of the rules which govern the interactions between metal ions and naturally occurring ligands. Many naturally occurring ligands are tremendously complicated molecules. The ligands whose complexes can be systematically studied are relatively simple. A metal-peptide complex may have some properties in common with a metal-protein complex. So far as these properties alone are concerned, the simple complex is a 'model compound' for the biological interaction. But we must remember that it is a model—not a replica."

A commonalty between plants and animals appears to exist in respect to the association of protein metabolism and mineral absorption and transfer. Thus, Sutcliffe (1962:165) concludes: "The remarkable ability of growing [plant] cells to attract solutes seems to be linked with a capacity to extract water and nutrients from the sieve tubes, *and may well depend on the intensity of protein synthesis or turnover*" (our italics).

In view of the intensity and sophistication of current research on the mineral metabolism of animals, the concluding remarks of Sutcliffe (1962:165–66) also seem apropos here: "Early inquirers embarked upon studies of salt absorption and transport in the belief that these processes could be explained largely without reference to other plant activities. It is now clear that to explain what seems to be a specific aspect of plant physiology one has almost to understand life itself. Thus the mechanism of an apparently simple function such as salt absorption proves to be immensely complex, and may tax the ultimate limits of our understanding.''

Transfer States of Macroelements

Calcium. Notable variations in calcium absorption rates in wild geese are indicated by the wide variations in the calcium content of feather keratin (tenfold to perhaps twentyfold [<1,000–10,200 ppm], as the lowest levels could not be determined by the analytical method that was routinely used). The question arises as to whether the various mechanisms of absorption held by current opinion to account for calcium absorption can encompass the variation in calcium flux indicated by such calcium-content variations in the feathers. Bray and Clark (1971) have reviewed the problems of equating *in vitro* experiments with *in vivo* studies, and Hazelwood (1972) has questioned the existence of cellular ''pumps'' as presently conceived. Various mechanisms of calcium transport have been recently reviewed by Wasserman (1972) and Simkiss (1974), but it seems clear from DeGrazia's (1971) earlier discussion that no unanimity exists on the subject as to the primary mechanism involved. Whether active calcium transport, calcium absorption into the mucosal cells via a calcium-binding protein, or calcium diffusion facilitated by electrochemical gradients and specialized subcellular structures such as microvilli most readily explains the considerable variations in calcium movement across the mucosal epithelium suggested by the feather data remains to be determined.

Egg laying increases calcium absorption in chickens, and this increase is correlated with an increase of calcium-binding protein (CaBP) in the duodenum (Wasserman et al. 1971). However, the short-term regulation of calcium absorption triggered by shell formation appears to involve another mechanism (Bar and Hurwitz 1971). In view of the fact that an osteoporotic condition of the tarsometatarsi (and presumably, a negative calcium balance) is associated with molt in geese, the CaBP content of the intestine of these birds at that time in relation to the CaBP content of the intestine in other periods of their life cycle would be of interest.

Wasserman (1972) has proposed an alternate route of calcium transfer that would involve the tight intercellular junctions and intercellular spaces. A passive, extracellular route for sodium and potassium has also been suggested (Jackson

and Smyth 1971). The importance of a ''shunt route'' has been emphasized by Ussing et al. (1974) and Schultz et al. (1974). The primary importance of tight intercellular junctions as a route for ions and small molecules across capillary walls and mesothelial cells appears to be well established (Michel 1970; Ussing 1971). Because of the function of tight junctions in the transport of water and solutes, Dibona and Civan (1973) have suggested the more appropriate term, ''limiting junction.''

In geese, because of the apparent association of high rates of calcium absorption with calcium turnover rates and the relation of the latter to sodium intake and absorption, intercellular junctions may be a calcium transfer route of primary importance. Perfusion studies of the duodenal area of human subjects have indicated that sodium chloride facilitated calcium absorption; the best absorption of calcium was achieved with isotonic solutions of sodium chloride and decreased rates of absorption were accompanied by hypotonic perfusions (Chapuy and Pansu 1971).

Birge et al. (1972) believe that the calcium efflux from the base of the mucosal cell is mediated by a specific enzyme, activated by sodium, on the serosal side of the cell. This calcium transport out of the mucosal cells is believed to be independent of bulk water flow and of net sodium transport and independent of a sodium-potassium exchange pump. These conclusions were, however, predicted by Schachter's (1963:173) experimental findings that indicated that ''sodium chloride may be required for energy-dependent, active transfer of calcium out of mucosal cells.''

Simkiss (1974) and his associates have determined that transported calcium in the cells of the chorioallantois never exists within the cell in ionic form. The main calcium carriers in human blood are proteins and chelates (Manery 1969:443), but 47.5 percent of the calcium in the plasma is in the ionic form (Walser 1969:242). This ionic fraction of the total calcium in the plasma is the most metabolically active form of calcium but the least variable quantitatively; the bound fraction, on the other hand, varies considerably in relation to total serum protein.

Albumins and globulins are the principal protein carriers of the calcium (particularly the former in man, Osterberg 1974) and in the chicken. In man the pre-albumin fraction showed the highest association with calcium (Bronner 1964:371). In this conjunction it is noteworthy that electrophoresis (Tiselius apparatus) revealed a particularly well-defined pre-albumin component in the plasma of two molting adult female Canada geese but not in wintering geese (Hanson 1958). The pre-albumin component of chicken plasma is normally found only in laying hens, but it has been produced in cockerels with diethylstilbestrol (Clegg et al. 1951). These investigators concluded that sex hormones, by increasing the amount of serum protein, increased the calcium binding capacity of the laying hens. Ericson et al. (1955), by means of ^{32}P studies, have shown that the pre-albumin component of laying hen sera is extremely rich in phosphorus. Because of the osteoporotic condition of the tarsometatarsi of geese during the molt, the pre-albumin fraction of the plasma may function as a carrier for both calcium and phosphorus, at least in adult females. Furthermore, the existence of a pre-albumin fraction during the molt is evidence that specialized protein molecules synthesized during, and as a result of, high estrogen levels during the breeding season persist for at least several weeks. (An increase in concentration and persistence into the molt period of another protein molecule, ceruloplasmin, can also be assumed to occur as a result of increased estrogen levels of the breeding season—at least in the case of birds. Such an increase would account for the inferred sex differential in the copper-binding capacity of the plasma of Ross' geese, as discussed earlier, and the transmittal of this increased copper load to the growing primaries. A basis for the belief is the finding that increases in concentrations of ceruloplasmin in sera of women administered contraceptive steroids are dose-dependent, Briggs and Briggs 1973).

Protein binding would not impede the transcapillary movement of calcium in the central pulp of the feather blastema, as the redistribution of calcium between the ionized and bound forms is rapid (Irving 1973:56). We assume that calcium (and other metal ions) is secondarily attached to the forming keratin in ionic form rather than incorporated into the keratin while bound to an amino acid.

Increases in protein intake increase the absorption of both calcium and magnesium (Tansy 1971:197). This increase is presumably the result of the enhancement of calcium absorption by various amino acids (Wasserman and Taylor 1969:385; Heinz 1972), an interaction suggesting that at least some calcium may be bound to amino acids during a stage of absorption.

If significant proportions of absorbed metals are transported in the plasma bound to both amino acids and amino acid complexes, as we suggest, it is noteworthy that concentrations of free amino acids are greatly increased in geese during the molt. Studies by Hanson and James D. Jones of the Mayo Clinic (unpublished) have revealed that the total concentrations of 16 free amino acids in the plasma of molting geese exceed spring and premolting levels by 30 and 31 percent respectively and concentrations in wintering geese by an average of 55 percent. These data are also noteworthy because the free amino acids are believed to be either the precursors or by-products of keratinization (Stettenheim 1972:29). Stettenheim (1972:6) states that keratinization "involves two processes, the synthesis of keratin protein and the breakdown of other elements in the epidermal cells by hydrolytic enzymes. In mammals, keratin protein synthesis, polymerization and initial keratin bonding is sequential; in birds it occurs simultaneously." The above findings are, therefore, coherent with the concept that the degradation of the breast muscles of geese and other waterfowl during the molt is an evolutionary development whereby free amino acids, particularly cystine and methionine, are made readily available for rapid growth of the primaries and secondaries (Hanson 1962).

Curan (1973) has shown that the transport of amino acids from the mucosal to the serosal side of *in vitro* preparation is almost completely dependent on the presence of sodium in the mucosal bathing solution. Conversely, the rate of active sodium transport in similar preparations is enhanced by the addition to the solution of alanine, an amino acid that is actively absorbed.

Magnesium. The mechanism of magnesium absorption is unknown, but only a small portion of the magnesium in the intestine is absorbed by simple diffusion. Both sodium and potassium have been reported to displace magnesium from the mucosal cells, the extent of which is believed to be "a function of monovalent cation entry into the epithelial cells. . . . It is suggested that a possible role of penetrating ions on the intestinal epithelium may be to loosen membrane-bound magnesium for participation in the regulation of passive permeability" (Tansy 1971:199). These conclusions seem to parallel the "shunt route" explanation discussed under calcium and sodium and may represent an alternate interpretation of the same phenomenon. Wholly in accord with our finding on the influence of sodium on magnesium absorption in geese and pheasants is the statement by Texter et al. (1971:392) that "sodium uptake by the intestinal segment increases the passage of magnesium into the bowel."

Of the magnesium in human plasma, 55 percent

exists as free ions, 32 percent is protein bound, and the remainder is variously complexed (Walser (1969:242). The ionic form is the metabolically active form of magnesium; because it is closely regulated, it is reasonable to conclude that the magnesium bound to keratin is derived chiefly from protein-bound forms, or, more likely, amino acids. The latter is conjecture, however, as amino acid-bound magnesium does not seem to have been reported.

Sodium. A satisfactorily complete model for sodium absorption has not been described. In general, sodium absorption has been associated with 1) concurrent absorption of a hexose or an amino acid and 2) a nonelectrogenic, or passive process associated with water transport (Jackson and Smyth 1971:147). Two of the most recent surveys of the problem are those of Ussing and Thorn (1973) and Ussing et al. (1974). Hyperosmotic solutions cause the opening of the tight junctures between cells and the dilation of the intercellular spaces during salt transport—even in the virtual absence of transepithelial water movement—thus providing an important route for the passive flow of water and solutes across the epithelium. According to Schultz, Frizzell, and Nellans (1974), this "shunt pathway" plays a major role in the passage of ions, small nonelectrolytes, and water across the small intestine. They estimate that sodium movement via the shunt pathway exceeds the flow via the transcellular route by a factor 7–10 to 1. Although Borlé (1974:367) states that "no definite conclusions can be presently drawn about the role of sodium in calcium absorption," the importance and nature of the mechanism of the shunt pathway that have been determined by many investigators suggest that a simple physical explanation, based on the extent of dilation of the tight junctures, might account for at least some of the increase in calcium and magnesium uptake that accompanies an increase of sodium in the diet and a corresponding increasing of levels of these elements in the feathers of geese and pheasants.

Phosphorus. Inorganic phosphorus in plasma is present as three ions due to the disassociation of orthophosphoric acid (H_3PO_4). In man the proportional concentration of the $H_2PO_4^-$ ion has been calculated to be 18.6 percent; HPO_4^{2-}, 81.4 percent; and PO_4^{3-}, 8×10^{-3} percent (Irving 1964:255). It is reasonable to conclude that phosphorus transmitted to the feather anlage is probably in these ionic forms and approximately in these proportions. Being negatively charged, these ions may find relatively fewer attachment sites on the

keratin being synthesized than do positively charged ions, thus accounting in part for the relatively low concentrations of phosphorus in feathers. And as mentioned earlier, phosphorus in the sera of laying hens has been associated with a pre-albumin fraction.

Transfer States of Trace Elements

The biologically active forms of trace elements are believed to be metal chelates. As metal ions, they may replace each other in biologically active compounds or interact indirectly in respect to absorption, transport, or physiological function (Matrone 1974; Riordan and Vallee 1974). If, however, what have been termed "mixed metal," metallothioneine-type, proteins (Mills 1974) can be shown to be important simultaneous carriers of two or more different transitional metal ions, perhaps these molecules would provide a partial explanation for some of the positive correlations found between transitional metals in primary feathers of geese.

Iron. The metabolic controls regulating iron transfer from the intestinal cell to the plasma is yet unknown. As appears to be the case regarding calcium absorption in geese, "there is a correlation between absorption and plasma iron turnover" (Turnbull 1974:394). However, iron absorption is depressed by inhibitors of protein synthesis presumably due to the lack of a protein carrier. The enhancement of iron absorption by histidine, cysteine, and lysine is believed to be related to the formation of tridentate iron chelates. "These studies provide strong support for the belief that iron taken up into the mucosal cell is partly bound to one or more specific carriers which appear to regulate its passage across the cell to the blood" (Turnbull 1974:388). But data for the Belcher Island Canada geese at least raises the question as to whether sodium enhances iron absorption and whether the "shunt pathway" may be involved. Perhaps it may be pertinent in this regard that iron absorbed from the gut follows a different pathway than that bound to mitochondria (Turnbull 1974:386).

Although absorbed as ferrous ions, systemic iron does not exist in the ionic form (Forth 1974:199). Iron is transported in the bloodstream as a ferritin mineral—presumably as a crystalline core surrounded by a protein shell—the combined components comprising transferrin. "Probably because of the similarities in size and shape, the distribution and rate of exchange of transferrin and albumin between the intravascular and extravascular compartments are almost identical" (Morgan

1974:46). In man, the transcapillary transfer rate of transferrin is estimated to be equal to 1.5 to 2.0 plasma pools per day and in the rabbit it is estimated to be as high as six pools per day. The rate of transferrin exchange with the interstitial fluids varies between tissues, depending on the permeability of the capillary wall (Morgan 1974). It is reasonable to assume that in the case of wild geese in which catabolism of breast muscle tissue, anabolism of the leg muscles and synthesis of feather keratin are all occurring simultaneously during the molt at very high rates, the rate of exchange of nutrients between the vascular and interstitial pools must be very high indeed. This high rate of turnover of nutrients between these pools accounts for the correlation between mineral levels in the environment, their absorption, availability in the interstitial fluid, and their incorporation into the feather keratin at the synthesis of the latter. Relatively small increases in mineral levels in the blood would be compounded each time (in a vernacular sense) there was a transfer of nutrients to the interstitial fluid equivalent to the plasma pool.

According to Frieden (1974:27), iron is reduced to the ferrous state before it is released to various cells and tissues, but Harrison and Hoy (1973:276) state that "the mechanism by which ferritin iron is mobilized *in vivo* is still controversial. It may be released from the intact molecules, from the cores after degradation of the protein, or both. It may be released by reduction to the more soluable ferrous ions, or by formation of soluable ferric chelates, or both." The mode of transfer of iron to the keratin-synthesizing cells is, presumably, unknown, but the transferrin molecule has been shown capable of penetrating into the interior of the reticulocyte. These four steps have been suggested to account for the exchange of iron with the recipient cell (Aisen 1973:300):

   1. Adsorption of transferrin to receptor sites on the reticulocyte.
   2. Formation of a firmer union between transferrin and the reticulocyte. This step may involve actual penetration of the protein into the interior of the cell.
   3. Transfer of iron from protein to cell.
   4. Release of protein.

The molt period in geese, as often pointed out, involves simultaneously catabolism and anabolism of breast muscle and leg muscle protein, respectively, followed by the sequence in reverse order (Hanson 1962). In view of the intense metabolic activity of keratin-synthesizing cells during molt and the extensive reorganization of protein at this time, presumably involving both deamination and transamination processes, it is perhaps reasonable to assume that when these cells transfer iron to within themselves, step 4 may be omitted, the protein shell is degraded, and the various amino acids are incorporated into the keratin.

Zinc. The mechanism of zinc absorption is little known (Becker and Hoekstra 1971:234): "About one-third of the plasma zinc is loosely associated with serum albumins and the remainder is more firmly bound to the globulins." A specific zinc-protein transport compound has not been identified in the plasma (Underwood 1971:214). According to Henkin (1974:301–2) 32 percent of the circulating zinc is bound to the alpha-2 globulin, 66 percent to albumin, and the remaining 2 percent mainly to the amino acids histidine and cysteine. The latter two forms of zinc constitute the exchangeable forms of this element in the blood.

The relative lability of zinc in respect to binding and its capacity to join subunits of proteins to form quaternary structures (Osterberg 1974) may account for its widespread distribution in the tissues of animals (Underwood 1971:210). Relatively large proportions of the total body content of zinc (in mammals) are found in the bone, skin, and other organs. This fact suggests that the effect of the high intake of zinc from plant foods and subsequent absorption pulses in geese are effectively dampened by the widespread binding of zinc to a variety of body tissues rather than its becoming concentrated primarily in feathers by selective incorporation into the synthesizing keratin. This relationship is further suggested by the finding that of the four metals studied by Sahagian et al. (1971)—zinc, mercury, calcium, and manganese—the metal saturation capacity of gut tissue is highest for zinc and lowest for manganese. It thus appears that the relatively small variability of the zinc content of feathers may be related, at least in part, to the lability of its binding as well as to homeostatic control of its absorption and excretion.

Manganese. Manganese, like calcium, is believed to be associated with the metabolism of amino acids with which it forms chelates. Pyridoxal, a derivative of pyridoxine and which has been shown to be an active form of vitamin B_6 in higher animals (West and Todd 1957:763), is believed to participate in this chelation. "These complexes of amino acid, pyridoxal, and manganese are transported in the body more rapidly than are amino acids alone. These very significant observations link protein metabolism to manganese turnover and establish the role of complexes in the metal transport (Pal and Christensen, 1959)'' (Cotzias 1962:431). Manganese in the blood is believed to be almost totally protein bound, but of the major plasma protein components, it has been found to be attached only to the β-globulin fraction (Cotzias and Foradori (1968:113).

Copper. Over 90 percent of the copper in mammalian plasma is bound in the liver to the alpha-2 globulin, ceruloplasmin (Adelstein and Vallee 1962:373); a second portion of the plasma copper is bound to serum albumin and is believed to be the exchangeable copper transferable across cell membranes (Van Campen 1971:220); a third small but significant portion bound to amino acids or complexes thereof has also been determined (Sarkar and Kruck 1965:183; Grassmann and Kirchgessner 1974:523; Margerum and Dukes 1974:205). Presumably, it is from the latter two fractions of plasma copper that copper is transferred to the feather follicle.

In man 90 percent of the plasma copper is bound to the macromolecule, ceruloplasmin, 9 percent to albumin, and the remaining 1 percent to the amino acids, histidine and cysteine. ''It is primarily these micromolecular liganded albumin-copper that comprise the major exchangeable or 'loosely bound' copper complexes in blood'' (Henkin 1974:299–301). The copper bound to a variety of amino acids is in equilibrium at physiological pH with serum-albumin copper. The latter is the major transporting agent of copper in the blood, but the copper that is transported across cell membranes is believed to be complexed with amino acids (Margerum and Dukes 1974:206). Copper ions, necessarily detached from amino acids prior to keratin synthesis, would presumably be available at the site in ionic form for binding to the newly synthesized keratin.

Hoffman (1964:47) believes that, because copper is not easily removed from alpha-2 globulin, the function of alpha-2 globulin is ''apparently not to transport copper but rather to limit the amount of copper deposited in the tissues [primarily liver and brain].'' In this respect feather keratin functions in a parallel, although perhaps adventitious, way. In the case of Wilson's disease, however, alterations in copper homeostasis are believed by Evans et al. (1973:1175) to ''result from the synthesis of an abnormal metal-binding protein with an increased affinity for copper.''

Silicon. Silicon is believed to be absorbed by the guinea pig as monosilicic acid (H_4SiO_4) (Underwood 1971) and is presumably carried in this form by the blood, but to our knowledge the transfer form of silicon and the nature of its binding with keratin substances have not been defined.

Aluminum. The prevalent form of aluminum in plasma and the exact nature of its carrier are unknown. Like other metals, it is believed to complex with proteins. The low level of aluminum generally found in feathers of geese (with certain notable exceptions; see appendix 1) is in agreement with generally low levels found in other animal tissues. These low levels of aluminum are believed due to the low solubility of aluminum compounds in the environment and the ability of aluminum to form insoluble complexes with other substances in the intestine. These factors relate to the physical and chemical properties of aluminum and its compounds (Ondreicka et al. 1971). Nevertheless, the frequency of correlations of aluminum with other elements was high as was also the statistical significance of the correlations (Tables 38–40).

Levels of elements in the primary feathers of geese reflect environmental levels sufficiently well that they can be used effectively to indicate the breeding (or molting) grounds of geese of unknown origin. However, due to many elemental interactions within and between various components of the environment and to metabolic controls and ion interactions (particularly those directly and indirectly related to sodium intake) within the body of the goose, element levels in the feathers are not wholly (in some cases only indirectly) related to environmental levels. We suspect that, with certain exceptions (sodium, potassium, boron, silicon, and aluminum), elemental levels in the feathers tend to most nearly reflect the proportionality of the ''mix'' of elements bound or chelated at some stage of absorption. Those metal ions which are attached to transporting molecules probably become detached just prior to their incorporation in the keratin structure, regardless of whether the carriers per se (amino acids) subsequently become part of the keratin structure.

A definition of the word *coherence* is ''to be logically connected, consistent.'' If a single word were chosen to summarize the relationships viewed here, it would be *coherence.* As a study of living organisms, this one is perhaps unique in that it is an attempt to provide an overview of the steps leading from ''the inert to the inert.'' Mineral levels in feather keratin can be almost as usefully interpreted from the most primary level—the characteristics of the atom as exemplified by the Irving-

Williams series of transition metals reordered in respect to their complexing power (Hughes 1974:76)—as from the mineral characteristics of each succeeding trophic level or explanations of the mechanisms of ion transfer through the nutrient chain of the ecosystem. The basic unity of the biosphere is scarcely news; we have merely had the privilege of tracing the mineral flow in the ecosystems of a significant portion of the Arctic avifauna and to speculate regarding a number of the unifying factors of life processes.

Appendixes
Literature Cited
Index

FEATHER mineral patterns of eight Canada geese shot in spring by Eskimos residing at Eskimo Harbor, Belcher Islands, N.W.T. (Figure *1*), are shown in Figure *266*. The basic similarities of these patterns (Figure *266*), which are believed to represent two rates of absorption of the same basic element "mix" in the nutrient chain, indicate that these geese molted in the same area the previous year. The full graphic extent of the high mean calcium value of the two-goose sample is shown for the sake of visual impact. Differences in consumption of brackish water seem most likely to account for these markedly different overall rates of mineral absorption. Because of the sinuous shape of the Belcher Islands, Canada geese on these islands live in close proximity to seawater. Very large salt glands characterize these geese—irrefutable evidence of their close association with saline environments—both on their breeding grounds and on their wintering grounds.

These unique feather mineral patterns constitute a classic record of an animal's metabolic responses to a mineral-rich environment. In addition to being a fitting appendix to this report, they are testimony to the soundness of the primary thesis of this report. We, therefore, decided they should be annotated in detail. The extremely high levels of the various elements in the primary feathers of the Belcher Islands geese can be accounted for in light of both the geology of these islands (Jackson 1960) and insights into mineral metabolism gained from study of the feather mineral patterns of other wild geese.

Limestone and dolomitic strata occur throughout the island system and account for the availability of calcium and magnesium, which were found in extreme concentrations in two geese. Exceptionally high intakes of sodium from brackish waters and from food plants (e.g., the halophyte *Triglochin maritima*) are believed to account for the notably high rates of turnover of calcium, magnesium, and potassium indicated by levels in the feathers. Clays enriched in potassium from exposure to seawater probably also explain the high levels of this element in these feathers. High phosphorus levels in these feathers can probably be related to a relatively high consumption of littoral invertebrates from adjacent Hudson Bay waters.

Soils developed from the Kipalu iron formation, which is from 200 to 410 feet (61 to 125 m) thick and is stratigraphically just below the thick and extensive basalt unit common to the islands, represent a likely source for the unusually high levels of iron found in the feathers of these geese. The major glacial movement was from northeast to southwest, and terranes to the lee of the elongate north-to-south exposures of the iron formation would be expected to be enriched with iron.

Zinc levels are excessively high in the feathers of these geese, possibly reflecting the approximately 20

Canada Geese of the Belcher Islands, N.W.T.

percent zinc content found in thin sections of one rock unit (not specified, by Richard Bell, Brock University, personal communication) on the Belcher islands.

Manganese is also unusually high in the feathers of two geese, indicating the presence of manganese-enriched organic soils derived from local igneous or silicic rocks. It is of interest, however, that Jackson (1960) found that manganese was the major mineral element in a single sample of graphitic coal ("anthraxolite") that he collected.

Sources of copper (chalcopyrite, malachite, bornite) found in veins in the extensive basalt unit could readily account for the above-normal copper levels of these feathers.

The source of boron, which in these feathers ranged as high as four times the normal levels in feathers of snow and Ross' geese, is assumed to be clays enriched by their previous exposure to seawater or ingestion of seawater.

Silicon levels were extremely high in the primaries of three geese, and in those of one (24,700 ppm) approached the highest level found in the feathers of all geese examined to date (28,000 ppm in the primaries of a Canada goose from Akimiski Island). Silicon associated with the widespread iron formations likely accounts for the high silicon levels in the feathers. Horsetail (*Equisetum*), a favored food of Canada geese, may be an important link in the nutrient chain accounting for the high levels of ingestion of both silicon and zinc (see discussion in chapter 7). However, as the Belcher Islands and Akimiski Island Canada geese are both "saltwater populations," we suspect that sodium intake may have had at least some effect on the level of silicon absorbed from ingested foods.

Levels of aluminum in the primaries of these Belcher Island geese can only be described as extraordinarily high. Its source is assumed to be the silts and clays of argillites that are rather common throughout the rock sequence of the islands.

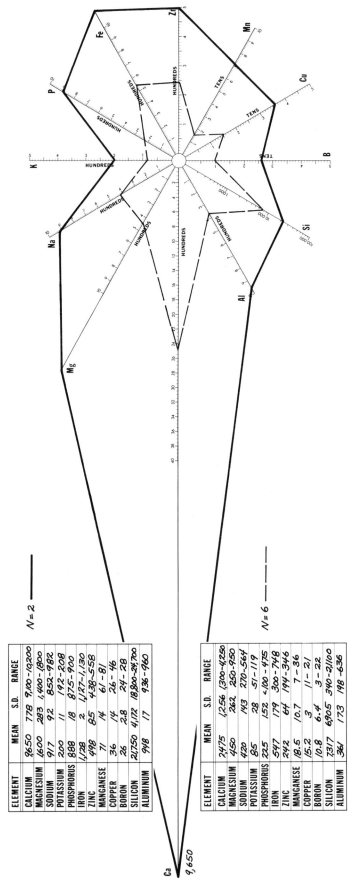

ELEMENT	MEAN	S.D.	RANGE
CALCIUM	9,650	778	9,100-10,200
MAGNESIUM	1,600	283	1,400-1,800
SODIUM	917	92	852-982
POTASSIUM	200	11	192-208
PHOSPHORUS	888	18	875-900
IRON	1,128	2	1,127-1,130
ZINC	498	85	438-558
MANGANESE	71	14	61-81
COPPER	36	14	26-46
BORON	26	2.8	24-28
SILICON	21,750	4,172	18,800-24,700
ALUMINUM	948	17	936-960

N = 2 ————

ELEMENT	MEAN	S.D.	RANGE
CALCIUM	2,475	1,256	1,300-4,250
MAGNESIUM	450	262	250-950
SODIUM	420	143	270-564
POTASSIUM	85	28	51-119
PHOSPHORUS	225	152	100-475
IRON	547	179	300-748
ZINC	242	64	194-346
MANGANESE	18.5	10.7	7-36
COPPER	15.2	3	11-21
BORON	10.8	6.4	3-22
SILICON	7,317	6,905	340-21,100
ALUMINUM	361	173	198-636

N = 6 ————

266. Feather mineral patterns representing eight Canada geese shot in spring by Eskimos residing at Eskimo Harbor, Belcher Islands, N.W.T.

THE PRECEDING discussions have dealt primarily with the movement of essental minerals through the ecosystem to their final incorporation in feather keratin. We have been heretofore concerned with neither the purely nutritional aspects of these elements nor the relation of their prevalence in the ecosystem to animal abundance, per se. In the case of blue, snow, and Ross' geese, which nest mainly in colonies, their presence in an area is indicative of at least a sufficiency of the essential elements in the environment for their particular needs.

To relate the intake of minerals to their prevalence in feather keratins, we necessarily have had to relate their absorption to some aspects of protein nutrition, particularly the role of the two sulfur-containing amino acids cysteine and methionine, in the metabolism of the molt. A consideration of these relationships led to consideration of the prevalence of sulfur in environments and of the role and relationships of sulfur compounds to the biotic world.

Despite the fact that sulfur is generally a ubiquitous element intimately tied in with protein production; that in some environments it is a limiting factor in the production of biomass; and that, present in reasonable abundance, it may play a synergistic role in protein production, a perusal of works that could be expected to deal with sulfur concentrations in the environment revealed that discussions of this element were either omitted or inadequately treated. In our review of sulfur in the environment we realized that there was frequent correspondence between high biological productivity and sulfur-rich environments.

It is generally conceded that studies of the sulfur metabolism of plants have lagged behind metabolic studies of carbon and nitrogen (Moir 1970:167). Oser (1965:557) states that sulfur does not come under the head of inorganic metabolism because only an insignificant part of it occurs in inorganic form. However, some vegetables probably contain sulfur in the form of sulfates in excess of that incorporated into the protein of the plant. Also, as our earlier reviews have shown, the separation of sulfur from the other macroelements of metabolism, because little of it occurs in inorganic form, is not valid.

Perhaps this somewhat dichotomous view of sulfur can be attributed to the fact that its relevance in the food chain has not been fully appreciated by ecologists and that nutritionists and physiologists seldom consider sulfur links in the nutrient chain below the level of proteins ingested or from the standpoint of its geochemical prevalence. However, it has been demonstrated that sulfur application to field crops is advantageous in large portions of the Pacific Coast states, east-central Nebraska, and areas of Mississippi, Alabama, Georgia, and Florida (Jordan and Reisenauer 1957:108). In short, deficiencies of sulfur with regard to agricultural and horticultural crops exist in the soils of these areas, and more recent studies (Robert Hoeft, personal communication) have shown that additional states (e.g., Minnesota, Wisconsin, Michigan, and Kansas) can be added to this list. It

Sulfur

was not without reason in the context of soil research in the early part of this century that McMurtrey and Robinson (1938:826) stated: "A general survey of the sulphur and phosphorus content of crop plants and soils reveals the fact that soils in general are more deficient in sulphur than in phosphorus. *It seems strange, then, that so little attention has been given to sulphur as a fertilizer constituent.*" The last sentence has been italicized because it offers such a parallel to Dequeker's (1972) expression of surprise (see chapter 7) that the sodium-calcium relationship has received so little attention in studies of human physiology. In respect to plant physiology, Allaway (1970:3) has expressed similar surprise that sulfur has not received "more massive scientific study."

Hanson quoted (1965:105) an observation made by Selwyn (the early director of the Geological Survey of Canada) in 1874 when crossing the relatively undisturbed Canadian prairie from Red River Settlement, Manitoba, to Rocky Mountain House, Alberta, because it seemed to have profound ecologic significance (although an explanation for the relationship Selwyn observed was not apparent at the time): "ducks of various kinds swarm upon nearly all the lakes and pools, and geese are frequently seen, *especially upon the saline lakes*" (our italics). In the process of concluding this report, a plausible explanation for the relation of saline lakes with high sulfate content to waterfowl abundance suddenly seemed self-evident. The saline, or "alkali" (as they are sometimes mistakenly referred to without regard to pH), potholes and lakes generally have a high content of sulfates—one of the sulfur sources for the syntheses of essential sulfur amino acids cysteine and methionine. And only microorganisms (bacteria, phytoplankton), fungi, and higher plants are capable of the synthesis of methionine. The pathway for the reduction of sulfate by fungi is believed to be: sulfate (SO_4^-)$\rightarrow$ sulfite ($SO_3^=$)$\rightarrow$ sulfide (S^-) or thiosulfate ($S_2O_3^=$)$\rightarrow$ cysteine$\rightarrow$methionine (Nicholas 1963:410). However, sulfite, sulfide, elemental sulfur, and thiosulfate through microbial mediation can also serve as sulfur sources for plants and microorganisms (Bandurski 1965:474).

The importance of sulfur amino acids in relation to the demands of egg laying and molt have been stressed earlier in this report and in Hanson 1962. Although sulfur, per se, has not generally received attention as one of the key elements in the nutrient chain, its differential abundance in the environment appears to be of such key importance that a brief discussion of it here may be helpful in understanding the abundance of waterfowl and other animals in relation to the overall content of other minerals in their environment.

ORIGIN OF SOIL AND WATER SULFATES

The sulfates present in the soils and waters of the Great Plains of the United States and the interior plains and lowlands east of the Rocky Mountains in Canada (Map 1254A, Geological Survey of Canada) are calcium sulfate dihydrate (gypsum), magnesium sulfate heptahydrate (epsomite), and sodium sulfate decahydrate (mirabilite or "salt cake" as it is known commercially). The white evaporite salts, or "alkali," that characterize the margins and, during droughts, the basins, of water areas on the plains are primarily sulfates of sodium and magnesium, with the former predominating. Calcium sulfate is a minor constituent of these evaporites in the southern sectors of the prairie provinces (Cole 1926:80).

According to Sloan (1972:22), in North Dakota, "the waters in the pothole environment, from fresh to saline, are from waters rich in calcium bicarbonate through those that are rich in magnesium carbonate to waters rich in calcium sulfate and magnesium sulfate. The most saline waters are enriched in sodium sulfate."

Gypsum beds, usually occurring as lens-like deposits, are associated with Paleozoic formations and are believed to be evaporites of seawater that covered midregions of the continent during that era. Magnesium sulfates are presumably similarly derived. Gypsum is soluble to the extent of several thousand parts per million and it weathers rapidly to yield sulfate ions (Sutcliffe 1962:141).

The probable origin and source of sodium sulfate deposits found in the western prairies is more uncertain and complex than in the case of gypsum due to the higher solubility of this salt. Cole (1926:77) favors bentonite clays as its probable source. Cole (1926:78) has postulated that calcium sulfate in solution through base exchange displaces the sodium from the bentonitic clays, thus creating the sodium sulfate. Grossman (1968) has postulated that east-flowing groundwaters from the Rocky Mountains dissolved evaporites associated with the Prairie formation of the Devonian age and moved them upward into the lakes situated in the Pleistocene deposits of the plains. Freezing concentrated the sodium sulfate, and residual brines drained into the Missouri River System (Sloan 1972:22–23). A similar thesis will be advanced below for a sulfate enrichment of the coastal marshes of Hudson and James bays.

According to Sloan (1972), "most theories for the accumulation of sodium sulfate salts in North Dakota require a bedrock source of highly charged brines or saline water." However, he points out that the salt content of the drift surrounding potholes is sufficient to account for their salinity. This drift is derived from the underlying bedrock—the Pierre Shale and Cretaceous and Tertiary rocks that are rich in gypsum and pyrite (FeS_2).

Whereas lower-lying basins in the landscape are enriched by drainage and leaching from waters in higher areas, these accumulated salts, when encrusted on mudflats adjacent to receding water lines, are "recirculated" by winds back on to the adjacent land. Cole (1926:77) records one instance in which a day-long wind whitened the vegetation in a direction downwind from the pond for a distance of one and a half to two miles.

The chemistry of the groundwater of the surficial deposits in the temperate areas of the Interior Plains of Canada is dependent on rainfall and temperature as well as the concentrations of mineral elements in the soils. If there is a net downward flow of water, dissolution processes dominate over precipitation, but in arid areas evaporation is more important in determining water chemistry due to capillary rise exceeding rainfall in effecting the movement of mineral ions (Brown 1967:146).

Because we believe the discussion in Brown (1967:146) is essential for an adequate understanding of the distribution of sulfates in the waters of the western plains we take the liberty of quoting this publication at some length (references cited are deleted for the sake of condensation).

"This macrozonation [of groundwater chemistry] is best expressed by the concentrations of the most soluble salts, principally $CaSO_4$ and $NaCl$. The effect is less noticeable for $CaCO_3$ whose solubility is determined by the amount of free CO_2, rather than by the amount of rainfall. Applying these general principles to the prairie environment, one may expect to find a hydrochemical reflection of the low-rainfall areas in southern Saskatchewan and Alberta (less than 16 inches), as compared to the regions of higher rainfall in western Alberta and east-central Manitoba (more than 18 inches).

"This was demonstrated . . . [for] groundwater in the till of western Alberta [which] is low in total dissolved solids (less than 800 ppm) and contains predominantly Ca-Mg bicarbonates. Eastward, the average amount of total dissolved solids increases to 2,500 ppm in east-central Alberta and central Saskatchewan . . . and sulphates become the predominant constituents—both sodium sulphate and calcium-magnesium sulphate. Farther east, toward southwestern Manitoba, the average amount of total dissolved solids diminishes to 2,000 ppm and to 1,100 ppm in southern Manitoba. . . . Calcium-magnesium sulphates are the predominant dissolved salts in both areas. The eastern boundary, between shallow sulphate water and bicarbonate water, was found [to be] in the Fannystelle area of central Manitoba. There, the average amount of total dissolved solids is 912 ppm, and calcium-magnesium bicarbonate dominates over calcium-magnesium sulphate in the proportion of about 2:1.

"On the whole, groundwater in surficial deposits *in areas with a total annual precipitation of more than 18 inches appears to be moderately mineralized* (less than 1,000 ppm) *and of the calcium-magnesium bicarbonate type. In areas with less than 16 inches annual precipitation groundwater in drift contains 1,000–2,500 ppm total dissolved solids and is of the calcium-magnesium sulphate or sodium sulphate type* [our italics]. The chemistry of shallow groundwater on the prairies can be interpreted so well in terms of Schoeller's macrozonation largely because of the uniformity of the surficial deposits. However, this simple hydrochemical zonation is modified once the drift water reaches the less uniform bedrock strata.''

RELATION OF AVAILABLE SULFUR TO PROTEIN CONTENTS OF PLANTS

As mentioned above, sulfur is absorbed by plants mainly as the sulfate ion. However, the proportion of N:S varies widely as ratios of from one to fifty-five to one have been documented, but generally the ratio in most plants is about fifteen to one (Moir 1970:167). It has been shown that the nitrogenous protein content of nutrient-grown perennial ryegrass (*Lolium perenne*) is directly proportional to the amounts of sulfate and nitrate in the nutrient solutions. On the average, there were thirty-six atoms of nitrogen for each atom of sulfur in the protein of this grass. If available, sulfate in excess of needs may be absorbed by plants; as a result, appreciable fractions of the total sulfur of plants may be in the form of sulfates (Epstein 1972:289–90; Nason and McElroy 1963:517). This excess sulfate would be converted into nonprotein structural components (Sutcliffe 1962:39). A notable example of sulfur ($SO_4^=$) absorption is found in stands of smooth cordgrass (*Spartina alterniflora* Loisel) in the coastal marshes of North Carolina. In this marsh grass the average sulfur content exceeded the total nitrogen (45.2 versus 43.3 kg/ha in the standing crop) (Broome et al. 1975:300).

Although the ratio of methionine to other amino acids does not vary significantly between normal and sulfur-deficient domesticated plants, the total protein in normal plants is higher. Allaway and Thompson (1966) state that ''S-adequate plants generally contain more S amino acids and are presumably of better nutritional quality for animals than are S-deficient plants.'' However, some plants that grow at optimum rates may contain N:S ratios which are deficient in sulfur for ruminants.

Sunflower and rape grown with sulfate-free nutrient solutions contain higher percentages of amino acids than do those raised in media containing sulfate. Because sulfur-containing amino acids could not be synthesized, nonsulfur-containing amino acids could not be condensed to form proteins (Eaton 1941; Eaton 1942).

The sulfur contents of plants vary appreciably. ''The amount of SO_4 in the ash of leaves is generally 3 to 6 percent, but values as low as 0.5 and as high as 18 percent are on record. Many cruciferae possess an exceptionally high content of sulfur'' (Stiles and Cocking 1969:293). According to Jordan and Reisenauer (1957:107), the sulfur content of plant proteins never exceeds 2 percent.

THE SULFUR CYCLE IN PONDS AND LAKES

The production of prairie potholes over any given period of years has been repeatedly linked to periodic droughts and the resultant oxidation and recycling of mineral nutrients on the return to ''normal'' water levels (Leitch 1964). Beule and Janisch (1974) found an 86-percent increase in the sulfates in the top two inches of sediments in Horicon Marsh, Wisconsin, following a drawdown. At the seven-inch level, the increase in sulfates was 368 percent. McKnight and Low (1969) have recorded approximately 60-percent increases in Corixidae and Chironomidae larvae following a drawdown of a spring-fed salt marsh in Utah.

Hutchinson (1957) has discussed the sulfur cycle in waters with authority and in detail. Only a few salient points need be cited here. Anaerobic and heterotrophic bacteria found in the sediments or at the water-mud interface of the hypolimnion of thermally stratified bodies of water reduce sulfates ($SO_4^=$) to hydrogen sulfide (H_2S). The rate of H_2S production has been found to be linearly related to sulfate concentrations. Part of the H_2S formed may be combined with any ferrous ions present, but much of it will be oxidized back to sulfates by sulfur bacteria.

In moderately deep potholes surrounded by high vegetation, the thermocline may be undisturbed for a considerable period in midsummer, and the potential for H_2S increases in these waters should be relatively great. Hutchinson (1957:760) has stated: ''In the waters of mineralized and often meromictic lakes, in which great quantities of the order of hundreds of milligrams per liter of H_2S are formed by the bacterial reduction of sulfates, very large quantities of gas may be expected in solution at quite high pH values, provided that the alkalinity is great. This hydrogen sulfide will be present almost exclusively as HS^-, and such alkaline waters will probably provide the optimum conditions for the development of sulfur bacteria.''

TOXICITY LEVELS OF DISSOLVED HYDROGEN SULFIDE

In certain waters of Indonesia, fish kills occur when lower waters rich in H_2S are overturned (Hutchinson 1957:773). Brigham and Gnilka (1973) have reported that phantom midge larvae (*Chaoborus*) succumbed to H_2S concentrations of from 0.64 to 0.32 mg per liter as sulfur. Hydrogen sulfide levels as low as 0.3 mg per liter are lethal to the beach flea *Gammarus* (Colby and Smith 1967), and minimum concentrations of 1.0 mg per liter are lethal for *Daphnia* and Ephemeroptera nymphs (Van Horn et al. 1949). Eggs and larvae of some fish are killed at H_2S levels as low as 0.076 to 0.009 mg per liter (Becker and Thatcher 1973:R5). Moyle (1956:311) reported that a lake in Hennepin County, Minnesota, which had 14.0 ppm of sulfate in the surface waters, contained none below the thermocline, and the lower waters smelled of hydrogen sul-

fide. He cites a paper by Fjerdingstad (1950) who found
that hydrogen sulfide had a marked effect on the plankton fauna and flora. These findings at least raise the
question as to what extent H₂S may be present in the
hypolimnion of some deeper, wind-sheltered potholes in
midsummer and to what degree invertebrate populations
may be affected, and therefore, to what degree H₂S may
be a factor in the senescence of potholes insofar as their
attractiveness to breeding waterfowl is concerned.

Hydrogen sulfide may also play an indirect role in
causing a pothole to become senescent in regard to waterfowl by depriving their food chain of essential trace
elements. As Hutchinson (1957:764) has stated, it is
"evident, however, that no limnologically possible concentrations of ionic ferrous iron could ever permit the
presence of analytically detectable amounts of copper or
silver in the presence of FeS" (and iron, manganese,
and zinc).

SULFATE CONCENTRATIONS IN SOME PRIMARY WATERFOWL PRODUCTION AREAS

Numerous studies have been made of duck production in
the region of prairie potholes, but evaluations of production, whether of waterfowl, their food plants, or the invertebrate fauna, that do not take into consideration the
important part sulfates play in the food chain will fall
short of their objectives. This viewpoint is affirmed by
Moyle and Hotchkiss (1945:19), who found that the concentration of sulfates in Minnesota waters exerted a
greater influence on aquatic plant distribution than did
their carbonate contents.

The sensitivity of plants to high concentrations of sulfate in the soil may relate to the fact that high sulfate
levels tend to limit the uptake of calcium (Jordan and
Reisenauer 1957:109). In such instances the role of calcium is reversed; more often calcium has been found to
be a depressant on the uptake of other macroelements
and trace elements.

Moyle (1956:304) stated that "the biological and
chemical processes, reactions and interrelationships in
waters are complicated and often not well understood."
Macan (1974) has echoed this assessment in respect to
invertebrates. Bolen (1964:154) has similarly noted that
"the ecological application of water chemistry to vascular plant life has not been clearly defined." Moyle
(1956:311) skirted this relationship when he commented: "The cause-and-effect relationship is not
known, but may be related to sulphur demands in plant
nutrition."

As mentioned before, many of the most productive
waterfowl areas of the North are coincident with areas of
post-glacial marine submergence. An enrichment in
minerals of the soils of these areas as a consequence of
their exposure to seawater has been assumed and often
alluded to by waterfowl biologists, but no specifics are
ever given. Overlooked is the fact that, next to the sodium and chloride ions, the sulfate ion is predominant in
seawater. The freeze-thaw-leach cycles and the integra-
tion of sulfate into organic matter insure retention of sulfur in these ecosystems and the reduction of the sodium
and chloride ions by leaching to tolerable ranges.

The importance of the coastal marshes of Hudson and
James bays in Manitoba and Ontario (the perimeter of a
former marine submergence area that extends inland for
100 to 200 miles) to the welfare of migrant blue and
lesser snow geese has been emphasized earlier. It is on
these marshes that the immature birds (the young of a
given year), particularly in years when hatches are delayed, must complete further physical maturation to enable them better to undertake the flight to midcontinent
areas and the Louisiana coast. For example, in some
phenologically late years a "fallout" of family flocks
with late-hatched broods is reported to occur in the
northern peninsula of Michigan after the migration
crosses Lake Superior. In other years when reproduction
is very poor, the flocks of geese—largely yearlings and
adults— do not halt their migration before reaching midcontinent areas.

Earlier we gave reasons for ruling out sodium chloride
as an important depressant factor on the ecology of these
coastal marshes in respect to waterfowl. A case can be
made, however, for sulfates as one of the key nutrients
that account for the attractiveness and importance of
these marshes to geese. Carried by wind-borne sprays as
well as high tides, the sulfates deposited in these cal-
cium-enriched coastal wetlands would be precipitated as
$CaSO_4 \cdot 2H_2O$ (Hutchinson 1957:570) when these
marshes freeze. Chlorides accumulated in the Hudson-
James bay marshes by these mechanisms would also, as
discussed earlier, be subject to constant seaward drainage and leaching. Thus, this sequence of events appears
to provide a valid model whereby sulfates are constantly
made available for plant growth, and sodium chloride
is held within levels tolerable to most vascular marsh
plants. The corollary assumption that can be made from
these events is that these coastal marsh plants are unusually high in protein content because of the availability of
sulfates and the interaction of nitrogen and sulfur in protein synthesis. Data on the nitrogen levels of plants in the
south coast marshes of Hudson Bay presented in chapter
3 under "Plant Relationships" are evidence in support
of this thesis.

Similar relationships between waterfowl densities and
sulfur abundance can be found elsewhere. The Canadian
Shield of Ontario supports a sparse duck population, but
according to Harry G. Lumsden (personal communication) there are two areas on the Shield that support
populations that in densities are comparable to breeding duck densities on the prairies. These areas are downwind (N.E.) from the smelters at Sudbury and Wawa.
Although the sulfur dioxide (SO_2) emissions have
drastically affected the seral stage of the vegetation in
the more immediate areas of these smelters, we strongly
suspect that the acidic sulfur contribution to the environment, when neutralized, acts as a significant fertilizer,
increasing the protein content of aquatic and emergent
plants as well as increasing plankton and bottom inver-

tebrate levels in these otherwise relatively infertile regional waters.

In Wisconsin there is an apparent relationship between the sulfate content of groundwaters used by municipalities (Baumeister 1972), the sulfate content of lake waters (Poff 1961), and breeding duck populations (March et al. 1973; March 1975, Figure 1). All values are highest in the southeast and east-central regions of the state areas represented by glacial drift and underlain chiefly by dolomitic limestones.

In Minnesota there is a northeast to southwest gradient in dissolved solids in lakes. Highest total alkalinity values were found for the shallow lakes of the waterfowl production areas of western and southwestern Minnesota. The average total sulfate content of the lakes of the southwestern portion of the state was fifteenfold higher than that of lakes in adjacent eastern areas; in contrast, total nitrogen and phosphorus contents were only twofold higher in lake waters of the southwest than in lakes to the east (Moyle 1956).

Relatively high salinities characterize most of the prime waterfowl areas of the Great Plains. The water of most ponds and lakes of the sandhills of Nebraska contain considerable concentrations of saline and alkaline compounds, but McAtee (1920) noted that the degree of alkalinity found was seldom a depressing factor for duck food plants. In South Dakota only extremely alkaline lakes are not used by ducks, but even these become attractive when rising waters reduce their alkalinity (Evans and Black 1956:38). In the midcontinent region, waterfowl water areas of North Dakota have received the most study in respect to their chemistry and invertebrate life. On the North Dakota prairies, the most heavily salted lakes are those that receive their drainage and seepage from lakes at slightly higher elevations. These lakes are generally shallow, and the dissolved salts are mainly sulfates of calcium and magnesium (Adomaitis and Shoesmith 1972:33; Swanson et al. 1972:44). Both sulfates and sodium may occasionally be present in toxic amounts (Adomaitis and Shoesmith 1972). Stewart and Kantrud (1972) have classified the vegetation and waters of prairie potholes in North Dakota in respect to water chemistry. Some of their data on ionic constituents are reproduced in Table 44; we have added the data for seawater for convenience of comparison.

The mixed-prairie areas of southern Alberta and Saskatchewan are recognized as one of the continent's most productive waterfowl breeding grounds (Keith 1961:31). Cole (1921) reported that alkali salts were widespread in the prairie provinces either as deposits in alkali lakes interbedded with peat or muck or in brine streams or springs. The composition of these salts was chiefly sulfates. In southeastern Alberta, sulfates and chlorides of calcium, magnesium, and sodium comprise the main salts of the alkali areas according to Keith (1961), but we suspect that as in the case of the more saline lakes of southern Saskatchewan (Rawson and Moore 1944), sulfate is the predominate anion. When artificial impoundments in this area are made, the local water table is

44. Some ionic constituents of surface water from prairie ponds and lakes in southeastern North Dakota

Relative Salinity*	Elements (ppm)				
	Ca	Mg	Na	SO$_4$	Cl
Fresh	25	15	9	32	14
Slightly brackish	86	62	42	200	19
Moderately brackish	99	150	235	775	33
Brackish	50	310	1,020	2,340	235
Subsaline	205	540	1,380	9,310	1,050
Saline	210	2,240	39,100	27,300	6,320
Seawater	400	1,272	10,561	2,590	18,980

Reference: Data derived from Stewart and Kantrud (1972, Table 3).

Notes: Figures represent mean values from four typical ponds or lakes in the indicated range of salinity. Data for seawater are added for comparative purposes.

* As indicated by plants.

raised and an upward movement of soluble salts into surface soils from lower levels of the glacial drift occurs (Keith 1961:25–26).

The chemical content of the rivers of the central plains of Canada and the United States dramatically reflects the high sulfate levels that characterize the soils and waters of this region. In comparison, major rivers of the Canadian Shield and Paleozoic Basin are extremely low in sulfates (Table 45).

An area with one of the highest densities of breeding ducks in North America is the Bear River marshes of Utah. These marshes are fed in the spring by the waters of the Bear River. The dissolved salts of this water and adjacent Great Salt Lake are given by Hutchinson (1957:569): chlorides 206 and 11,289, sulfates 52 and 1359, respectively. The ratio of abundance of these ions (Cl$^-$/SO$_4^=$) is of interest: Bear River, 3.97; Great Salt Lake, 8.31. Thus, in respect to sulfates in water, water going into the Bear River marshes would be classified as nearly fresh in terms of Stewart and Kantrud's classification based on vegetation (Table 44). In respect to chlorides, the water would be classified as brackish. However, in the marsh itself, the concentration of these anions is undoubtedly far higher due to evapotranspiration and the release of waters from Great Salt Lake into the marsh in summer.

The sulfates have nutritional advantages for plants short of toxic levels; the chlorides, on the other hand, tend to be a depressant, except at low levels, on the growth of most aquatic and wet-soil plants except halophytes. For these reasons it seems doubtful that the practice of evaluating water areas used by waterfowl on the basis of electrical conductance as it reflects the full complement of dissolved solids has adequate ecological relevance. The reverse side of the coin is studies that have considered only the sodium chloride content of waters in respect to productivity of potholes and marshes. The point in both instances, of course, is that the contribution

45. Median concentrations (ppm) of some anions in rivers draining three geologic areas of Canada west of Hudson and James bays and mean concentrations for the Missouri and James rivers in the United States

Geologic Province, River, and Sampling Site	Age of Rocks in River Basin	Latitude and Longitude of Sampling Site	n	$SO_4^= - S$	n	$NO_3^= + NO_2^= - N$	n	$PO_4^= - P$
CANADIAN SHIELD								
Seal River (below Great Island)	Precambrian	58°53′ 96°16′	6	1.6	8	.010	3	.002
Hanbury River (above Hoare Lake)	Precambrian	63°55′ 105°09′	6	1.5	6	.010	4	.003
Lockhart River (below Artillary Lake)	Precambrian	62°53′ 108°28′	10	1.3	10	.008	9	.002
Snare River (outlet of Big Spruce Lake)	Precambrian	63°30′ 116°00′	21	1.8	21	.045	6	.003
PALEOZOIC BASIN								
Churchill River (at Fort Churchill intake)	Mostly Precambrian; lower portion, Ordovician and Silurian	58°40′ 94°10′	54	3.4	28	.070	29	.003
Hayes River (below Gods River)	Ordovician and Silurian	54°31′ 92°47′	1	2.8	1	.080	—	—
Winisk River (below Ashewig River)	Mostly Silurian	54°31′ 87°14′	16	2.5	13	.029	—	—
Albany River (at Hat Island)	Precambrian and Silurian; lower third, Devonian	51°19′ 83°52′	29	4.3	25	.011	—	—
INTERIOR PLAINS—NORTH								
Athabasca River (at McMurray)	Mostly Upper Cretaceous	56°46′ 111°24′	63	26.7	58	.055	14	.007
Peace River (at Peace Point)	Upper four-fifths Lower Cretaceous; Lower fifth Upper and Middle Devonian	59°06′ 112°25′	70	24.0	56	.060	13	.003
Salt River (5 mi. N of Ft. Resolution)	Middle Devonian	60°02′ 112°22′	2	1232.0	1	.030	2	.004
Slave River (5 mi. NE of Ft. Resolution)	Precambrian (L. Athabasca watershed) and Quarternary plus waters of Athabasca and Peace rivers	61°15′ 113°40′	7	18.1	3	.020	2	.005
Little Buffalo River (at Hywy 5)	Middle Devonian	60°02′ 112°46′	14	635.0	10	.013	6	.002
Little Buffalo River (east of Pine Point)	Middle Devonian	60°59′ 113°45′	9	985.0	4	.065	5	.002
Buffalo River (at Hywy 5)	Middle and Upper Devonian	60°42′ 114°55′	23	29.0	16	.031	13	.003
Hay River (at Hywy 5)	Lower two-thirds Cretaceous, and Upper and Middle Devonian	60°44′ 115°51′	33	69.1	23	.090	18	.008
Kakisa River (at outlet Kakisa Lake)	Lower Cretaceous and Upper Devonian	60°55′ 117°25′	27	31.3	17	.020	14	.003
Horn River (north mouth)	Middle and Upper Devonian	61°32′ 117°57′	22	72.1	17	.030	15	.002
Mackenzie River (near Ft. Simpson)	Upper Devonian	61°52′ 121°10′	52	24.3	38	.055	21	.002
Mackenzie River (above Arctic Red River)	Lower Cretaceous, Middle and Upper Devonian	67°27′ 113°42′	95	31.2	54	.072	23	.002

252

Table 45. *Continued*

Geologic Province, River, and Sampling Site	Age of Rocks in River Basin	Latitude and Longitude of Sampling Site	Anion Concentrations [a] (ppm)					
			n	$SO_4^=-S$	n	$NO_3^=+$ $NO_2^=-N$	n	$PO_4^=-P$
INTERIOR PLAINS—SOUTH								
North Saskatchewan River (east of Prince Albert)	Upper Cretaceous up to sampling sites	53°12′ 105°20′	52	60.6	44	.489	12	.021
South Saskatchewan River (at St. Louis)	Upper Cretaceous up to sampling sites	52°55′ 105°48′	98	64.6	90	.230	7	.007
James River (near Scotland)	Cretaceous	43°11′ 97°38′	5	776.0[b]	5	<.01[b]	5	.09[b]
Missouri River (at Yankton)	Cretaceous	42°52′ 97°24′	5	212.0[b]	5	.01[b]	5	.01[b]

References: Data for Canada courtesy of the Inland Waters Directorate, Environment Canada. Data for the James and Missouri rivers from Larimer and Leibbrand (1974).

[a] Data are mainly for the ice-free period of the year.

[b] Data are for the May–September period; values are arithmetic means.

that sulfates make to plant growth and invertebrate populations has not been assessed.

The foregoing discussions in this section have related primarily to duck populations. It is perhaps significant that the largest examples of the largest race of Canada geese (*Branta canadensis maxima*) seem to have originated in the Dakotas (Hanson 1965) and that the northern Great Plains are fall and spring feeding-staging areas for well over a dozen races of Canada geese comprising hundreds of thousands of individuals and, according to Harry G. Lumsden (personal communication), 1.8 million blue and lesser snow geese.

Although the evidence is not complete, the hypothesis can be advanced that the sites of blue and lesser snow goose colonies can be related to *relatively* high sulfate environments: 1) coastal marshes of areas of former marine submergence that may be enriched with sulfates as a result of the freeze-precipitation-thaw cycle discussed earlier; 2) the deltas of sulfate-rich waters (e.g., Kendall Island in the Mackenzie River mouth—see Table 45; the Anderson River, which in its lower portion flows through sulfate-rich terrane; the range of the Wrangel Island colony which includes areas of solonchak soils).

Other species of geese that nest in high densities in the Arctic (brant—*Branta bernicla;* cackling geese—*Branta canadensis minima*) also have coastal ranges. The greater snow geese, on the other hand, nest in small colonies or at scattered points on inland or upland areas. At least a partial explanation for the capacity of some Arctic areas to support dense populations of geese is the fertility of the soils concerned, of which the sulfate content may be an important factor.

SULFATES AND WATERFOWL FOODS

Various investigators, McAtee (1920), Moyle and Hotchkiss (1945), and Keith (1961), are consistent in

their findings on which species of aquatic plants are tolerant of saline waters. Two species are notable in this respect—sago pondweed (*Potamogeton pectinatus*) and widgeon grass (*Ruppia occidentalis*). Both are duck foods of primary importance. According to Moyle and Hotchkiss (1945:30), widgeon grass makes its best growth in waters with a sulfate ion content greater than 125 ppm. Stewart and Kantrud (1972) have written the classic, definitive paper describing the tolerances of the various aquatic and wet-soil plant species of potholes in North Dakota to total salinities. In the Bear River marshes, musk grass (*Chara*) is found in very saline waters but is a favored food of ducks (Wetmore 1921).

Major halophytic wetland plants of southeastern Alberta are wild oats (*Hordeum jubatum*) and salt grass (*Distichlis stricta*) (Keith 1961:26). It appears significant, therefore, that five Canada geese from this area examined for stomach contents had fed principally on the leaves of *Hordeum jubatum* (Keith 1961:34). Keith also cites a study of the food habits of 122 molting pintail ducks from Pel Lake, Saskatchewan, in which 32 percent of the foods consisted of the seeds of alkali bulrush (*Scirpus maritimus*), a species associated with moderate to strongly alkaline waters.

We earlier cited the high protein content of the halophyte *Triglochin maritima* and its high palatability for Canada geese. In a related context, it is of interest to note that *Equisetum arvense* (a species of a family of plants (Equisetaceae) that is equally sought after by Canada geese) had the highest content of sulfur (0.74 percent) of 158 species of plants in Wisconsin analyzed for their mineral content (Gerloff et al. 1964).

The classical limnological study of saline lakes was made in Saskatchewan by Rawson and Moore (1944). Most of the lakes they studied contained between 300 and 30,000 ppm of dissolved solids. In waters containing over 700 ppm of dissolved solids, the sulfate ion was the dominant ion in the negative group. The greatest number of species of rooted aquatic plants were recorded from relatively fresh water (171 ppm of dissolved solids), but *Potamogeton richardsoni* and *P. pectinatus*, both important duck foods, were present in waters having as high as 2,550 and 14,200 ppm of dissolved solids, respectively. More importantly, the populations of bottom invertebrates peaked in two lakes with salinities of 1,300 and 2,300 ppm, and highest populations of *Entomostraca* were found in waters of one lake having 8,000 ppm of dissolved solids. Judging from Rawson's and Moore's (1944:Fig 4) summary data, sulfates in Saskatchewan lakes of moderate to high salinity make up about 90 percent of the total anions. The biomass of plankton life in the seas of high latitudes is renown; it may be relevant in conjunction with evaluating Rawson and Moore's data that the sulfate content of seawater is about 2,600 ppm (Table 44). In the Bear River marshes, as elsewhere, alkali flies (*Ephydra*) and brine shrimp (*Artemia fertilis*) are important foods for ducks and shore birds (Wetmore 1921:15).

RELATION OF SULFATE-RICH ENVIRONMENTS TO SOME PLANTS AND OTHER VERTEBRATE POPULATIONS

There is always the temptation—and danger—of attempting to make an attractive hypothesis apply too widely. Sulfates are but one dimension of the chemical environment, and the latter must be considered in perspective along with such other major aspects of the environment as climate, topography, and vegetation; but the influences of all are obviously interwoven in their relationships with plant and animal distributions and population densities. Nevertheless, it may be stimulating to call to attention some plants and animal populations whose presence or densities appear to be related importantly to sulfate concentrations in soils and waters.

Crop and Range Plants of the Great Plains

The most important cereal crop of the northern Great Plains is, of course, wheat. Gypsum has been found to be a very satisfactory source of sulfur for wheat (Rasmussen et al. 1975), there having been substantial increases in the sulfur content of the straw and grain portions when gypsum was incorporated in sulfur-deficient soil. Different levels of sulfur applications did not raise the level of sulfur in the grain above 0.16 percent although increases of sulfur in straw for all treatments were noted (perhaps largely as accumulated sulfate, which the authors did not differentiate). Also sulfur had a tendency to increase protein content although the effect was not as pronounced as that of nitrogen.

Other domestic crop plants of the plains that are notably high in sulfur are the sunflowers (*Helianthus*) and the mustards (Cruciferae rape [*Brassica napus* L. and *B. campestris* L.]). Sunflower seed contains the highest percentage of sulfur amino acids of all feed stocks. Forty-seven percent of the oil meal consists of crude protein, as compared with 11 percent for corn meal. In re-

spect to methionine, sunflower meal contains 1.6 percent, corn meal only 0.3 percent (Boucher 1956). Rape, like other mustards, is noted for its high content of sulfur compounds. Whole seeds of *Brassica* contain 24 to 28 percent protein; the extracted meal contains 42 to 44 percent protein (Appelquist 1972:166).

Sunflowers, domestic stocks and wild, are grown or found in greatest profusion in the prairies and on the Great Plains (Weibel 1951), where they also serve as a valuable forage plant for open land grazing by cattle. Rape is grown in northwestern Minnesota and is extensively planted as a crop in the prairie provinces, its range extending in an arc from the Winnipeg, Manitoba, area northwestward to Edmonton, Alberta (Ohlson 1972:30). Cold-hardiness characterizes both. Interestingly, in this conjunction, it has been shown that alfalfa will winter better when provided adequate sulfur (Robert Hoeft, personal communication). Although other factors also determine their use as crops, the sulfur content of these two plants and the generally relatively high sulfate content of the soils and waters of the areas in which they are most favorably grown can hardly be fortuitous.

The protein content of native grasses in North Dakota is relatively high. It is well above minimal protein requirements for cattle in spring and early summer, but is reported to be below minimum protein requirements for cattle in late summer and fall due to drying and weathering of the leaves (Whitman et al. 1951). However, this somewhat theoretical appraisal, based on Morrison's *Feeds and Feeding* (20th edition), did not take into consideration nonprotein sulfate that doubtless exists in western range grasses or the sulfate that may be adventitiously attached on prairie grasses or that sulfate present in the drinking water that would be available to the bacteria of the rumen. Yet Whitman et al. (1951:34) point out that cattle (as did, of course, the bison; see discussion below) at the Livestock Experiment Station near Miles City, east-central Montana, did well the entire year on native grass range forage with minimal supplements of salt, bone meal, and hay during the most severe winter weather. Although it may not have been an important factor in the study at Miles City, overlooked is the fact that bone is the major reservoir of sulfur in vertebrates; therefore cattle given rations supplemented with bone meal for its calcium and phosphorus content, in effect, were also given sulfur which can be utilized by the bacteria of the rumen to produce protein, thereby supplementing the dietary intake of protein contained in the forage.

Ring-necked Pheasants

Ring-necked pheasants probably attained their historic high densities in North America in the eastern half of South Dakota in the early 1940s. War, physiography, weather, nesting cover, land use, and crop prices all helped to set the stage for the tremendous densities that pheasants achieved in this area and to a lesser extent in some adjacent portions of North Dakota, Minnesota, and Iowa (Kimball et al. 1956). But these factors are of no avail unless soil nutrients are present as the underlying base. Kimball et al. (1956:212) state that soil fertility was the factor directly limiting the abundance of the pheasant on the Great Plains; we suggest that the high sulfate content of the soils and waters of the region may be one of the more important aspects of the environment that made it possible for pheasants to achieve those historic high densities. (Compare pheasant-range maps given by Kimball et al. [1951] with maps showing sulfate content of waters and gypsiferous areas as indicated by commercial deposits, outcrops, and evaporite basins, pages 81 and 125 in National Atlas of The United States of America, 1970). In 1945, Rasmussen and McKean reported that the Uinta valley of northwestern Utah held "an abundance of pheasants." This abundance may have been related in part to interspersions of excellent cover associated with a patchwork of cultivated fields—the results of uneven terrain and stony soil—but it is of interest that the authors also remarked upon the rankness of the vegetation and the "alkali" present in the soils.

One of the important plant links in the nutrient chain for pheasants in the plains states was wild sunflowers. The abundance of these plants in east-central South Dakota during the years of the record pheasant population (the late 1930s and the early 1940s) was such that, in bloom, their yellow blossoms characterized for miles the extensive tracts of uncultivated land that existed then (Ronald Labisky, personal communication). Seeds of wild sunflowers constituted a major food for pheasants in the Dakotas (Errington 1945:193), the seeds of sunflowers and other Compositae were the dominant foods of pheasants in Nebraska during the same period (Sharp and McClure 1945:207–18).

Bison

The fertility of prairie soils is probably equaled only by that of certain deltaic soils. The density and diversity of mammalian populations they supported prior to settlement were correspondingly great. One has only to read Audubon's Missouri River journal and his accounts of the bison (*Bison bison bison*) herds to appreciate this point—which need not be belabored further here.

If it can be conceded that the ability of the prairies to sustain the enormous herds of bison that once existed there may be partly related to a sulfate-rich environment, what of the northern wood bison (*Bison bison athabascae*)? This larger and darker race once ranged over most of the northern third of Alberta (Alberta Lowland) and northeastward from the valley of the Liard River to Horn Plateau and Lac la Martre. The western limits of the Canadian Shield formed the eastern limits of its range (Soper 1941:360). However, greatest densities of the wood bison were apparently attained in the region west of the Slave River and north of the Peace River and westward to the valley of the Hay River (Soper 1941); indeed, the Buffalo River arises and flows northward from this area. Preble (1908:144) notes that the Indians reported to Mackenzie in the late 1700s that bison abounded on the plains along the Horn River, which

originates along the north slope of the Horn Plateau and empties into the northeast corner of Mills Lake, an expansion of the Mackenzie River (Table 45). The Horn Plateau is composed of gypsiferous shales of the Cretaceous period (Day 1968:15). It is not surprising, therefore, that the dry playa basins east of the Horn Plateau (seen by Hanson in 1975) are salt encrusted. The wood bison were exterminated from this area, but have been reestablished there in recent years by transplants from Wood Buffalo Park.

Charles Camsell was one of the first geologists actively to explore the region. In 1903 he reported cliffs of gypsum ten feet high along the Little Buffalo River. In 1917, in his report on the salt and gypsum deposits between the Peace and Slave rivers, Camsell (1917) stated that at almost all outcrops of the Paleozoic strata important deposits of gypsum occurred. Along the lower Peace River, outcrops of the Middle Devonian rocks are exposed for eighteen miles. The lowest bed is composed of gypsum that attains a maximum thickness of fifty feet. Outcrops of gypsum were also found along Slave River in the vicinity of La Butte, at Ennuyeaux, at Bell Rock below Fort Smith, and at the forks of the Salt River. Waters from springs four miles south of Brine Springs are reported to be milky white in color from suspended sulfates. Twenty-seven miles north of McMurray, at La Saline near the former southern limits of the wood bison, springs near the base of the Devonian escarpment produce a brine rich in sulfates (Allen 1929). The Middle Devonian rock formations of the Great Slave Lake area east of the Canadian Shield (Little Buffalo Formation, the Nyarling Formation and Lonely Bay Formation) are either underlain by a gypsiferous evaporite unit (the Chinchaga Formation) or the uppermost strata themselves consist of interbedded gypsum and carbonate rocks (Nyarling Formation) (Norris 1965).

The Salt Plains adjacent to the Salt River are visited by bison (Soper 1941), presumably for the available salt, but the boggy "muskeg" on the higher ground to the southwest of the Slave River and to the north of the Peace River constitutes the chief winter range (Soper 1941:369). The summer range of these bison is the area of prairie interspersions on the eastern portions of the Alberta Plateau (Soper 1941), a summering area referred to by Fuller (1961:10) as "the gypsiferous terrain," a relationship the significance of which is only now apparent. Both Raup (1935:47) and Soper (1941:370–72) in listing the plants of this area allude to their luxurious growth. The area is pitted with sinkholes as a result of the dissolving away of pockets of gypsum (Norris 1965); the sinkholes that hold water contain an abundant plankton of green and blue-green algae and "minute" invertebrates (Raup 1935:31).

In recent years, 4,000 to 5,000 bison have occupied the combined deltas of the Peace and Athabasca rivers. Sulfate enrichment of these alluvial soils and of the forage produced on them can be assumed from the foregoing discussions.

In June 1974 Hanson was privileged to accompany

Roy Jacobson of the Canadian Wildlife Service on a flight over the Great Slave Plain between the west shores of Great Slave Lake and Horn Plateau. New growth had not yet appeared, and the dried vegetation in combination with salt (sulfate) encrusted playa lake basins made it a desolate-appearing terrain. Nevertheless, the bison thrive there now as in early times. The parallels between the gypsiferous range of the wood bison and the saline ("alkali") conditions of the midcontinent range of the prairie bison seem to be too close to be coincidental. Ruminants, however, share in the best of two worlds, as they can derive sulfur amino acids both from plants and from the proteins synthesized by the bacteria of their rumina, which can utilize excess sulfates contained in plant foods (Allaway and Thompson 1966). The reduction of sulfates to sulfides in the rumen is well established, and the latter form of sulfur is believed to be the primary source of sulfur for microbial protein synthesis. Experiments by Mitchell (1963:24) with cows given intravenous injections of C^{14}-labeled nutrients to bypass normal food intake have shown that essential amino acids found in the milk could only have been produced by the bacteria of the rumen. In other experiments, sulfate-supplemented feed given to cattle and sheep has been shown to increase the amount of amino acids nine to twenty times in the rumen contents. Increased milk production in cows fed silage supplemented with 2 to 3 kg of sulfate per metric ton has also been demonstrated (Garrigus 1970:136). Obviously, ruminants such as bison are eminently suited to capitalize on sulfate-rich environments.

Moose

The relation of moose populations to levels of sulfate in their primary habitats and in their foods needs study. In the winter of 1771–72, Samuel Hearne found moose (*Alces americanus*) "very plentiful" on the south side of Great Slave Lake east of the mouth (delta) of the Slave River (Preble 1908:134). Many other examples of high moose densities coinciding with high sulfate environments can doubtless be given.

Moose tend to be primarily associated with riparian habitats and early forest successional stages. Their primary foods are willows (*Salix*) and poplar (*Populus*) (Cowan et al. 1950; Peterson 1953). Cowan et al. (1950) found the bark of *Populus tremuloides* particularly high in protein. It would, therefore, appear significant that Gerloff's et al. (1964) data for the sulfur content of 2 species of *Salix* and 3 species of *Populus* in Wisconsin average far higher for sulfur (hence protein) than for any of the plant families represented by the 158 species they analyzed for mineral content.

Viewed in the context of the present essay, it is apparent that the metabolic significance of "salt licks" for ruminants is still poorly understood. Sodium chloride is not necessarily the sought-after salt. Murie (1951:309) states that the need of wild game for salt (sodium chloride) is open to question and that "it is necessary to approach the salting question with caution," noting that salt is not required for keeping cattle on *alkali* ranges in southeastern Montana (our italics). He points out that licks rarely consist mainly of sodium chloride, "and in fact contain very little of it."

Peterson (1955:120–21) has evaluated salt licks from the standpoint of moose populations. A study in Yellowstone National Park indicated that soil-eating in one area was related to its content of chlorides and *sulfates* of sodium and calcium (our italics). A study in the Rocky Mountain parks of Canada concluded that neither sodium chloride nor phosphorus was the essential element sought at licks by big game but that trace elements may have been the critical components. On the Kenai Peninsula of Alaska, the soil of a salt-lick area in an open field has been eaten down several feet below the surrounding level. Although a large part of the mineral concentration was sodium chloride, *sulfuric acid* and *sulfides* were also present (Hosley 1949; our italics). On Isle Royale, water and soil from four licks contained .10 to .25 percent salt. "The content of these samples was *chiefly calcium sulfate* and traces of sodium chloride. The single mud sample analyzed contained much more of the latter than the liquid samples" (Peterson 1955:121; our italics).

The American bison is reported to make use of salt licks—especially in the past in eastern portions of the United States (where sulfate levels in the environment are low in comparison to the Great Plains). A portion of Roe's (1972:842–43) review of the literature on the American bison is therefore significant in relation to our thesis.

"Neither of these witnesses alludes to them rubbing down trees. Other witnesses do so, however; and their observations pertain to the same general types of country. . . . 'The open space around and near the *sulphur or salt springs,* instead of being an old field, as had been supposed by Mr. Mausker [Mansker], at his visit here in 1769, was thus freed from trees and underbrush by the innumerable herds of buffalo and deer and elk that came to these waters. . . .' [Our italics.] John Filson describes the Blue Licks, Kentucky, 1784 (in part) as follows: '. . . the vast space of land around these springs desolated as if by a ravaging enemy, and hills reduced to plains; for the land near these springs is chiefly hilly. . . .' Fortescue Cuming's 'old-timer' host, Captain Waller, with reference to the same approximate period said that 'about the salt licks and springs they frequented, they pressed down and destroyed the soil to a depth of three or four feet, as was conspicuous yet in the neighbourhood of the Blue Lick, where all the old trees have their roots bare of soil to that depth . . .' (1789–1809)."

Realizing the nutritional benefits that ruminants achieve by supplementing their diet (more specifically, with supplying the bacteria of their rumina) with sulfates, the sulfate content of "salt licks" used by ungulates should be examined in light of their special needs.

Diving Ducks

An event of perhaps comparable ecologic significance to the bison observed on this flight was the spectacular concentration of male diving ducks—goldeneyes, scaups, canvasbacks, redheads, ringnecks, and scoters (all three species)—observed in the north channel of the Mackenzie River north of Big Island. Their total numbers were in the tens of thousands. The food resources (obviously mostly animal and nonmotile) which attract these ducks can be linked to the Slave River, which drains about one-ninth of the land area of Canada and which is estimated to empty 60,000 tons of dissolved minerals daily into Great Slave Lake (Rawson 1951:672). These minerals provide the nutrients for a plankton base (Rawson 1956) and for a relatively rich bottom fauna (Rawson 1953) which in turn is reflected in the well-known fishery of Great Slave Lake. These plankton-rich waters, moving with the current in the Mackenzie River, in turn provide the food base for the invertebrate bottom fauna on which diving ducks feed.

Muskrats

A mammal eminently suited to be a nutritional barometer of the wetland environment is the muskrat (*Ondatra zibethicus*). Its range extends over almost all of North America north of Mexico and south of the tree line, excepting most of California and Texas. Fifteen subspecies are currently recognized (Hall and Kelson 1959:756). Their response to the nutritional quality of their environment can be judged on the basis of three primary factors: 1) size of the animal, 2) quality of the fur, and 3) density of the populations. The factor of body size can be partly ruled out; it tends to be the converse of Bergmann's rule, the smallest animals occurring in the north and the largest in the south. The length of the growing season for plants as it affects plant biomass production would appear to be a factor affecting, through selection, size attainable within genetic control.

According to Charles Wilson of the Hudson's Bay Company (Winnipeg office), the best muskrat pelts in North America are taken in western New York (e.g., in the Finger Lake area), and the best quality Canadian pelts are taken within a fifty-mile radius of Montreal, Quebec. Pelts of eastern muskrats (Quebec and Ontario south of the Canadian Shield) are preferred to those of the western muskrat, being darker and larger. In western Canada three areas stand out in respect to production per unit area and quality of fur: 1) the marshes of the Saskatchewan River delta of Manitoba, 2) the Athabasca-Peace and Slave river deltas, and 3) the Mackenzie River delta. All three areas are noted for the tremendous sizes of the muskrat populations they have been capable of supporting.

The one environmental factor most of these areas have in common is sulfate-rich waters (Table 45). A sulfate evaporite basin underlies the Devonian strata of western New York (*National Atlas of the United States of America:* 125 and 181). The North and South Saskatchewan rivers flow for 2,000 miles through the sulfate-rich soils of the prairie provinces, terminating in the famed deltaic marshes near the northwest end of Lake Winnipeg. The sulfate-rich waters of the Peace and Athabasca rivers discharge part of their dissolved mineral load in their joint delta; the remainder is carried to the Slave River delta and into Great Slave Lake. This flow, augmented by waters draining the northern portion of the country underlain by Devonian rocks and associated gypsum formations—between Great Slave Lake and Lac la Martre—contributes additional sulfates to the Mackenzie River. The last 200 miles of the Mackenzie River again traverse Upper and Middle Devonian rocks, as do the lower portions of a major tributary, the Peel River (where Preble [1908:192] also found muskrats excessively abundant).

How do muskrats from the three western areas compare in pelt quality, which is so acutely dependent on nutrition? Of the muskrats of the Athabasca-Peace and Slave river deltas, (*Ondatra z. spatula*) Fuller (1951:362) states: "The northwestern muskrat produces pelts with heavy leather and dense underwool, which compares favorably in size with those of most of the other races."

Stevens (1953), who studied the Mackenzie River delta population of the same race, states that "in regions of *favorable habitat* [our italics] the typical *spatula* has short, heavy underfur and is valued in the fur trade for certain superior qualities of its pelt."

However, for a definitive, comparative assessment of the relative quality of the fur of these three muskrat populations, no one can speak with greater authority than the chief fur graders of the Hudson's Bay Company. According to Charles Wilson (personal communication), the muskrat pelts of the Saskatchewan River delta (and associated marshes) rate as the best, followed by those from the Athabasca-Peace and Slave river deltas. The Mackenzie River muskrats are rated third because of the woolly nature of their underfur. Although the latter may rate lower as a fashion fur, the dense woolly underfur is likely, because of the greater insulative demands made upon it by the long cold periods of the subarctic habitat, to weigh more on an area basis than underfur consisting of straight fibers. The metabolic and nutritional demands on these subarctic muskrats, especially in respect to cysteine requirements, would therefore be greater than the nutritional requirements that challenge more southerly populations.

The underlying assumption of this discussion is, of course, that muskrats, through high intake of protein-rich, sulfate-containing vegetative foods, are able to grow better fur in high-sulfate environments. There are significant experimental findings to support this hypothesis. Ewes given supplemental sodium sulfate produced a higher percentage of first-class karakul pelts than those that were not fed supplemental sulfate (Garrigus 1970:129).

Although muskrats and waterfowl alike benefit from high sulfate environments, it seems that they also can be adversely affected, even if indirectly, by marsh senescence related in part to the reduction of sulfate content. McLeod (Errington 1963:604) in noting that muskrat populations of the Saskatchewan River delta underwent ebbs and flows in numbers, each lasting about five years, theorized that: "The general opinion is that the land (wetland and shallow waters areas) 'sours' in time and its productivity is reduced. The one soil element very essential for plant growth and the one most likely to be absent in soils of low fertility is nitrogen. It was thought that the so-called 'sour' condition of the soil might refer to a nitrogen deficiency. There are two possibilities here: either the total nitrogen content of the soil is low, or if nitrogen is present in reasonable quantity it may all be taken up during the first four or five years of growth and become bound up in organic form. In the absence of complete decomposition of plant material in the water and mud by bacteria and animal organisms the nitrogen would be found in no simpler form than amino acids. *It might be noted that there was an almost complete absence in the water of animal forms such as Cladocera and Copepoda, and bottom dwelling forms such as chironomid larvae were quite rare"* (our italics).

Errington (1963:604) also cites the finding of Wilde, Youngberg, and Hovinds (1950) that the bottom soils of beaver ponds become saturated with hydrogen sulfide. They also make the significant point that hydrogen sulfide is directly injurious to plant roots.

Cutthroat Trout

The sport fishery for cutthroat trout (*Salmo clarki*) in the rivers and streams of Yellowstone National Park is spectacular by all accounts. The overwhelming source of mineral input into the streams of the region is the multitude of mineral springs (3,000) with high sulfur content that are found throughout the park. Notable in this respect are the Yellowstone and Firehole rivers which are "distinguished by a sulphury taste and odor the year round. . . . Practically all other streams in the park carry considerable amounts of sulphur in suspension during the flood period. . . . But as the waters grow clear, that is, as the sulphur sands have silted out, they improve in potability. Moreover, *the fish then lose the sulphury taste they all seem to have during the flood period"* (Muttkowski 1929:175; our italics). The average annual sulfate content of the Yellowstone River near the Wyoming border in Montana is about 30 ppm, a concentration equal to half of that carried by the North Saskatchewan River and equal to or exceeding the sulfate load of the major rivers of the northern plains of Canada (Table 44). It would, therefore, appear that the cutthroat trout and the invertebrates upon which they feed are both physiologically tolerant of periodic exposures to high concentrations of sulfur and sulfur compounds and that populations of both groups benefit from moderately high sulfate levels in the waters.

Salamanders

Most of the small water bodies of the plains lack fish populations, but their predatory niche in the ecosystem is replaced by the tiger salamander (*Ambystoma tigrinum*). In many of the western areas they do not progress beyond the gilled, larval, or axolotl stage. In the high saline (sulfate) lakes in which they occur, high densities apparently characterize their populations. According to W. E. LaBerge (personal communication), these salamanders are so abundant in the small saline lakes of the sandhill country of Nebraska that they constitute an important part of the diet of the white pelicans (*Pelecanus erythrorhynchos*) that nest in the region. In North Dakota, the tiger salamander appears *"often in astonishing numbers"* after the first warm spring rains (Wheeler and Wheeler 1966:31; our italics).

Wetmore (1920) studied the birdlife of Lake Burford, New Mexico. This lake was once known as Stinking Lake from the input of a spring of sulfur-laden waters that it receives in addition to meltwater, rainwater and other springs. Although acceptable to livestock, the waters of this lake are decidedly "alkaline," but aquatic plants typically associated with high sulfate waters attained luxuriant growth in it, particularly sago pond weed which was enveloped by large masses of green algae. There were no fish in the lake but larval salamanders, which fed on chironomids, were abundant—so much so that when large numbers of these salamanders died in June, their bodies produced an "effluvium" in early morning, as Wetmore (1920) delicately expressed it. (The numbers of wild ducks [twelve species] Wetmore recorded around the lake calculates out to about one nesting pair per hundred yards of shoreline, a substantial density considering the fact that the lake is isolated in desert country.)

Clostridium botulinum

To conclude this discussion of some of the relationships of sulfate concentrations in the environment to the biotic world, some searching questions should be asked regarding the bête noire of western waterfowl populations—*Clostridium botulinum,* the organism producing the extraordinarily toxic by-product of their metabolism. It is, on the face of it, significant that severe botulism outbreaks are confined to the more arid regions of the country. Outbreaks of botulism have, in the past before the real cause of die-offs of ducks was known, been called "western duck sickness" or "alkali poisoning," the latter term because of the association of dead ducks with alkali (sulfate) encrusted flats. The more typical outbreak area is indeed associated with alkaline pH. Summer temperatures, falling water levels resulting in extended exposures of shoreline and an anaerobic intermix of soil, dead insect bodies, or other organic matter set the stage for outbreaks (Jensen and Williams 1964).

The question is, why are outbreaks confined largely to the West? As we have repeatedly discovered in the course of this study, many biochemical and physiological studies, largely of necessity, are surprisingly parochial in scope and outlook. Apparently in all the years *Clostridium botulinum* has been studied, the ecology of the organism in natural environments has received relatively scant attention. Fortunately, as we completed this report, our attention was called to a paper in press by Louis D. S. Smith (1975) in which he reports on twenty-five species of *Clostridium* found in twenty-one soil samples from Virginia, Iowa, South Dakota, Idaho, Washington, and Costa Rica. Smith found the restriction of *C. botulinum* type A in six soil samples and proteolytic type B in neutral or alkaline soils significant at the .001 level. Types E and F were found in one acid soil sample. Judging from the pH range of the other types he studied, it seems safe to conclude that the geographic range of Type C, the serotype producing toxins causing waterfowl mortality, is largely related to alkaline soils—and in the West, these are soils high in sulfates. Conversely, acidity is recognized as a limiting factor for *Clostridium botulinum* in food (Alexander 1971:132).

The rather primitive state of knowledge of the ecology of *Clostridium* is indicated by Smith's assessment of his findings: "The reasons for the association of certain clostridia, such as *C. botulinum* type A and *C. mangenoti* with soil of certain characteristics are not known. Although the association with alkaline and acid soils, respectively, seems clear, this may be more than the simple effect of the level of hydrogen ion concentrations. The complexity of soil structure allows the establishment of a multitude of different microenvironments in a small volume and, consequently, the existence of a complex microbial population. Nevertheless, the restriction of these two species to soils of different levels of activity indicates *that some simple parameters* may be of importance in determining the make-up of microbial populations at least for the anaerobes" (our italics).

Botulism in ducks occurs frequently in Wisconsin in only one area—along the lower Fox River and at its mouth at the south end of Green Bay (Richard A. Hunt, personal communication). Ducks that have died from botulism are found along the southwest shore of the bay. It is probably immensely significant that at least six paper pulp mills discharge sulfurous and high BOD effluents into the Fox River—which varies in sulfate content from 9–48 and averages 24 ppm (James H. Wiersma, unpublished), depending on the time of year. However, it should also be pointed out that lake and ground waters used for drinking water by municipalities are highest in sulfate content in this sector of the state.

The question is what role do sulfates play in the metabolism of Clostridia that permits their populations to irrupt? The relationship appears to be an indirect one as *Clostridium botulinum* is reported to be unable to reduce sulfates in a dissimilatory manner. It would seem that a possible answer lies in a symbiotic relationship of the Clostridia with other biota such as dissimilatory sulfate-reducing bacteria and possibly other organisms more directly affected by high sulfur concentrations (M. P. Bryant, personal communication; Alexander 1971:225).

Klein and Cronquist (1967:183–84) have concluded from morphological and physiological data that sulfate-reducing bacteria are present-day representatives of organisms which are one evolutionary stage removed from fermentative Clostridia (the most primitive microbes). Clostridia (some species) "can also reduce and assimilate sulfur through another route with energy supplied by ATP. Not so simple, but probably almost as primitive, is dissimilatory sulfate reduction in which APS [adenosine-5'-phosphosulfate] is directly reduced to sulfate and hence to sulfide. Both of these reactions are also found in *Clostridium* species, as well as in *Desulfovibrio*. The difference, of course, is that *Desulfovibrio*, since it is autotrophic, is more advanced than *Clostridium*, which cannot utilize efficiently the energy available from these reactions."

AVAILABLE PROTEIN AND SULFATES—SYNERGISTIC DENOMINATORS OF PRODUCTIVITY?

The association of breeding waterfowl with the various limnologically classified water areas, particularly the prairie waterfowl with eutrophic lakes and rich invertebrate populations, was pointed out by Hanson et al. (1949:201–2). The high protein requirements of game birds are well known. Cook (1964:577) points out, for example, that wood ducks deprived of a high protein diet fail to reproduce. Thus, breeding ducks select lakes rich in free-swimming invertebrates. Significantly, marsh soils found to be poor for plants were also noted to be unproductive in animal foods (Cook 1964:577). Correspondingly, the largest standing crops of invertebrates are associated with rich growths of aquatic plants. The dependence of nesting female ducks and their young broods on small invertebrates as a source of vitally needed protein is being clearly shown by investigations of Swanson and his coworkers at the Northern Prairie Wildlife Research Station, Jamestown, North Dakota (Swanson et al. 1974) and in Canada by Sugden (1973).

It has long been known that periodic droughts revitalize the prairie potholes and the food chains upon which waterfowl are dependent. However, aside from the overall explanation that nutrients bound up in the organic matter are oxidized as a result of desiccation and again are made available for plant and animal life (Leitch 1964), the key nutrients and mechanisms involved in this sequence of events have not been elucidated. Aquatic and wetland plants and dense populations of small aquatic invertebrates are but links in the productivity of prairie potholes and lakes. It is suggested here that the wide prevalence of sulfates in the prairies and high plains—as well as other sulfate-rich environments—is a key factor that accounts for the high productivity often associated with these areas. The primary benefit of sulfates in the environment in excess of minimal metabolic needs is that they apparently maximize protein production by plants and animals as well as having an amino-acid-sparing action in birds and mammals and doubtless also in other vertebrates. Almquist (1970:206), for example, has shown that chicks given supplemental sodium sulfate make higher weight gains and more efficient feed conversions.

Our discussion has focused on the importance of sulfur in the distribution of certain fauna and flora and in the unusual productive capacity of these sulfur-sufficient environments, but we do not mean to denigrate the importance of nitrogen and phosphorus. As Moyle (1956:311) notes, "waters [in Minnesota] high in sulfates are usually high in carbonates, phosphorus, and nitrogen, but there are exceptions." Nevertheless, we detect that Hutchinson (1966:684) was not fully satisfied with current concepts of what constitutes fertility in waters: "Although it is widely believed that the biomass of phytoplankton is likely to be determined by the concentrations of combined dissolved nitrogen and dissolved phosphate, the evidence that this is so is confined largely to a few field experiments in lake fertilization and some enrichment experiments with lake water in bottles. Experimental studies on lakes (e.g., Stross and Hasler 1960) seem to indicate that lime, presumably largely as a source of bicarbonate, may be a limiting factor."

In a similar vein, Almquist (1970:204) points out that most studies of the nutrition and physiology of sulfur amino acids have disregarded sulfate intake, although considerable amounts of sulfate may be incorporated in experimental diets in their preparation. Applied in the context of biomass production in the wild, we echo Almquist's musing: "I have wondered how often sulfate has been the actual identity of somebody's 'unidentified growth factor.' "

There are no in-depth discussions even in recent texts of the ecological relationships of sulfur (e.g., Frey 1966; Rorison 1969; Alexander 1971; Hasler 1975); it is equally apparent that much research and a definitive treatment of data existing in the literature of the relationship of environmental sulfur to food chains is badly needed if, in light of the evidence presented, we are to understand adequately animal populations and the productivity of natural environments.

Literature Cited

Adelstein, S. J., and B. L. Vallee. 1962. Copper. Pages 371–401 in Comar, C. L., and Felix Bronner. 1962.

Adomaitis, V. A., and J. A. Shoesmith. 1973. The ecological chemistry of waterfowl habitat. U.S. Bur. Sport Fish. and Wildl. Northern Prairie Wildl. Res. Center. Ann. Prog. Rept. 65 pp.

Aisen, P. 1973. The transferrins (siderophilins). Pages 280–305 in Eichhorn, Gunther L., Vol. 1. 1973.

Albanese, Anthony A. 1959. Protein and amino acid nutrition. Academic Press, New York. 604 pp.

———. 1972. Newer methods of nutritional biochemistry. Academic Press, New York. 5:252 pp.

Alexander, Martin. 1971. Microbial ecology. John Wiley & Sons, Inc., New York. 511 pp.

Allan, J. A., and R. L. Rutherford. 1929. Salt and gypsum in Alberta, Canada. Inst. Min. and Metal. Trans. 32:232–54.

Allaway, W. H. 1970. The scope of the symposium: outline of current problems related to sulfur nitrition. Pages 1–5 in Muth, O. H., and J. E. Oldfield. 1970.

———, and J. F. Thompson. 1966. Sulfur in the nutrition of plants and animals. Soil Sci. 101(4):240–47.

Allen, Durward L. 1956. Pheasants in North America. The Stackpole Co. and Wildl. Mgt. Inst. 490 pp.

Almquist, H. J. 1970. Sulfur nutrition of nonruminant species. Pages 196–208 in Muth, O. H., and J. E. Oldfield. 1970.

Altman, Phillip L., and Dorothy S. Dittmer (Ed.). 1968. Metabolism. Biological Handbook. Fed. Amer. Soc. Exper. Biol. 737 pp.

Anderson, William L. 1964. Survival and reproduction of pheasants released in southern Illinois. Jour. Wildl. Mgt. 28(2):254–64.

———. 1969. Condition parameters and organ measurements of pheasants from good, fair, and poor range in Illinois. Jour. Wildl. Mgt. 33(4):979–87.

———. 1972. Dynamics of condition parameters and organ measurements in pheasants. Ill. Nat. Hist. Surv. Bull. 30(8):453–98.

———, and Peggy L. Stewart. 1969. Relationship between inorganic ions and distribution of pheasants in Illinois. Jour. Wildl. Mgt. 33(2):254–70.

———, and ———. 1970. Concentrations of chemical elements in pheasant tissues. Ill. Nat. Hist. Surv. Biol. Note No. 67:15 pp.

———, and ———. 1973. Chemical elements and the distribution of pheasants in Illinois. Jour. Wildl. Mgt. 37(2):142–53.

Anke, Manfred. 1965. Mineral and trace element content of cattle hair as indicators of Ca, Mg, P, K, Na, Fe, Zn, Mn, Cu, Mo, and Co Nutrition. II. Relationship to cutting depth, hair type, hair color, hair age, animal age, lactation stage, and pregnancy. Arch. Tierernähr. 15(6):469–85 (Ger.). (Biol. Abs. 47:92059)

———. 1967. Mineral and trace element content of cattle hair as indicators of Ca, Mg, P, K, Na, Fe, Zn, Mn, Cu, Mo, and Co nutrition. V. The mineral supply of dairy cows on soils of various geographical origin measured by the mineral content of black cattle cover hair and of red clover. Arch. Tierernähr. 17(1–2):1–26 (Ger.) (Biol. Abs. 48:1–2197).

Anonymous. 1969. Boron residue tolerance approved for citrus fruits. Oil, Paints, Drug Reporter, June 9:40.

Appelquist, L.-A. 1972. Chemical constituents of rape seed. Pages 123–73 in Appelquist, L.-A., and R. Ohlson. 1972.

———, and R. Ohlson. 1972. Rape seed. Cultivation, composition, processing and utilization. Amer. Elsevier Publishing Co., New York. 391 pp.

Armands, Gösta. 1967. Geochemical prospecting of a uraniferous bog deposit at Masugnsbyn, northern Sweden. Pages 127–54 in Kvalheim, Aslak. 1967.

Armstrong, J. E. 1947. The Arctic archipelago. Pages 311–24 in "Officers of the Geological Survey" [of Canada]. 1947. Geology and economic minerals of Canada (Third edition) Canada Dept. Mines and Resources Econ. Geol. Ser. No. 1, Publ. No. 2478. Ottawa. 357pp.

Art, Henry W., F. Herbert Bormann, Garth K. Voigt, and George M. Woodwell. 1974. Barrier Island ecosystem: role of meteorologic nutrient inputs. Science 184(4132):60–62.

Arthur, David. 1965. Interrelations of molybdenum and copper in the diet of the guinea pig. Jour. Nutr. 87(1):69–76.

Baker, G., L. H. P. Jones, and I. D. Wardrop. 1959. Cause of wear in sheep's teeth. Nature 184(4698):1583–84.

———, ———, and Angela A. Milne. 1961. Opal uroliths from a ram. Aust. Jour. Agric. Res. 12(3):473–82.

———, and Hans Joachine Schneider. 1966. Dependence of inorganic components in human hair on sex, age, hair color, and type of hair. Z. Gesamte Inn. Med. Thre Grenzgeb. 21(24):794–801 (Ger.) (Chem. Abs. 66(8):63204c).

Bandurski, Robert S. 1965. Biological reduction of sulfate and nitrate. Pages 467–90 in Bonner, James, and J. E. Varner. 1965.

Bar, Arie, and Shmuel Hurwitz. 1971. Relationship of calcium-binding protein to calcium absorption in the fowl. Pages 50–51 in Menczel, J., and A. Harell. 1971.

Barry, T. W. 1960. Waterfowl reconnaissance in the western Arctic. Arctic Circ. 13:51–58.

Barzel, Uriel S. 1970. Osteoporosis. Grune and Stratton, New York. 290 pp.

Bate, L. C., and Frank F. Dyer. 1965. Trace elements in human hair. Nucleonics 25(10):74–81.

Bauer, Göran C. H., Arvid Carlsson, and Bertil Lindquist. 1961. Metabolism and homeostatic function of bone. Pages 609–76 in Comar, C. L., and Felix Bronner. 1961.

Baumeister, Robert. 1972. Chemical analyses of selected public drinking water supplies (including trace metals). Wis. Dept. Nat. Res. Tech. Bull. 63:16 pp.

Bayrock, L. A., and S. Pawluk. 1967. Trace elements in tills of Alberta. Can. Jour. Earth Sci. 4:597–607.

Beals, C. S. (Ed.). 1968. Science, history and Hudson Bay. Vol. II. Can. Dept. Energy, Mines and Res., Ottawa. 503–1057.

Becker, C. D., and T. O. Thatcher. 1973. Toxicity of power plant chemicals to aquatic life. Battelle Pacific Northwest Laboratories, Richland, Wash. 1249 UC-VI.

Becker, W. M., and W. G. Hoekstra. 1971. The intestinal absorption of zinc. Pages 229–56 in Skoryna, S. C., and D. Waldron-Edward. 1971.

Bell, D. J., and B. M. Freeman. 1971. Physiology and biochemistry of the domestic fowl. Academic Press, New York. 1:601 pp.

———, and ———. 1971. Physiology and biochemistry of the domestic fowl. Academic Press, New York. 3:1153–488.

Benedict, Francis G., and Robert C. Lee. 1937. Lipogenesis in the animal body with special reference to the physiology of the goose. Carnegie Inst. Wash. Publ. No. 489:232 pp.

Bent, Arthur Cleveland. 1951. Life histories of North American waterfowl. II. 316 pp., 60 pl. Dover Publishing, Inc., New York. Pages 225, 229–30.

Bernstein, Leon. 1970. Calcium and salt tolerance of plants. Science 167(3923):1387.

Beule, John, and Thomas Janisch. 1974. Soil sediment survey. Wis. Dept. Nat. Res. Proj.:W-141-R-9, Prog. Rept. Mar. 1, 1973–Feb. 28, 1974. 13 pp.

Birge, Stanley J., Helen R. Gilbert, and Louis V. Avioli. 1972. Intestinal calcium transport: the role of sodium. Science 176(4031):168–70.

Bittar, E. Edward, and Neville Bittar. 1968. The biological basis of medicine. Academic Press, New York. 1:590 pp.

———, and ———. 1969. The biological basis of medicine. Academic Press, New York. 3:493 pp.

Blackadar, R. G. 1966. Geological reconnaissance, southern Baffin Island, District of Franklin. Geol. Surv. Can. Paper 66-47:32 pp.

Blackburn, S. 1961. The chemical composition of keratins and skin proteins. Pages 701–5 in Long, Cyril. 1961.

Bloom, W., A. V. Nalbandov, and M. A. Bloom. 1960. Parathyroid enlargements of laying hens on a calcium-deficient diet. Clin. Orthopoedics 17:206–9.

Bolen, Eric G. 1964. Plant ecology of spring fed salt marshes in western Utah. Ecol. Monog. 34(2):143–66.

Bonner, James, and J. E. Varner. 1965. Plant biochemistry. Academic Press, New.York. 1954 pp.

Borle, André. 1974. Calcium and phosphate metabolism. Pages 361–90 in Comroe, Julius H., Jr. et al. 1974.

Boucher, R. V. (Chairman of Committee). 1956. Composition of concentrate by-product feeding stuffs. Nat. Acad. Sci., Nat. Res. Coun. Publ. 449, Washington, D.C. 126 pp.

Bould, C. 1963. Mineral nutrition of plants in soils. Pages 16–96 in Steward, F. C. 1963.

Bourne, Geoffrey. 1971. The biochemistry and physiology of bones. 2nd Ed. Academic Press, New York. 3:584 pp.

Bowen, H. J. M. 1966. Trace elements in biochemistry. Academic Press, New York. 241 pp.

Bradfield, Robert B., Theresa Yee, and Juan M. Baertl. 1969. Hair zinc levels of Andean Indian children during protein calorie malnutrition. Amer. Jour. Clin. Nutr. 22(10):1349–53.

Bray, R. 1943. Notes on the birds of Southampton Island, Baffin Island, and Melville Peninsula. Auk 60(4):504–36.

Bray, Richard C., and Irwin Clark. 1971. Vagaries in the use of isolated intestinal mucosal preparations with particular emphasis on calcium uptake. Pages 313–37 in Nichols, George Jr., and R. H. Wasserman (Ed.). 1971.

Brickman, Arnold S., Shaul G. Massry, and Jack W. Coburn. 1971. Calcium deprivation and renal handling of calcium during saline infusions. Amer. Jour. Physiol. 220(1):44–48.

Briggs, Michael and Maxine Briggs. 1973. Effects of some contraceptive steroids on serum proteins of women. Biochem. Pharmacol. 22(18) 2277–81.

Brigham, Warren U., and Sarah R. Gnilka. 1973. Sulfur budget of Lake Shelbyville, Illinois, and the effects of sulfides upon *Chaoborus*. Univ. of Ill. Water Resources Center, Res. Rept. 66:59 pp.

Britton, A. A., and G. C. Cotzias. 1966. Dependence of manganese turnover on intake. Amer. Jour. Physiol. 211(1):203–6.

Bronner, Felix. 1964. Dynamics and function of calcium. Pages 341–44 in Comar, H. L., and Felix Bronner. 1964.

Broome, S. W., W. W. Woodhouse, Jr., and E. D. Seneca. 1975. The relationship of mineral nutrients to growth of *Spartina alterniflora* in North Carolina. I. Nutrient status of plants and soils in natural stands. Soil Sci. Soc. Amer. Proc. 29(2):295–301.

Brown, I. C. (Ed.). 1967. Groundwater in Canada. Geol. Surv. Can. Econ. Geol. Rept. 24:228 pp.

Buck, W. Keith, and Amil Dubnie. 1968. Economic possibilities. Pages 935–43 in Beals, C. S., Vol. 2. 1968.

Burger, J. Wendell, and Walter N. Hess. 1960. Function of the rectal gland in the spiny dogfish. Science 131(3401):670–71.

Camsell, Charles. 1903. The region southwest of Fort Smith, Slave River, N.W.T. Geol. Surv. Can. Summ. Rept. 1902. Ann. Rept. P & A. 15:151–69.

———. 1917. Salt and gypsum deposits of the region between Peace and Slave Rivers, northern Alberta. Geol. Surv. Can. Summ. Rept. 1916(26):134–45.

Cannon, H. L. 1960. Botanical prospecting for ore deposits. Science 132(3427):591–98.

Carlisle, E. M. 1974. Essentiality and function of silicon. Pages 407–23 in Hoekstra, W. G., et al. 1974.

Chapin, F. Stuart. 1974. Phosphate absorption capacity and acclimation potential in plants along a latitudinal gradient. Science 183 (4124):521–23.

Chapman, Charles. 1968. Practices affecting south Atlantic and gulf coast marshes and estuaries. Pages 93–106 in Newsom, J. D. 1968.

Chapman, L. J., and M. K. Thomas. 1968. The climate of northern Ontario. Can. Dept. Trans. Met. Br. Clim. Stud. 6:58 pp.

Chapman, V. J. 1960. Salt marshes and salt deserts of the world. Leonard Hill (Books), Ltd., London. Intersciences Publishers, Inc., New York. 392 pp.

Chapuy, M. C., and D. Pansu. 1971. Influence of sodium, glucose and xylose on calcium absorption. Pages 54–56 in Menczel, J., and A. Harell. 1971.

Chauvaux, G., F. Lomba, I. Fumiere, and V. Bienfet. 1965. Copper and manganese in cattle, methods of determination and biological significance of copper

and manganese in hair. Ann. Med. Veter. 109(3):174–226 (Fr.). (Biol. Abs. 47:63546).

Christensen, F. W., T. H. Hopper, and O. A. Stevens. 1947. Feeding slough hays and oat straw. N. D. Agric. Exper. Sta. Bull. 349:20 pp.

Christie, George A. 1972. Nutritional and metabolic aspects of circadian rhythms. Pages 1–32 in Albanese, Anthony A. 1972.

Christopher, John P., John C. Heganauer, and Paul D. Saltman. 1974. Iron metabolism as a function of chelation. Pages 133–45 in Hoekstra, W. G. et al. 1974.

Clegg, R. E., P. E. Sanford, R. E. Hein, A. C. Andrews, J. S. Hughes, and C. D. Mueller. 1951. Electrophoretic comparisons of the serum proteins of normal and diethylstilbesterol-treated cockerels. Science 114(2965):437–38.

Colby, P. J., and L. L. Smith, Jr. 1967. Survival of walleye eggs and fry on paper fiber sludge deposits in Prairie River, Minnesota. Trans. Amer. Fish Soc. 96(3):278–96.

Cole, L. H. 1926. Sodium sulphate of western Canada, occurrence, uses and technology. Can. Mines Br. Pub. 646:160 pp.

Collins, Douglas H. 1966. Pathology of bone. Butterworths, London. 254 pp.

Comar, C. L., and Felix Bronner (Eds.). 1960. Mineral metabolism. Academic Press, New York. 1(A):386 pp.

———, and ——— (Eds.). 1962. Mineral metabolism. Academic Press, New York. 2(B):623 pp.

———, and ——— (Eds.). 1964. Mineral metabolism. Academic Press, New York. 2(A):649 pp.

———, and ——— (Eds.). 1969. Mineral metabolism. Academic Press, New York. 3:548 pp.

Comroe, Julius H., Jr., Ralph R. Sonnenschein, and Kenneth L. Zierler. 1974. Ann. Rev. of Physiol. Ann. Rev., Inc., Palo Alto, Calif. 36:583 pp.

Cooch, F. G. 1961. Ecological aspects of the blue-snow goose complex. Auk 78(1):72–89.

Cook, Arthur H. 1964. Better living for ducks—through chemistry. Pages 569–678 in Linduska, Joseph P., and Arnold L. Nelson. 1964.

Copeland, Eugene. 1966. Salt transport organelle in *Artemia salensis* (Brine shrimp). Science 151(3709):470–71.

Copp, D. Harold. 1969. Parathormone, calcitonin and calcium homeostasis. Pages 453–513 in Comar, C. L., and Felix Bronner. 1969.

Cotzias, George C. 1962. Manganese. Pages 403–42 in Comar, C. L., and Felix Bronner. 1962.

———, and Arnaldo C. Foradori. 1968. Trace metal metabolism. Pages 105–21 in Bittar, E. Edward, and Neville Bittar. Vol. 1. 1968.

Cowan, I. McT., W. S. Hoar, and J. Hatter. 1950. The effect of forest succession upon the quantity and upon the nutritive values of woody plants used as foods by moose. Can. Jour. Res. D. 28:249–71.

Craig, B. G. 1961. Surficial geology of northern District of Keewatin. Geol. Surv. Can. Dept. of Mines and Tech. Surv. Paper 61-65:8 pp.

Crone, Christian, and Niels A. Lassen. 1970. Capillary permeability. Academic Press, New York. 681 pp.

Curan, Peter F. 1973. Amino acid transport in intestines. Pages 298–314 in Ussing, H. H., and N. A. Thorn. 1973.

Davis, George K. 1972. Competition among mineral elements relating to absorption by animals. Pages 62–69 in Hopps, Howard C., and Helen L. Cannon. 1972.

Day, J. H. 1968. Soils of the upper Mackenzie River area, Northwest Territories. Can. Dept. Agric. Res. Br. 77 pp.

DeGrazia, Joseph A. 1971. The intestinal absorption of calcium. Pages 151–71 in Skoryna, S. C., and D. Waldron-Edward. 1971.

Delacour, Jean. 1954. The waterfowl of the world. County Life, Ltd., London. Vol. 1. 284 pp.

Dementiev, G. P., and N. A. Gladkov (Eds.). 1967. Birds of the Soviet Union. Translated from Russian. U.S. Dept. of Interior, Nat. Sci. Found. and Israel Prog. for Sci. Trans. 4:683 pp.

Dent, C. E., and L. Watson. 1966. Osteoporosis. Postgrad. Med. Jour. 42:583–608.

Dequeker, Jan. 1972. Bone loss in normal and pathological conditions. Leuven Univ. Press. 214 pp.

Dibona, Donald R., and Mortimer M. Civan. 1973. Intercellular pathways for water and solute movement across the toad bladder. Pages 161–68 in Ussing, H. H., and N. A. Thorn. 1973.

Douglas, R. J. W. (Ed.). 1970. Geology and economic minerals of Canada. Geol. Surv. Can. Econ. Geol. Rept. No. 1. 838 pp.

Dowben, Robert M. 1969. Biological membranes. Little, Brown and Co., Boston. 303 pp.

Dunson, William A., and Robert D. Weymouth. 1965. Active uptake of sodium by soft shell turtles (*Trionyx spinifer*). Science 149(3679):67–69.

Dunson, William A., Randall K. Packer, and Margaret K. Dunson. 1971. Sea snakes: an unusual salt gland under the tongue. Science 173(3995):437–41.

Dzubin, A., H. Boyd, and W. J. D. Stephen. 1973. Blue and snow goose distribution in the Mississippi and central flyways. Can. Wildl. Serv., Ottawa. 81 pp. mimeo.

Eaton, Scott V. 1941. Influence of sulphur deficiency on the metabolism of the sunflower. Bot. Gaz. 102(3):536–56.

———. 1942. Influence of sulphur deficiency on the metabolism of black mustard. Bot. Gaz. 104(2):306–15.

Eatough, D. J., W. A. Mineer, J. J. Christensen, R. M. Izatt, and N. F. Mangelson. 1974. Ferric fructose and treatment of anemia in pregnant women. Pages 659–63 in Hoekstra, W. G. et al. 1974.

Eichhorn, Gunther L. 1973. Inorganic biochemistry. Elesvier Scientific Publishing Co., New York. 1:607 pp.

Ellis, J. H. 1960. The call of the land. Agric. Inst. Rev. 15(2):10–12.

Ellis, Richard A., and John H. Abel, Jr. 1964. Intercellular channels in the salt-secreting glands of marine turtles. Science 144(3624):1340–42.

Epstein, Emanual. 1972. Mineral nutrition of plants: principles and perspectives. John Wiley and Sons, Inc., New York. 412 pp.

Ericson, A. T., R. E. Clegg, and R. E. Hein. 1955. Influence of calcium on mobility of the electrophoretic components of chicken serum bloods. Science 122(3161):199–200.

Errington, Paul L. 1945. The pheasant in the northern prairie states. Pages 190–202 in McAtee, W. L. 1945.

———. 1963. Muskrat populations. Iowa State Univ. Press, Ames, Iowa. 665 pp.

Evans, Charles D., and Kenneth E. Black. 1956. Duck production studies on the prairie potholes of South Dakota. U.S. Fish and Wildl. Serv. Spec. Sci. Rept. Wildl. 32:59 pp.

Evans, Gary W., and Carole J. Hahn. 1974. Copper- and zinc-binding components in rat intestine. Pages 285–97 in Friedman, Mendel. 1974.

Evans, G. W., R. S. Dubois, and K. M. Hambridge. 1973. Wilson's disease: identification of an abnormal copper-binding protein. Science 181(4105):1175–76.

Evans, Harold J., and George J. Sorger. 1966. Role of mineral elements with emphasis on the univalent cations. Ann. Rev. Plant Physiol. 17:47–76.

Eyubov, I. A. 1968. Effect of supplementary feeding of copper, cobalt, and manganese salts on their level in lamb wool in the zone of nutritional anemia. Mater. Respub. Konf. Probl. "Mikroelem. Med. Zhivotnovod.," 1st 1968, pp. 172–74 (Russ.). (Chem. Abs. 74:29496.)

Farner, Donald S., and James R. King (Eds.). 1972. Avian Biology. Academic Press, New York. Vol. 2. 612 pp.

Fehrenbacher, J. B., J. L. White, H. P. Ulrich, and R. T. Odell. 1965. Loess distribution in southeastern Illinois and southwestern Indiana. Soil Sci. Soc. Amer. Proc. 29(5):566–72.

Fjerdingstad, E. 1950. The microphyte communities of two stagnant freshwater ditches rich in H_2S. Dansk Botanisk Arki. 14(3):44 pp.

Fleischer, Michael. 1972. An overview of distribution patterns of trace elements in rocks. Pages 6–16 in Hopps, Howard C., and Helen L. Cannon. 1972.

Florkin, Marcel, and Howard S. Mason. 1962. Comparative biochemistry. A comprehensive treatise. Vol. 4. Pt. B. 841 pp.

Flynn, Arthur, and Albert W. Franzmann. 1974. Seasonal variations in hair mineral levels of Alaskan moose. Pages 444–47 in Hoekstra, W. G. et al. 1974.

Forth, W. 1974. Iron absorption, a mediated transport across the mucosal epithelium. Pages 199–215 in Hoekstra, W. G. et al. 1974.

———, and W. Rummel. 1971. Absorption of iron and chemically related metals in *vitro* and in *vivo:* specificity of the iron binding system in the mucosal of the jejunum. Pages 173–91 in Skoryna, S. C., and D. Waldron-Edward. 1971.

Fourmann, Paul, and Pierre Royer. 1968. Calcium metabolism and the bone. Blackwell Scientific Publ. Oxford. 656 pp.

Fraser, J. A. 1964. Geological notes on northeastern district of Mackenzie, Northwest Territories. Geol. Surv. Can. Paper 63-40:20 pp.

Freeman, Hans C. 1973. Metal complexes of amino acids and peptides. Pages 121–66 in Eichhorn, Gunther L. 1973.

Frey, David G. 1966. Limmology in North America. Univ. Wisc. Press, Madison. 734 pp.

Frieden, Earl. 1974. The evolution of metals as essential elements (with special reference to iron and copper). Pages 1–31 in Friedman, Mendel. 1974.

Friedman, Mendel. 1974. Protein-metal interactions. Plenum Press, New York. 692 pp.

Frost, Harold M. 1966. The bone dynamics in osteoporosis and osteomalacia. Charles C Thomas, Springfield. 176 pp.

Fuller, William A. 1951. Measurements and weights of northern muskrats. Jour. Mam. 32(3):360–62.

———. 1962. The biology and management of the bison of Wood Buffalo National Park. Can. Wildl. Serv. Wildl. Mgt. Bull. 1(16):52 pp.

Fyles, J. G. 1962. Physiography. Pages 8–17 in Thorsteinsson, R. and E. T. Tozer. 1962.

———, J. A. Heginbottom, and V. N. Rampton. 1972. Quaternary geology and geomorphology, Mackenzie Delta to Hudson Bay. Guidebook, excursion A-30, Internat. Geol. Cong., 24th, Montreal. 23 pp.

Garby, Lars, and Areekul Suvit. 1971. A method for estimation of the leak to pore number ratio in capillary membranes. Pages 306–9 in Crone, Christian, and Niels A. Larsen. 1970.

Garn, Stanley M., Christabel G. Rohmann, and Betty Wagner. 1967. Bone loss as a general phenomenon in man. Fed. Proc., Fed. Amer. Soc. Exp. Biol. 26(6):1729–36.

Garrigus, U. S. 1970. The need for sulfur in the diet of ruminants. Pages 126–52 in Muth, O. H., and J. E. Oldfield. 1970.

Gauch, Hugh G. 1972. Inorganic plant nutrition. Dowden, Hutchinson, and Ross, Inc., Stroudsburg, Pa. 488 pp.

Geis, James W., and Robert L. Jones. 1971. Ecological significance of biogenic opaline silica. Soil Microcommunities Conf. USAEC, First, Syracuse. pp. 74–85.

Gerloff, G. C., D. D. Moore, and J. T. Curtis. 1964. Mineral content of native plants of Wisconsin. Univ. Wisc. Coll. Agric. Exp. Sta. Res. Rept. 14:27 pp.

Gibbs, Ronald J. 1973. Mechanisms of trace metal transport in rivers. Science 180(4081):71–73.

Gilbert, A. 1971. The egg: its physical and chemical aspects. Pages 1379–99 in Bell, D. J., and B. M. Freeman. Vol. 3. 1971.

Gnibidenko, G. S. 1968. New Data on the Paleozoic stratigraphy of Wrangel Island. Doklady Akad. Nauk SSSR. 179(2):407–9.

Goodman, Donald C., and Harvey I. Fisher. 1962. Functional anatomy of the feeding apparatus in waterfowl. Aves:Anatidae. Southern Illinois Univ. Press, Carbondale, 193 pp.

Gordus, Adon A. 1973. Environmental aspects of trace metals on human hair. Mimeo Rpt. No. 2, Nat. Sci. Found. Grant G1-35116. 52 pp.

Goss, Richard J. 1964. Adaptive growth. Logos Press, London. Academic Press (Distributor), New York. 360 pp.

Graf, Donald L. 1960. Geochemistry of carbonate sediments and sedimentary carbonate rocks. Part III. Minor element distribution. Ill. State Geol. Surv. Circ. 30:5–31.

Grassman, E., and M. Kirchgessner. 1974. On the metabolic availability of absorbed copper and iron. Pages 523–26 in Hoekstra, W. G. et al. 1974.

Grossman, I. G. 1968. Origin of sodium sulfate deposits of the northern Great Plains of Canada and the United States. U.S. Geol. Surv. Prof. Paper 600-B:B104–9.

Hall, E. Raymond, and Keith R. Kelson. 1959. The mammals of North America. The Ronald Press Co., New York. 2:547–1083 and index. 79 pp.

Hammer, Douglas I., John F. Finklea, Russell H. Hendricks, Thomas A. Hinners, Wilson B. Riggan, and

Carl M. Shy. 1971. Trace metals in human hair as a simple epidemiologic monitor of environmental exposure. Trace substances in Env. Health—V. Proc. 5th Conf., Columbia, Mo. pp. 25–38.

Hanson, Harold C. 1958. Studies on the physiology of wintering and of molting Canada geese (*Branta canadensis interior*). Ph.D. thesis, Univ. of Ill., Urbana. 125 pp.

———. 1959. The incubation patch of wild geese: its recognition and significance. Arctic 12(3):139–50.

———. 1962. The dynamics of conditions factors in Canada geese and their relation to seasonal stresses. Arctic Inst. N. Amer. Tech. Paper. 12:68 pp.

———. 1965. The giant Canada goose. Southern Ill. Univ. Press, Carbondale, 226 pp.

———, and Robert L. Jones. 1968. Use of feather minerals as biological tracers to determine the breeding and molting grounds of wild geese. Ill. Nat. Hist. Surv. Biol. Notes No. 60:8 pp.

———, and ———. 1974. An inferred sex differential in copper metabolism in Ross' geese (*Anser rossii*): biogeochemical and physiological considerations. Arctic 27(2):111–20.

———, Paul Queneau, and Peter Scott. 1956. The geography, birds and mammals of the Perry River region. Arctic Inst. of N. Amer. Spec. Publ. No. 3:98 pp.

———, Murray Rogers, and Edward S. Rogers. 1949. Waterfowl of the forested portions of the Canadian Pre-Cambrian Shield and the Palaeozoic Basin. Can. Field-Nat. 63(5):183–204.

———, Harry G. Lumsden, John J. Lynch, and Horace W. Norton. 1972. Population characteristics of three mainland colonies of blue and lesser snow geese nesting in the southern Hudson Bay region. Ontario Ministry of Nat. Res. Res. Rept. (Wildl.) No. 92:38 pp.

Harper, James A., and Ronald F. Labisky. 1964. The influence of calcium on the distribution of pheasants in Illinois. Jour. Wildl. Mgt. 28(4):722–31.

Harris, William H., and Robert P. Heaney. 1970. Skeletal renewals and metabolic bone disease. Little, Brown and Co., Boston. 89 pp.

Harrison, Pauline M., and T. G. Hoy. 1973. Ferritin. Pages 253–79 in Eickhorn, Gunther L. Vol. 1. 1973.

Hartmans, J. 1967. Some critical comments on the use of hair samples for mineral research (cattle). Inst. Biol. Sheik. Onderg. Landbouwgewassen Wageningen JAARB. 1967. p. 113–18. (Biol. Abs. 50:7576).

———. 1974. Tracing and treating mineral disorders in cattle under field conditions. Pages 261–71 in Hoekstra, W. G. et al. 1974.

Hasler, Arthur D. (Ed.). 1975. Coupling of land and water systems. Springer-Verlag, New York. 309 pp.

Hawkins, A. S., E. G. Wellwin, and W. F. Crissey. 1950. A waterfowl reconnaissance of the James Bay-Hudson Bay area. 9 pp. (unpublished).

Hazelwood, Robert L. 1972. The intermediary metabolism of birds. Pages 471–526 in Farner, Donald S., and James R. King. 1972.

Healy, W. B. 1974. Ingested soil as a source of elements to grazing animals. Pages 448–50 in Hoekstra, W. G. et al. 1974.

Heinz, Erich. 1972. Transport of amino acids by animal cells. Pages 455–501 in Hokin, Lowell E. 1972.

Hemphill, Delbert D. 1972. Availability of trace elements to plants with respect to soil-plant interaction. Pages 46–51 in Hopps, Howard C., and Helen L. Cannon. 1972.

Henkin, R. I. 1974. Metal-albumin-amino acid interactions: chemical and physiological interrelationships. Pages 299–328 in Friedman, Mendel. 1974.

Herring, G. M. 1973. The mucosubstances of bone. Pages 75–94 in Zipkin, Isador. 1973.

Hewitt, E. J. 1963. The essential nutrient elements: requirements and interactions in plants. Pages 137–360 in Steward, F. C. 1963.

Heyland, J. D., and H. Boyd. 1970. Greater snow geese (*Anser caerulescens atlanticus* Kennard) in northwest Greenland. Dansk Ornithologisk Forenings Tidsskrift 64:198–99.

Heywood, W. W. 1961. Geological notes, northern district of Keewatin. Geol. Surv. Can. Dept. Mines and Tech. Surv. Paper 61–1:9 pp.

Hill, K. J. 1971. The physiology of digestion. Pages 25–49 in Bell, D. J., and B. M. Freeman. Vol. 1. 1971.

Hill, R. 1974. Changes in circulating copper, manganese, and zinc with the onset of lay in the pullet. Pages 632–34 in Hoekstra, W. G. et al. 1974.

Hoekstra, W. G., J. W. Suttie, H. E. Ganther, and Walter Mertz (Eds.). 1974. Trace element metabolism in animals. Univ. Park Press, Baltimore. 775 pp.

Hoffman, L. 1964. Project Mar. The conservation and management of temperate marshes, bogs and other wetlands. International Union for the Conservation of Nature and Natural Resources. Publ. 3. New Series. 475 pp.

Hoffman, William S. 1970. The biochemistry of clinical medicine. 4th Ed. Year Book Medical Publishers, Inc., Chicago. 856 pp.

Hokin, Lowell E. 1972. Metabolic pathways. 3rd Ed. Vol. VI. Metabolic Transport. Academic Press, New York. 704 pp.

Hood, Peter J. (Ed.). 1969. Earth science symposium on Hudson Bay. Geol. Surv. Can. Paper 68–53:385 pp.

Hopps, Howard C., and Helen L. Cannon. 1972. Geochemical environment in relation to health and disease. N.Y. Acad. Sci. Annals 199:352 pp.

Hosley, N. W. 1949. The moose and its ecology. U.S. Fish and Wildl. Serv. Wildl. Leaflet 312:51 pp.

Hudson, D. A., R. J. Levin, and D. H. Smith. 1971. Absorption from the alimentary tract. Pages 51–71 in Bell, D. J., and B. M. Freeman. Vol. 1. 1971.

Hughes, M. N. 1974. The inorganic chemistry of biological processes. John Wiley and Sons, New York, 304 pp.

Hutchinson, G. Evelyn. 1957. A treatise on liminology. Vol. 1. Geography, physics and chemistry. John Wiley & Sons, Inc., New York. 1015 pp.

Hutchinson, G. E. 1966. The prospect before us. Pages 683–90 in Frey, David G. 1966.

Irving, James T. 1964. Dynamics and function of phosphorus. Pages 249–313 in Comar, C. L., and Felix Bronner. II. Pt. A. 1964.

———. 1973. Calcium and phosphorus metabolism. Academic Press, New York. 246 pp.

Ivanov, O. N. 1973. Stratigraphy of Wrangel Island. Izvestiia Akad. Nauk USSR, Ser. Geol., No. 5, pp. 104–15.

Jackson, G. D. 1960. Belcher Islands, Northwest Territories. Geol. Surv. Can. Dept. Mines and Tech. Surv., Paper 60–20:13 pp.

Jackson, Michael J., and D. H. Smyth. 1971. Intestinal absorption of sodium and potassium. Pages 137–50 in Skoryna, S. C., and D. Waldron-Edward. 1971.

Jackson, W. P. U. 1967. Calcium metabolism and bone disease. Edward Arnold, Ltd., London. 134 pp.

Jacobs, A., and M. Worwood. 1974. Iron in biochemistry and medicine. Academic Press, New York. 769 pp.

Jaffe, Henry L. 1972. Metabolic, degenerative, and inflammatory diseases of bones and joints. Lea and Febiger, Phila. 1101 pp.

Jaworowski, Z., J. Bilkiewicz, and W. Kostanecki. 1966. The uptake of 210 Pb by resting and growing hair. Int. Jour. Radiat. Biol. 11(6):563–66.

Jensen, Wayne I., and Cecil Williams. 1964. Botulism and fowl cholera. Pages 333–41 in Linduska, Joseph P., and Arnold L. Nelson. 1964.

Johnels, A. G., and T. Westermark. 1969. Mercury contamination of the environment in Sweden. Pages 221–41 in Miller, Morton W., and George G. Berg. 1969.

Jones, M. M., and J. E. Hix, Jr. 1973. Metal induced ligand reactions involving small molecules. Pages 361–437 in Eichhorn, Gunther L. 1973.

Jones, Robert L., and A. H. Beavers. 1963. Sponge spicules in Illinois soils. Soil Sci. Soc. Amer. Proc. 27(4):438–40.

———, ———, and J. D. Alexander. 1966. Mineralogical and physical characteristics of till in moraines of La Salle County, Illinois. Ohio Jour. of Sci. 66(4):359–68.

———, Ronald F. Labisky, and William L. Anderson. 1968. Selected minerals in soils, plants, and pheasants. Ill. Nat. Hist. Surv. Biol. Note 63:8 pp.

Jordan, Howard V., and H. M. Reisenauer. 1957. Sulfur and soil fertility. Pages 107–11 in U.S.D.A. Yearbook of Agric. 1957.

Jowsey, Jenifer, and Gilbert Gordan. 1970. Bone turnover and osteoporosis. Pages 201–38 in Bourne, Goeffrey. Vol. 3. 1970.

Kastelic, Joseph, Harold H. Draper, and Harry P. Broguist. 1963. Proceedings of symposium on protein nutrition and metabolism. Univ. Ill. Coll. Agric. Spec. Publ. 4:154 pp.

Kauranne, L. K. 1967. Trace element concentration in layers of glacial drift at Kolina, central Finland. Pages 181–91 in Kvalheim, Aslak. 1967.

Keith, Lloyd B. 1961. A study of waterfowl ecology on small inpoundments in southeastern Alberta. The Wildl. Soc. Wildl. Monog. 6:88 pp.

Kelsall, John P. 1970. Comparative analysis of feather parts from wild mallards. Can. Wildl. Serv. Prog. Notes 18:6 pp.

———, and J. R. Calaprice. 1972. Chemical content of waterfowl plumage as a potential characteristic tool. Jour. Wildl. Mgt. 36(4):1088–97.

Kerbes, R. H. 1969. Biology and distribution of nesting blue geese on Koukdjuak plain. N.W.T. Proj. Can. Wildl. Serv. Compl. Rept. and M.Sc. thesis. univ. of Western Ontario. 122 pp., mimeo.

———. In press. The nesting population of lesser snow geese in the eastern Canadian Arctic: a photographic inventory of June 1973. Can. Wildl. Serv., Ottawa. Rept. Series, No. 35.

Kimball, James W., Edward L. Kozicky, and Bernard A. Nelson. 1956. Pheasants of the plains and prairies. Pages 204–63 in Allen, Durward L. 1956.

Klein, Richard M., and Arthur Cronquist. 1967. A consideration of the evolutionary and taxonomic significance of some biochemical and micromorphological and physiological characters in the thallophytes. Quart. Rev. Biol. 42(2)105–296.

Kovda, V. 1956. Composition minérale des plantes et formation des sols. Congress of Soil Sci. 6th (Paris), Commission E, pp. 201–6.

Krista, L. M., C. W. Carlson, and O. E. Olson. 1961. Some effects of saline waters on chicks, laying hens, poults, and ducklings. Poul. Sci. 40(4):938–44.

Krolak, Marian. 1968. Manganese content in cattle hair and in hay from different regions of the Glansk Province. Pol. Arch. Wet. 11(1):159–70 (Pol.).

Kvalheim, Aslak (Ed.) 1967. Geochemical prospecting in Fennoscandia. Interscience Publishers. John Wiley & Sons. New York 349 pp.

Labisky, Ronald F. 1968. Ecology of pheasant populations in Illinois. Ph.D. thesis. Univ. of Wisc., Madison. 511 pp.

———, James A. Harper, and Frederick Greeley. 1964. Influence of land use, calcium, and weather on the distribution and abundance of pheasants in Illinois. Ill. Nat. Hist. Surv. Biol. Note 51:19 pp.

Låg, J. 1967. Soils of Fennoscandia and some remarks on the influence of the soils in geochemical prospecting. Pages 85–95 in Kvalheim, Aslak. 1967.

LaHaye, P. A., and Emanuel Epstein. 1969. Salt toleration by plants: enhancement with calcium. Science 166(3903):395–96.

———, and ———. 1970. Calcium and salt tolerance of plants. Science 167(3932):1388.

Lang, A. 1970. Prospecting in Canada. 4th Ed. Geol. Surv. Can., Dept. Energy, Mines and Resources. 380 pp.

———, A. M. Goodwin, R. Mulligan, D. R. E. Whitmore, G. A. Gross, R. W. Boyle, A. G. Johnston, J. A. Chamberlain, and E. R. Rose. 1970. Economic minerals of the Canadian Shield. Pages 153–226 in Douglas, R. J. W. 1970.

Larimer, O. J., and N. F. Leibbsand. 1974. Water resources data for South Dakota, 1973. Pt. 2 water quality records. U.S. Geol. Surv. 150 pp.

Lassiter, J. W., W. J. Miller, M. W. Neathery, R. P. Gentry, E. Abrams, J. C. Carter, Jr., and P. E. Stake. 1974. Manganese metabolism and homeostasis in calves and rats. Pages 557–59 in Hoekstra, W. G. et al. 1974.

Leach, R. M., Jr. 1974. Biochemical role of manganese. Pages 57–59 in Hoekstra, W. G. et al. 1974.

Leitch, William G. 1964. Water. Pages 273–81 in Linduska, Joseph P., and Arnold L. Nelson. 1964.

Lemieux, Louis. 1959. The breeding biology of the greater snow goose on Bylot Island, Northwest Territories. The Can. Field-Nat. 73(2):117–28.

Lemieux, Louis, and Joan M. Heyland. 1967. Fall migration of blue geese Chen caerulescens and lesser snow geese Chen hyperborea hyperborea from the Koukdjuak River, Baffin Island, Northwest Territories. Naturaliste Can. 94:677–94.

Leopold, A. 1931. Report on a game survey of the North Central states. Sport. Arms and Amm. Manufacturers' Inst., Madison, Wis. 299 pp.

Leslie, R. J. 1964. Sedimentology of Hudson Bay, District of Keewatin, 34, 44, and 54. Geol. Surv. Can. Paper 63–48:31 pp.

Linduska, Joseph P., and Arnold L. Nelson. 1964. Waterfowl tomorrow. U.S. Dept. Interior Bur. Sport Fish. and Wildl. 770 pp.

Long, Cyril (Ed.). 1961. Biochemists' handbook. D. Van Nostrand Co., Inc. Princeton, N.J. 1192 pp.

Longley, Richmond W. 1972. The climate of the prairie provinces. Environment Canada. Clim. Stud. 13:79 pp.

Lord, C. S. 1953. Geological notes on southern district of Keewatin, Northwest Territories. Geol. Surv. Can. Paper 52–22:11 pp. and 2 maps.

Lounamaa, J. 1967. Trace elements in trees and shrubs growing on different rocks in Finland. Pages 287–317 in Kvalheim, Aslak. 1967.

Lovering, T. S., and Celeste Engel. 1967. Translocation of silica and other elements from rock into Equisetum and three grasses. U.S. Geol. Surv., Prof. Paper No. 5948:16 pp.

Lynch, John J. 1967. Values of the south Atlantic and gulf coast marshes and estuaries to waterfowl. Pages 51–63 in Newson, John D. 1967.

Lynch, John J. 1973. 1972 productivity and mortality among geese, swans and brant. U.S. Bureau of Sport Fish. and Wildl. Prog. Rept. Albuquerque, New Mexico. 42 pp., mimeo.

———, Ted O'Neil, and Daniel W. Lay. 1947. Management significance of damage by geese and muskrats to Gulf Coast marshes. Jour. Wildl. Mgt. 11(1):50–76.

Macan, T. T. 1974. Freshwater ecology. John Wiley & Sons, New York. 343 pp.

McAtee, W. L. 1911. Winter ranges of geese on the Gulf Coast; notable bird records for the same region. Auk 28(2):272–74.

———. 1920. Wild-duck foods of the sandhill region of Nebraska. U.S. Dept. Agric. Bull. 794(2):37–77.

———. (Ed.). 1945. The ring-necked pheasant. Amer. Wildl. Inst. 320 pp.

McBean, Louis D., Mohsen Mahloudji, John G. Reinhold, and James A. Halsted. 1971. Correlation of zinc concentrations in human plasma and hair. Amer. Jour. Clin. Nutr. 24(5):506–9.

McCullough, Robert A., and C. L. Grant. 1952–53. Unpublished studies with Pittman-Robertson-Dingell-Johnson projects in New Hampshire involving analyses of fish and game.

Macdonald, I., and P. J. Warren. 1961. The copper content of the liver and hair in kwashiorkor. Abs. of Commun., 142 Meet. Nutr. Soc., Roy. Soc. Med., London. 20:36.

McEwen, E. H. 1958. Observations on the lesser snow goose nesting grounds. Egg River, Banks Island Can. Field–Nat. 72(3):122–27.

McIlhenny, E. A. 1932. The blue goose in its winter home. Auk 49(3):229–306.

McKnight, Donald E., and Jessop B. Low. 1969. Factors affecting waterfowl production on a spring-fed salt marsh in Utah. N. Amer. Wildl. and Nat. Res. Conf. Trans. 34:307–14.

McLean, Franklin C., and Marshall R. Urist. 1961. Bone. An introduction to the physiology of skeletal tissue. Univ. of Chicago Press. 2nd Ed. 261 pp.

McMurtrey, Jr., J. E., and W. O. Robinson. 1938. Neglected soil constituents that affect plant and animal development. Pages 807–34 in U.S. Dept. Agric. 1938.

Manery, J. F. 1969. Calcium and membranes. Pages 405–52 in Comar, C. L., and Felix Bronner. 1969.

Manning, T. D. 1942. Blue and lesser snow geese on Southampton and Baffin Islands. Auk 59(2):158–75.

March, James R. 1975. Mallard population and harvest dynamics in Wisconsin. Univ. Wis. Dept. Wildl. Ecol. Ph.D. thesis.

———, Gerald F. Martz, and Richard A. Hunt. 1973. Breeding duck populations and habitat in Wisconsin. Wis. Dept. Nat. Res. Tech. Bull. 68:36 pp.

Margerum, Dale W., and Gary R. Dukes. 1974. Kinetics and mechanisms of metal-ion and proton-transfer reactions of oligopeptide complexes. Pages 157–212 in Siegel, Helmut. Vol. 1. 1974.

Mason, Brian. 1966. Principles of geochemistry. John Wiley and Sons, Inc., New York. 329 pp.

Masoro, Edward J. 1973. Editor's forward. Page vii in Schepartz, Bernard. 1973.

Masri, Merle Sid, and Mendel Friedman. 1974. Interactions of keratins with metal ions: uptake profiles, mode of binding, and effects on properties of wool. Pages 551–87 in Friedman, Mendel (Ed.). 1974.

Matrone, Gennard. 1974. Chemical parameters in trace-element antagonisms. Pages 91–102 in Hoekstra, W. G. et al. 1974.

Meister, Waldemar. 1951. Changes in histological structure of the long bones of birds during the molt. Anat. Rec. 111(1):1–21.

Menczel, J., and A. Harell. 1971. Calcified tissue. Structural, functional and metabolic aspects. Academic Press, New York. 214 pp.

Michel, C. C. 1970. Direct observations of sites of permeability for ions and small molecules in mesothelium and endothelium. Pages 628–42 in Crone, Christian, and Niels A. Lassen. 1970.

Miller, Morton W., and George G. Berg (Eds.). 1969. Chemical fallout: current research on persistent pesticides. Rochester Conf. on Toxicity. Proc. 531 pp.

Miller, W. J., G. W. Powell, W. J. Pitts, and H. F. Perkins. 1965. Factors affecting zinc content of bovine hair. Jour. Dairy Sci., 48(8):1091–95.

Mills, C. F. 1974. Trace-element interactions: Effects of dietary composition on the development of imbalance and toxicity. Pages 79–89 in Hoekstra, W. G. et al. 1974.

Mitchell, H. H. 1959. Some species and age differences in amino acid requirements. Pages 11–43 in Albanese, Anthony A. 1959.

———. 1963. Some comparative studies in protein nutrition. Pages 21–22 in Kastelic, J. et al. 1963.

———. 1964. Comparative nutrition of man and domestic animals. Academic Press, New York. 1:701 pp.

Moir, R. J. 1970. Implications of the N/S ratio and differential recycling. Pages 165–81 in Muth, O. H., and J. E. Oldfield. 1970.

Moore, Carl V. and Reubenia Dubach. 1962. Iron. Pages 287–348 in Comar, C. L., and Felix Bronner. 1962.

Moo-Young, A. J., H. Schraer and R. Schraer. 1970. The copper content of the isthmus mucosa and certain organs of the domestic fowl. Proc. Soc. Exper. Biol. Med. 133:497–99.

Morgan, E. H. 1974. Transferrin and transferrin iron. Pages 29–71 in Jacobs, A., and M. Worwood. 1974.

Morgan, James P. 1970. Deltas—a resumé. Jour. Geol. Educ. 18(3):107–17.

Mortvedt, J. J., P. M. Giordano, and W. L. Lindsay (Eds.). 1972. Micronutrients in agriculture. Soil Sci. Soc. Amer., Madison, Wis. 666 pp.

Moyle, John B. 1956. Relationships between the chemistry of Minnesota surface waters and wildlife management. Jour. Wildl. Mgt. 20(3):303–20.

———, and Neil Hotchkiss. 1945. The aquatic and marsh vegetation of Minnesota and its value to waterfowl. Minn. Dept. Cons. Tech. Bull. 3:122 pp.

Murie, Olaus J. 1951. The elk of North America. The Stackpole Co., Harrisburg and The Wildl. Mgt. Inst., Washington. 376 pp.

Muth, O. H., and Oldfield, J. E. 1970. Symposium: sulfur in nutrition. The Avi Publ. Co., Inc., Westport, Conn. 251 pp.

Muttkowski, Richard A. 1929. The ecology of trout streams in Yellowstone National Park. Roosevelt Wildl. Ann. 2(2):147–263.

Nagel, John E. 1969. Migration patterns and general habits of the snow goose in Utah. Div. of Fish and Game. Utah Dept. Nat. Res. Publ. 69(6):74 pp., mimeo.

Nason, Alvin, and William D. McElroy. 1963. Modes of action of the essential mineral elements. Pages 451–536 in Steward, F. C. 1963.

Newsom, John D. 1967. Proceedings of the marsh and estuary management symposium. Louisiana State Univ. Div. of Contin. Educ., Baton Rouge, La. 250 pp.

Nicholas, D. V. D. 1963. Inorganic nutrient nutrition of microorganisms. Pages 363–447 in Steward, F. C., Vol. 3. 1963.

Nichols, George Jr., and R. H. Wasserman (Eds.). 1971. Cellular mechanisms for calcium transfer and homeostasis. Academic Press, New York. 513 pp.

Norris, W. A. 1965. Stratigraphy of middle Devonian and older Paleozoic rocks of the Great Slave region, Northwest territories. Geol. Surv. Can. Dept. Mines Tech. Surv. Mem. 322:180 pp., 9 maps.

Odum, Eugene P. 1964. The role of tidal marshes in estuarine production. Pages 70–79 in L. Hoffmann.

Officers Geological Survey of Canada. 1947. Geology and economic minerals of Canada. 3rd Ed. Econ. Geol. Ser. 1:357 pp.

Ohlson, R. 1972. Production of and trade in rape seed. Pages 9–35 in Appelquist, L. A., and R. Ohlson. 1972.

O'Mary, C. C., M. C. Bell, N. N. Sneed, and W. T. Butts. 1970. Influence of ration copper on minerals in the hair of Hereford and Holstein calves. Jour. Anim. Sci. 31(3):626–30.

Ondreicka, Rudolf, Josef Kortus, and Emil Ginter. 1971. Aluminum, its absorption, distribution, and effects on phosphorus metabolism. Pages 293–305 in Skoryna, S. C., and D. Waldron-Edward. 1971.

Orten, James M. 1966. Biochemical aspects of zinc metabolism. Pages 38–47 in Prasad, Ananda S. 1966.

Oser, Bernard L. (Ed.). 1965. Hawk's physiological chemistry. 14th Ed. McGraw-Hill Book Co., New York. 1472 pp.

Österberg, Ragnar. 1974. Metal ion-protein interactions in solution. Pages 45–88 in Siegel, Helmut. Vol. 3. 1974.

Pal, Prabhot R., and Halvor N. Christensen. 1952. Interrelationships in cellular uptake of amino acids and metals. Jour. Biol. Chem. 2343:613–17.

Palmer, Ralph S. 1972. Patterns of molting. Pages 65–102 in Farner, Donald S., and James R. King. 1972.

Panić, B., L. J. Bezbradica, N. Nedeljkov, and A. G. Istwani. 1974. Some characteristics of trace-element metabolism in poultry. Pages 635–43 in Hoekstra, W. G. et al. 1974.

Park, Charles F., Jr. and Roy A. MacDiarmid. 1964. Ore deposits. W. H. Freeman and Co., San Francisco. 475 p.

Parmalee, D. F., H. A. Stephens, and R. H. Schmidt. 1967. The birds of southeastern Victoria Island and adjacent small islands. Nat. Mus. Can. Bull. 222:229 pp.

Peacor, D. R. 1972. Manganese. Crystal Chemistry. Pages 25-A-1-A-12 in Wedepohl, K. H. 1972.

Pearson, G. A., and L. Bernstein. 1958. Influence of exchangeable sodium on yield and chemical composition of plants. II. Wheat, barley, oats, rice, tall fescue and tall wheat grass. Soil Sci. 86(5):254–61.

Peisach, Jack, Philip Hisen, and William E. Blumberg. 1966. The biochemistry of copper. Academic Press, New York. 588 pp.

Pelletier, B. R. 1969. Submarine physiography, bottom sediments, and models of sediment transport in Hudson Bay. Pages 100–135 in Hood, Peter J. 1969.

———, F. J. E. Wagner, and A. C. Grant. 1968. Marine geology. Pages 557–613 in Beals, C. S. 1968.

Perris, A. D. 1971. The calcium homeostatic system as a physiological regulator of cell proliferation in mammalian tissues. Pages 101–26 in Nichols, George Jr., and R. H. Wasserman. 1971.

Petering, Harold G., David W. Yeager, and Sylvan O. Witherup. 1971. Trace element content of hair. I. Zinc and copper content of human hair in relation to age and sex. Arch. Env. Health 23(9):202–7.

Peterson, A. J., and R. H. Common. 1972. Estrone and estradiol concentrations in peripheral plasma of laying hens as determined by RIA. Can. Jour. Zool. 50(4):395–404.

Peterson, Randolph L. 1953. Studies of the food habits and the habitat of moose in Ontario. Contrib. Roy. Ont. Mus. Zool. and Paleol. 36:49 pp.

———. 1955. North American moose. Univ. Toronto Press, Toronto. 280 pp.

Poff, Ronald. 1961. Ionic composition of Wisconsin lake waters. Wis. Con. Dept. Fish Mgt. Div. Misc. Rept. No. 4:15 pp., mimeo.

Potter, J. G. 1965. Snow cover. Can. Dept. Trans. Met. Br. Clim. Stud. 3:69 pp.

Prasad, Ananda S. (Ed.). 1966. Zinc metabolism. Charles C. Thomas. Springfield, Ill. 465 pp.

Preble, Edward A. 1908. A biological investigation of the Athabaska-Mackenzie region. U.S. Bur. Biol. Surv. N. Amer. Fauna 27:574 pp.

Rajaratnam, V. A., J. B. Lowry, P. N. Avadhani, and R. H. V. Corley. 1971. Boron: possible role in plant metabolism. Science 172(3988):1142.

Rampton, V. N., and J. Ross Mackay. 1971. Massive ice and icy sediments throughout the Tuktoyaktuk Peninsula, Richards Island, and nearby areas, District of Mackenzie. Geol. Surv. Can. Paper 71–21:16 pp.

Rasmussen, D. Irwin, and William T. McKeen. 1945. The pheasant in the intermountain irrigated region. Pages 234–53 in McAtee, W. L. 1945.

Rasmussen, P. E., R. E. Ramig, R. R. Allmaras, and C. M. Smith. 1975. Nitrogen-sulfur relations in soft white winter wheat. II. Initial and residual effects of sulfur application on nutrient concentration, uptake, and N/S ratio. Agron. Jour. 67(2):224–28.

Raup, Hugh M. 1935. Botanical investigations in Wood Buffalo Park. Nat. Mus. Can. Bull. No. 74:174 pp.

Rawson, D. S. 1951. The total mineral content of lake waters. Ecology 32(4):669–72.

————. 1953. The bottom fauna of Great Slave Lake. Jour. Fish. Res. Bd. Can. 10(8):486–522.

————. 1956. The net plankton of Great Slave Lake, Jour. Fish. Res. Bd. Can. 13(1):53–127.

————, and J. E. Moore. 1944. The saline lakes of Saskatchewan. Can. Jour. Res. D. 22(6):141–201.

Reinhold, John G., Georga A. Keoury, and Teresa A. Thomas. 1967. Zinc, copper and iron concentrations in hair and other tissues: effects of low zinc and low protein intakes in rats. Jour. Nutr. 92(2):173–82.

Riordan, J. F., and B. L. Vallee. 1974. The functional roles of metals in metalloenzymes. Pages 33–57 in Friedman, Mendel. 1974.

Ritchie, J. C. 1962. A geobotanical survey of northern Manitoba. Arctic Instit. N. Amer. Tech. Paper 9:47 pp.

Rodan, Gideon A. 1973. Cellular functions of calcium. Pages 187–206 in Irving, James T. 1973.

Roe, Frank Gilbert. 1970. The North American buffalo. A critical study of the species in its wild state. Univ. Toronto Press, Toronto. 2nd Ed. 991 pp.

Romanoff, Alexis L., and Anastasia J. Romanoff. 1949. The avian egg. John Wiley & Sons, Inc., New York. 918 pp.

Rorison, I. H. 1969. Ecological aspects of the mineral nutrition of plants. Blackwell Scientific Publ., Oxford. 484 pp.

Rösler, H. J., and H. Lange. 1972. Geochemical tables. Elsevier Publishing Co., New York. 468 pp.

Russell, Richard Joel. 1936. Physiography of lower Mississippi River delta. Pages 3–199 in Russell, Richard Joel et al. 1936.

————, Henry V. Howe, James H. McGuirt, Christian F. Dohm, Wade Hadley, Fred B. Kniffen, and Clair A. Brown. 1936. Lower Mississippi river delta; reports on the geology of Plaquemines and St. Bernard parishes. La. Dept. Cons. Bull. No. 8:454 pp.

Ryder, John Pemberton. 1967. The breeding biology of Ross' goose in the Perry River region, Northwest Territories. Can. Wildl. Serv. Rept. Serv. No. 3:56 pp.

————. 1969. Nesting colonies of Ross' goose. Auk 86(2):282–92.

————. 1971. Distribution and breeding biology of the lesser snow goose in central Arctic Canada. Wildfowl. 22:18–28.

Sahagian, Benjamin M., I. Harding-Barlow, and H. Mitchell Perry, Jr. 1971. Effects of inhibitors and accelerators on intestinal absorption of divalent trace metals. Pages 321–37 in Skoryna, S. C., and D. Waldron-Edward. 1971.

Sarker, Bibudhendra, and Theo P. Kruck. 1965. Copper-amino acid complexes in human serum. Pages 183–96 in Peisach, Jack, Philip Hisen, and William E. Blumberg. 1965.

Sauchelli, Vincent. 1969. Trace elements in agriculture. Van Nostrand Reinhold Co., New York. 248 pp.

Saville, P. D. 1965. Changes in bone mass with age and alcoholism. Jour. Bone Joint Surg. 47 A, pp. 492–99.

Schachter, David. 1969. Toward a molecular description of active transport. Pages 157–76 in Dowben, Robert M. 1969.

Schellner, G. 1971. Sodium, zinc, and manganese supply to dairy cows grazing on weathered soils of different geological origins as measured by the black hairs of the animals as well as hops and purple clover. Monatsh. Veterinaermed. 26(7):249–55 (Ger.). (Chem. Abs. 75(9):60585).

Schepartz, Bernard. 1973. Regulation of amino acid metabolism in mammals. W. B. Saunders., Phila. 205 pp.

Schmidt-Nielsen, Knut, and Ragnär Fange. 1958. Salt glands in marine reptiles. Nature 182:783–85.

Schrenk, W. G. 1964. Minerals in wheat grain. Kansas Agric. Exp. Sta. Tech. Bull. 136:23 pp.

Schroeder, Henry A., and Nason, Alexis P. 1969. Trace elements in human hair. Jour. Invest. Dermatol. 53(1):71–78.

Schroeder, W. A., Lois M. Kay, Marry Lewis, and Nancy Munger. 1955. The amino acid composition of certain morphologically distinct parts of white turkey feathers, and of goose feather barbs, and goose down. Jour. Amer. Chem. Soc. 77(14):3901–8.

Schultz, Stanley G., Raymond A. Frizzell, and Hugh N. Nellans. 1974. Iron transport by mammalian small intestine. Pages 51–91 in Comroe, Julian H. et al. 1974.

Schütte, Karl H. 1964. The biology of the trace elements. J. B. Lippincott Co., Philadelphia. 228 pp.

Schwartz, Irving L. 1960. Extrarenal regulation with special reference to the sweat glands. Pages 337–86 in Comar, C. L., and Felix Bronner. 1960.

Schwarz, Klaus. 1974. New essential trace elements

(Sn, V, F, Si): progress report and outlook. Pages 355–80 in Hoekstra, W. G. et al. 1974.

Selwyn, A. R. C. 1874. Notes on a journey through the Northwest Territory from Manitoba to Rocky Mountain House. Can. Nat. New Ser. 7:193–216.

Shacklette, Hansford T., J. C. Hamilton, Josephine G. Boerngen, and Jessie M. Bowles. 1971. Elemental composition of surficial materials in the United States. U.S. Geol. Surv. Prof. Paper 574-D:71 pp.

Sharp, Ward M., and H. Elliott McClure. 1945. The pheasant in the sandhill region of Nebraska. Pages 203–33 in McAtee, W. L. 1945.

Shear, C. B. 1970. Calcium and salt tolerance of plants. Science. 167(3923):1387–88.

Sherratt, H. S. A. 1961. Anthocyanidins, flavones, flavonols and flavanones. Pages 1027–29 in Long, Cyril. 1961.

Shoemaker, Vaughan H. 1972. Osmoregulation and excretion in birds. Pages 527–74 in Farner, Donald S., and James R. King, Vol. II. 1972.

Shorland, F. B. 1961. Chemical composition of animal depot fats. Pages 691–701 in Long, Cyril. 1961.

Shorrocks, Victor M. 1974. Boron deficiency: its prevention and cure. Borax Consolidated, Ltd., London. 55 pp.

Siegel, Helmut. 1974. Metal ions in biological systems. Marcel Dekker, Inc. 1:267 pp.; 3:289 pp.

Simkiss, K. 1974. Calcium translocation by cells. Endeavor 33(120):119–23.

Skoryna, S. C., and D. Waldron-Edward (Eds.). 1971. Intestinal absorption of metal ions, trace elements and radionuclides. Pergamon Press, Oxford. 431 pp.

Sloan, Charles E. 1972. Ground-water hydrology of prairie potholes in North Dakota. U.S. Geol. Surv. Prof. Paper 585C:28 pp.

Slusher, David F. 1968. Tour guide. Soil Sci. Soc. Amer. Proc., Division S-5, New Orleans, La. 26 pp., mimeo.

Smith, Louis D. S. 1975. Common mesophilic anaerobes, including *Clostridium botulinium* and *Clostridium tetani* in 21 soil specimens. Appl. Microbiol. 29(5). In press.

Sognnaes, Reidar F. (Ed.). 1960. Calcification in biological systems. Amer. Assoc. Adv. Sci. Publ. No. 64:511 pp.

Soper, J. Dewey. 1941. History, range and home life of the northern bison. Ecol. Monog. 11(4):348–412.

———. 1942. Life history of the blue goose, *Chen caerulescens (Linnaeus)*. Boston Soc. Nat. His. Bull. 42(2):121–225.

Spencer, Herta, Joan A. Friedland, and Vernice Ferguson. 1973. Human balance studies in mineral metabolism. Pages 689–727 in Zipkin, Isadore. 1973.

Steinbach, H. B. 1962. Comparative biochemistry of the alkali metals. Pages 677–720 in Florkin, Marcel, and Howard S. Mason, Vol. 4. Pt. B. 1962.

Stettenheim, Peter. 1972. The integument of birds. Pages 1–63 in Farner, Donald S., and James B. King, Vol. II. 1972.

Stevens, W. E. 1953. The northwestern muskrat of the Mackenzie delta, Northwest Territories. Can. Wildl. Serv. Wildl. Mgt. Bull. Ser. 1, No. 8:40 pp.

Stevenson, F. J., and M. S. Ardakani. 1972. Organic matter reactions involving micronutrients in soils. Pages 79–114 in Mortvedt, J. J., P. M. Giordano, and W. L. Lindsay (Eds.). 1972.

Steward, F. C. (Ed.). 1963. Plant physiology. Vol. III. Inorganic nutrition of plants. Academic Press, New York. 811 pp.

Stewart, Robert E., and Harold A. Kantrud. 1971. Classification of natural ponds and lakes in the glaciated prairie region. U.S. Bur. Sport Fish. and Wildl. Resource Publ. 92:57 pp.

———, and ———. 1972. Vegetation of prairie potholes, North Dakota, in relation to quality of water and other environmental factors. U.S. Geol. Surv. Prof. Paper 585D:36 pp.

Stiles, Walter, and E. C. Cocking. 1969. An introduction to the principles of plant physiology. Methuen and Co., Ltd., London. 633 pp.

Strain, William H., and Walter J. Pories. 1966. Zinc levels of hair as tools in zinc metabolism. Pages 363–77 in Prasad, Ananda S. 1966.

———, Walter J. Pories, Arthur Flynn, and Orville A. Hill, Jr. 1971. Trace element nutriture and metabolism through head hair analysis. Trace Substances in Env. Health—V. Proc. 5th Conf., Columbia, Mo. pp. 383–97.

———, Lucille T. Steadman, Charles A. Lankau, Jr., William P. Berliner, and Walter J. Pories. 1966. Analysis of zinc levels in hair for the diagnosis of zinc deficiency in man. Jour. Lab. Clin. Med. 68(2):244–49.

Stresemann, E., and V. Stresemann. 1966. Die Manser der Vogel. Jour. fur Ornithologie (Sonderheft) 107:1–447.

Stross, R. G., and A. D. Hasler. 1960. Some lime induced changes in lake metabolism. Limnol. Oceanogr. 5:265–72.

Suelter, C. H. 1970. Enzymes activated by monovalent cations. Science 168(3933):789–95.

Sugden, Lawson G. 1973. Feeding ecology of the pintail, gadwall, American widgeon and lesser scaup ducklings in southern Alberta. Can. Wildl. Serv. Rept. Ser. 24:45 pp.

Sutcliffe, J. F. 1962. Mineral salts absorption in plants. Pergamon Press, New York. 194 pp.

Svatkov, N. M. 1958. Soils of Wrangel Island. Sov. Soil Sci., No. 1:80–87 (trans. edition).

Swanson, George A., Gary L. Krapu, Jerome R. Serie, and Mavis I. Meyer. 1972. Relationships between the liminology of prairie potholes and the feeding ecology of ducks. U.S. Bur. Sport Fish. and Wildl., Northern Prairie Wildl. Res. Center, Ann. Prog. Rept. 70 pp.

————, Vyto A. Adomaitis, John A. Shoesmith, and Mavis I. Meyer. 1972. Water quality characteristics of North Dakota prairie wetlands and their influence on the aquatic ecosystem. U.S. Bureau Sport Fish. and Wildl., Northern Prairie Wildl. Res. Center, Ann. Prog. Rept. 70 pp.

————, Morris I. Meyer, and Jerome R. Serie. 1974. Feeding ecology of breeding blue-winged teals. Jour. Wildl. Mgt. 38(3):396–407.

Sykes, A. M. 1971. Formation and composition of urine. Pages 233–78 in Bell, D. J., and B. M. Freeman. Vol. 1. 1971.

Tansy, Martin Francis. 1971. Intestinal absorption of magnesium. Pages 193–210 in Skoryna, S. C., and D. Waldron-Edward. 1971.

Taucins, E., and A. Svilane. 1965. Content of major and trace elements in sheep wool as a function of their levels in the ration. Tr. Lab. Biokhim. Fiziol. Zhivotn., Inst. Biol. Akad. Hank Latv. SSR 4:247–50 (Russ.). (Chem. Abs. 66:35784).

Tedrow, J. C. F., and L. A. Douglas. 1964. Soil investigations on Banks Island. Soil Sci. 98(1):53–65.

Texter, E. Clinton, Jr., Higino C. Laureta, and Ching-Chung Chou. 1971. Vascular factors in metal ion absorption. Pages 383–94 in Skoryna, S. C., and D. Waldron-Edward. 1971.

Thompson, John F., Iran K. Smith, and David P. Moore. 1970. Sulfur in plant nutrition. Pages 80–96 in Muth, O. H., and J. E. Oldfield. 1970.

Thornton, Iain. 1974. Biogeochemical and soil ingestion studies in relation to the trace-element nutrition of livestock. Pages 451–54 in Hoekstra, W. G. et al. 1974.

Thorsteinsson, R., and E. T. Tozer. 1962. Banks, Victoria, and Steffansson Islands, Arctic Archipelago. Geol. Surv. Can. Mem. 330:85 pp., 1 map.

Tilman, S. M., S. G. Byalobzheskiy, A. D. Chekhov, and O. M. Kuvayev. 1964. The geologic structure of Wrangel Island: pp. 53–98 in Tektonika i glubinnoye stroyeniye severo-vostoka SSSR; Trudy severo-vostochnogo kompleksnogo n-isl. inst., Akad. nauk SSSR (Magadan) vyp. 11.

Turnbull, Adam. 1974. Iron absorption. Pages 369–403 in Jacobs, A., and M. Worwood. 1974.

Umarji, G. M., and R. A. Bellare. 1966. Hair manganese in normal subjects. Indian Jour. Exp. Biol. 4(4):212–14. (Chem. Abs. 66(9):73854d).

Underwood, E. V. 1971. Trace elements in human and animal nutrition. 3rd Ed. Academic Press, New York. 543 pp.

U.S. Department of Agriculture. 1938. Soils and men. Yearbook of Agric., U.S. Govt. Printing Office, Washington, D.C. 1232 pp.

————. 1957. Soil. Yearbook of Agric., U.S. Govt. Printing Office, Washington, D.C. 784 pp.

U.S. Geological Survey. 1974. Water resources data for Wyoming, 1973. U.S. Dept. of Interior. 237 pp.

Upenski, S. M. 1965. The geese of Wrangel Island. The Wildfowl Trust Ann. Rept. 16:126–29.

Urist, Marshall R. 1969. Rarefying diseases of bone. With a comment on translocation of remodelled bone in osteoporosis. Pages 425–52 in Bittar, E. Edward, and Neville Bittar, Vol. 3. 1969.

Usik, Lily. 1969. Review of geochemical and geobotanical prospecting methods in peatland. Geol. Surv. Can. Paper 68–66:43 pp.

Ussing, H. H. 1971. Tracer studies and membrane structure. Pages 654–56 in Crone, Christian, and Niels A. Lassen. 1971.

————, and N. A. Thorn (Eds.). 1973. Transport mechanisms in epithelia. Academic Press, New York. 627 pp.

————, David Erlij, and Ulrik Lassen. 1974. Transport pathways in biological membranes. Pages 17–49 in Comroe, Julius H., Jr. et al. 1974.

Van Campen, Darrell R. 1971. Absorption of copper from the gastrointestinal tract. Pages 211–27 in Skoryna, W. C., and D. Waldron-Edward. 1971.

Van Horn, W. M., J. B. Anderson, and M. Katz. 1949. The effect of Kraft pulp mill wastes on some aquatic organisms. Trans. Amer. Fish. Soc. 79:55–63.

Vaughan, Janet M. 1970. The physiology of bone. Clarendon Press, Oxford. 325 pp.

Vinogradov, A. P. 1959. The geochemistry of rare and dispersed chemical elements in soils. Consultants Bureau Inc., New York. 2nd Ed. 209 pp.

Wacker, Warren E. C., and Bert L. Vallee. 1964. Magnesium. Pages 438–521 in Comar, C. L., and Felix Bronner. 1964.

Walser, Mackenzie. 1969. Renal excretion of alkaline earths. Pages 235–320 in Comar, C. L., and Felix Bronner. 1969.

Warren, Harry V., Robert E. Devault, and Christine H. Cross. 1967. Possible correlations between geology and some disease patterns. Ann. N.Y. Acad. Sci. 136(Art. 22):657–710.

Wasserman, R. H. 1972. Calcium transport by selected animal cells and tissues. Pages 351–84 in Hokin, Lowell E. 1972.

————, and A. N. Taylor. 1969. Some aspects of intestinal absorption of calcium with special reference to vitamin D. Pages 321–403 in Comar, C. L., and Felix Bronner. 1969.

————, R. L. Morrissey, A. N. Taylor, and R. A. Corradino. 1971. Biochemical and physiological aspects of adaptation. Pages 49–50 in Menczel, J., and A. Harell. 1971.

Webber, P. J., J. W. Richardson, and J. T. Andrews. 1970. Post glacial uplift and substrate age at Cape Henrietta Maria, southeastern Hudson Bay, Canada. Can. Jour. Earth Sci. 7:317–25.

Wedepohl, K. H. 1972. Handbook of geochemistry. Vol. 11/3. SpringerVerlag, New York. 92 Chap. (indiv. paged).

Weibel, R. O. 1951. Sunflowers as a seed and oil crop for Illinois. Univ. of Ill. Agric. Exp. Sta. Circ. 681:16 pp.

Werner, A., and M. Anke. 1960. Trace element content of beef-cattle hair as an aid in recognizing deficiency phenomena. Arch. Tierernähr 10:142–53 (Chem. Ab. 54:16574).

West, Edward Staunton, and Wilbert R. Todd. 1951. Textbook of biochemistry. The Macmillan Co., New York. 1345 pp.

Wetmore, Alexander. 1920. Observations on the habits of birds at Lake Burford. New Mexico. Auk 37(2):221–47.

————. 1921. Wild ducks and duck foods of the Bear River marshes, Utah. U.S. Dept. Agric. Bull. 936:20 pp.

Wheeler, George C., and Jeanette Wheeler. 1966. The amphibians and reptiles of North Dakota. Univ. of N. Dak. Press, Grand Forks. 104 pp.

White, W. Arthur. 1959. Chemical and spectrochemical analyses of Illinois clay materials. Ill. State Geol. Surv. Circ. 282:1–53.

Whitman, Warren C., D. W. Bolin, E. W. Klosterman, H. V. Klosterman, K. D. Ford, L. Mooman, D. G. Hoog, and M. L. Buchanan. 1951. Carotene, protein, and phosphorus in range and tame grasses of western North Dakota. N. Dak. Agric. Exp. Sta. Bull. 370:55 pp.

Widdowson, Elsie M., and J. W. T. Dickerson. 1964. Chemical composition of the body. Pages 1–247 in Comar, C. L., and Felix Bronner. 1964.

Wilde, S. A., C. T. Youngberg, and J. H. Hovind. 1950. Changes in composition of ground water, soil fertility, and forest growth produced by the construction and removal of beaver dams. Jour. Wildl. Mgt. 14(2):123–28.

Williams, R. J. P. 1967. Heavy metals in biological systems. Endeavor 26(98):96–100.

Wolstenholme, G. E. W., and Cecilia M. O'Connor (Eds.). 1956. Ciba foundation symposium on bone structure and metabolism. Little, Brown and Co., Boston. 299 pp.

Wood, Harold Bacon. 1950. Growth bars in feathers. Auk 67(4):486–91.

Yorath, C. J., and H. R. Balkwill. 1970. Stanton map-area, Northwest Territories (107D). Geol. Surv. Can. Paper 69-9:7 pp.

Zipkin, Isadore (Ed.). 1973. Biological mineralization. John Wiley and Sons, New York. 899 pp.

Index

An unusual number of illustrations characterizes this study. It has, therefore, often been necessary to place graphs and maps many pages beyond the corresponding text material. To aid the reader in finding the respective maps and graphs, figure numbers and page numbers have been cited, enclosed in parentheses, wherever pertinent. Some geographical place names may also be unfamiliar to many readers; for this reason, figure numbers for pertinent maps and their respective page numbers have also been given, enclosed by parentheses, following the initial citations.

"Geese" in the index refers to blue and lesser snow geese. Citations referring to greater snow geese, Ross' geese, Canada geese, and other species are so designated.